AF291386

QUANTUM ERA WARFARE

QUANTUM ERA WARFARE

Atul Tripathi

Cmde Manish Sinha (Retd.)

First published in 2026 by
PENTAGON PRESS LLP
206, Peacock Lane, Shahpur Jat
New Delhi-110049, India
Contact: 011-26490600

Typeset in AGaramond, 11.5 Point
Printed at Aegean Offset Printers, Greater Noida

ISBN 978-81-994764-3-1

www.pentagonpress.in

Contents

FOREWORD

There are moments in history when the nature of human conflict transforms so fundamentally that generations of strategists, scientists, and soldiers must reimagine the art of warfare. The arrival of gunpowder, flight, nuclear weapons, and cyberspace shifted how nations understood power and security. We now face another such moment, marked by the quiet hum of quantum computing and the subtle signals of entangled photons.

Quantum Era Warfare by Atul Tripathi and Commodore Manish Sinha addresses this transformative juncture—a change as inevitable as it is revolutionary. Commissioned as a research project by the National Maritime Foundation, this book is the product of relentless research and a remarkable ability to envision future conflict through the lens of quantum technology. The authors approach this complex subject pragmatically, translating abstract quantum concepts into the practical language of strategy and warfare.

This book arrives at a time when militaries worldwide are navigating an expanding battlespace—one extending beyond geography into computation and perception. It seeks to bridge two universes that rarely converse: the theoretical physicist's laboratory and the commander's operations room. The authors have commendably mediated this dialogue with authority and balance. Atul Tripathi's expertise in information security and technology governance provides insight into how scientific breakthroughs enter national systems. On the other hand, Commodore Sinha's operational experience deepens understanding of their practical impact across land, sea, air, space, and cyber domains. Together, they offer a narrative that is both rigorous and accessible.

Quantum theory's intrinsic probabilistic nature is turning uncertainty into a strategic instrument. It promises faster computing, unbreakable encryption, and near-instant secure communication—all capable of compressing decision cycles, strengthening deterrence, and maintaining operational advantage. However, as the authors caution, these technologies also bring new vulnerabilities. For India, whose security landscape demands agility and innovation, this work is especially timely because our investments in quantum computing, cryptography, communications satellites, and defence R&D would underpin our success in future conflicts.

This book speaks directly to the new generation of military professionals who will lead through the quantum transition. It is my hope that this book will inspire innovative thinking, foster cooperation, and nurture the confidence that the future of technology, like that of warfare, is not predetermined, but shaped by wisdom, purpose, and preparedness.

New Delhi
October 2025

Admiral Karambir Singh
PVSM, AVSM (Retd)
Chairman
National Maritime Foundation

List of Figures

Chapter One

Quantum Shift: Preparing for the Future of Warfare

"Quantum Shift: Preparing for the Future of Warfare," orients the readers to the strategic significance of quantum technologies and why they matter now. The chapter explains, in clear operational terms, the basic building blocks qubits, superposition, and entanglement and links each concept to concrete defense outcomes such as faster decision making, resilient communications, and next generation artificial intelligence. Without assuming prior scientific training, it shows how these effects convert into battlefield advantage by enabling superior sensing, computation, and secure information exchange across the full spectrum of conflict.

The chapter then translates science into force protection and mission assurance through the "Quantum Fortress" lens. It outlines how quantum secure communications especially Quantum Key Distribution (QKD) and the transition to post quantum cryptography harden command networks against interception and "harvest now, decrypt later" threats. Officers are introduced to quantum enhanced navigation, sensing, and radar concepts that improve detection, reduce susceptibility to stealth and GPS denial, and elevate real time situational awareness in contested electromagnetic environments.

Beyond immediate protection, the chapter addresses the wider transformation of warfare driven by quantum enabled autonomy and precision. It examines how quantum powered algorithms can guide autonomous platforms, optimise logistics and fires, and compress the sensor to shooter timeline, while also surfacing the ethical, legal, and arms control dilemmas that will accompany these capabilities. Leaders are equipped to recognise where doctrine, rules of engagement, and governance must adapt as quantum effects change the speed, transparency, and escalation dynamics of conflict.

Finally, the chapter situates these developments within India's emerging quantum defence doctrine and the accelerating global competition. It offers practical direction on securing digital battlefields, employing quantum intelligence in cyber operations, and reshaping information warfare, alongside training and acquisition steps for joint quantum readiness. Throughout, it emphasizes interoperable, scalable adoption ensuring Indian forces can integrate maturing quantum technologies responsibly and decisively from laboratory prototypes to operational advantage across land, sea, air, space, and cyber.

1.1 QUANTUM FRONTIERS: THE NEXT ERA OF MILITARY TECH

Qubits in Combat: The Science Behind the Strategy

Demystifying quantum mechanics through a war fighter's lens; understanding the building blocks of tomorrow's military systems

Qubits: The New Ammunition of Computation

In the rapidly evolving theatre of modern warfare, qubits, the foundational units of quantum information are emerging as the next-generation computational ammunition, poised to revolutionize how militaries think, sense, decide, and act. Quantum systems will perform complex calculations at speeds and scales unimaginable with today's classical architectures, offering a transformative leap in processing power. In classical computing, the fundamental unit of information is a bit, which can exist in one of two states, either 0 or 1, much like a light switch that is either ON or OFF. In contrast, a qubit (quantum bit), the foundational element of quantum computing, can exist in both states simultaneously due to a property called superposition. You can think of a qubit as a switch that is ON and OFF at the same time until it is observed. This unique characteristic allows quantum computers to process vast and complex possibilities in parallel.

For a military commander planning a strike multiple variables must be considered since multiple targets, various weather conditions, different enemy responses, and limited resources millions of combinations need to be considered. A traditional computer checks one scenario at a time, like flipping through a playbook page by page. A qubit changes the game.

A qubit can hold many possible values at once, thanks to a quantum property called superposition. Instead of checking scenarios one by one, a quantum computer with multiple qubits can evaluate many scenarios simultaneously. Let's think of it like this:

- A traditional soldier scouting terrain walks one path at a time.
- A quantum-enabled scout sees all paths at once, instantly identifying the best one.

This has profound implications for modern warfare. Qubits enable rapid intelligence processing, allowing analysis of thousands of tactical outcomes in parallel. They improve target prediction by managing uncertainty and partial information an everyday reality in combat and surveillance operations. Furthermore, quantum encryption powered by qubits offers next-generation cyber security, protecting military communication channels from even the most sophisticated cyber threats.

In essence, the qubit is far more than a high-speed computing unit. It is a strategic enabler that brings multi-dimensional thinking into military decision-making, with potential applications in missile defence, ISR (Intelligence, Surveillance, and Reconnaissance), and operational logistics.

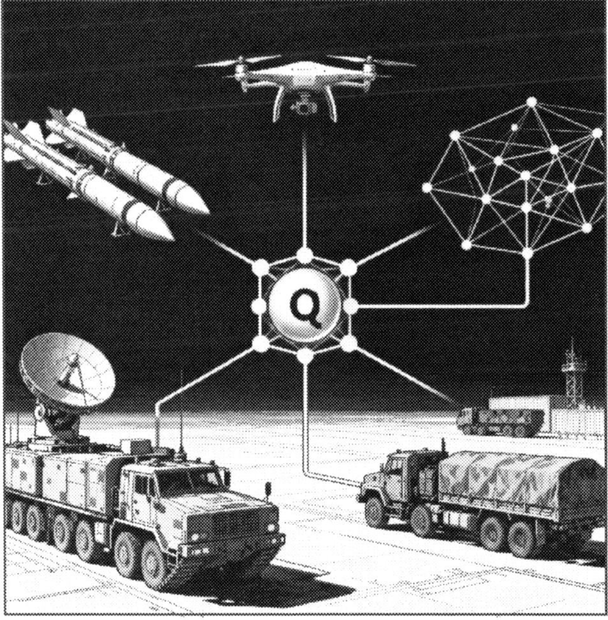

Fig. 1.1: Quantum Command: The Qubit Edge in Defence Decision-Making

Physically realised through highly controlled systems such as trapped ions, superconducting circuits, or single photons, qubits require an unprecedented level of precision and environmental stability. Their fragility and sensitivity are not weaknesses, but rather indicators of their raw potential; they are like high-value, mission-critical assets. In military terms, if classical computing is akin to deploying large infantry units to carry out broad, linear operations, qubits are the equivalent of elite special forces, capable of executing high-risk, high-reward missions with precision, speed, and stealth. They require fewer resources, but deliver disproportionately higher strategic value.

From an operational perspective, qubits offer transformative capabilities across multiple defence domains. Quantum computers powered by qubits can model complex environments such as multi-domain battlefields, logistics under duress, or adversary behaviour in real time, enhancing decision superiority. Quantum sensors leveraging qubit entanglement and coherence properties can provide ultra-sensitive detection of submarines, stealth aircraft, and nuclear materials, even in electromagnetically contested or GPS-denied environments. Perhaps most critically, qubits form the foundation of quantum communication systems, offering unbreakable encryption through quantum key distribution (QKD), effectively nullifying the threat posed by both traditional and quantum-enabled cyber adversaries.

Furthermore, as quantum hardware and software ecosystems evolve, qubits will integrate into a broader military digital infrastructure, linking with classical systems via hybrid computing architectures and enhancing the performance of AI, ISR (Intelligence, Surveillance, Reconnaissance), and autonomous systems. Their impact will be particularly significant in cyber-electromagnetic activities (CEMA), space-based assets, and strategic command and control (C2), where speed, reliability, and resilience are mission-critical.

Ultimately, as quantum technologies mature and operationalise, qubits will not merely support military operations, but will also redefine warfare and fighting doctrines. Whether it is in enabling decision advantage, securing communications, navigating in denied environments, or unlocking new forms of situational awareness, qubits represent a decisive edge. Just as gunpowder and satellites once redefined approaches to war fighting, quantum bits may become the standard-bearers of 21st-century military supremacy, ushering in a new era of computational warfare.

Superposition: Unlocking Parallel Possibilities for Battlefield Superiority

In classical computing, information is stored and processed using binary digits, or bits, which can be in one of two definite states 0 or 1. Every operation follows a linear, deterministic path executing one calculation at a time. This approach has served well across decades of digital innovation, from logistics to signal processing. However, when faced with the sheer scale, complexity, and speed required in modern military operations such as evaluating thousands of potential enemy manoeuvres, integrating cross-domain intelligence, or simulating real-time battle scenarios, classical systems start to show their limitations. Their linear architecture simply cannot cope with the exponential data explosion and fast-paced nature of the 21st-century warfare.

Quantum computing, through the concept of superposition, offers a radical departure from this model. As mentioned earlier, a qubit, can exist not just in the state of 0 or 1, but in a superimposed state of both simultaneously. This means that a quantum system can process an entire spectrum of possibilities in parallel, rather than step-by-step.

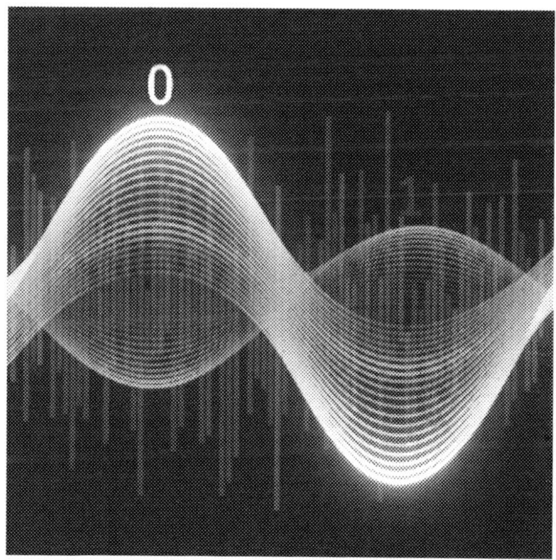

Fig. 1.2: Superposition: The Power of Parallel Possibilities

To visualise this advantage in military terms: Imagine a drone that is not pre-assigned to a single target but is dynamically poised to strike multiple high-priority targets simultaneously, reacting in real-time to changing tactical conditions. That is the operational power superposition brings to quantum systems; it enables real-time, multidimensional situational awareness and response.

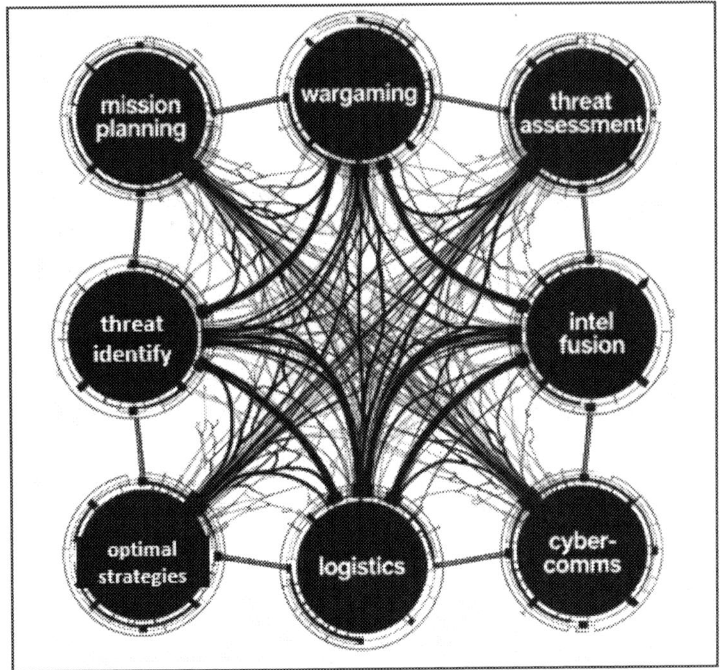

Fig. 1.3: Superposition to Command Advantage

For the Indian Armed Forces, this opens a new frontier in strategic and operational capabilities:

- **Mission Planning & War Gaming**: Superposition allows quantum systems to simulate thousands of battle plan permutations concurrently, enabling commanders to test and select optimal strategies faster than ever before.

- **Dynamic Threat Assessment**: When analysing live satellite feeds, intercepted signals, or troop movement patterns, quantum computers can instantly explore vast permutations to identify probable threat vectors or infiltration routes.

- **Intelligence Fusion**: Superposition helps correlate fragmented inputs human intelligence, UAV imagery, cyber alerts into a unified threat picture, empowering better decision-making under pressure.

- **Operational Logistics**: In high-stress conditions say, resupplying forward posts under fire quantum models can optimise supply chains by simulating countless delivery permutations across terrain, weather, and threat variables.

- **Cyber and Communications Superiority**: In cyber defence and encryption, quantum systems using superposition can rapidly analyse potential vulnerabilities or generate uncrackable encryption patterns on the fly.

In essence, superposition doesn't just offer computational speed it enables strategic foresight. It transforms how information is processed, how decisions are modelled, and how quickly responses are activated. In theatre-level operations, where milliseconds can determine mission success or failure, the ability to evaluate multiple scenarios in parallel is nothing short of transformative.

For India's military leadership, integrating quantum capabilities driven by superposition could significantly shorten the sensor-to-shooter loop, enhance multi-domain operations, and deliver a decisive edge in environments where speed, complexity, and uncertainty dominate.

Entanglement: Instantaneous, Encrypted Connectivity for the Modern Battle Space

One of the most counter-intuitive yet powerful principles of quantum mechanics is entanglement, a phenomenon in which two or more quantum particles become intrinsically linked, such that a change in the state of one particle is reflected *instantaneously* in its entangled partner. This occurs regardless of the distance separating them. To those used to conventional communication systems, this might appear to violate the very notion of signal propagation. But this is not about the speed-of-signal transmission, but about *instantaneous correlation*, enabled by the laws of quantum physics.

In the classical world, secure RF communication between a forward operating post and a command centre depends on transmissions that can be delayed, jammed, or intercepted. In contrast, entangled systems offer a secure, real-time, and tamper-proof link, immune to traditional electronic warfare threats. Imagine a Para SF team operating near the Line of Actual Control (LAC), deep in high-altitude, electromagnetically contested terrain. Conventional SATCOM or VHF/UHF channels are either not available, degraded due to terrain, or actively jammed. Now imagine a communications system based on entangled qubits. Once entanglement is established prior to deployment, the team can remain linked to command elements with instantaneous quantum state correlations that cannot be intercepted, jammed, or spoofed. No signals are broadcast, no frequencies are intercepted, just secure, stealthy state correlation that makes the communication undetectable and unbreakable.

Fig. 1.4: Entanglement: Instant Connection Across Distance

Quantum entanglement offers a foundation for next-generation C4ISR (Command, Control, Communications, Computers, Intelligence, Surveillance, and Reconnaissance). In the context of India's joint operations be it Army-Air coordination in Arunachal Pradesh, maritime surveillance in the Indian Ocean Region (IOR), or tri-service information fusion at Andaman & Nicobar Command, entangled communication channels can significantly reduce latency and enhance survivability.

Key advantages include:

- **Anti-jamming Capability**: Unlike classical RF-based networks that can be targeted by adversary EW systems (such as those deployed along the Chinese border or in hostile naval encounters), quantum links do not emit detectable signals.

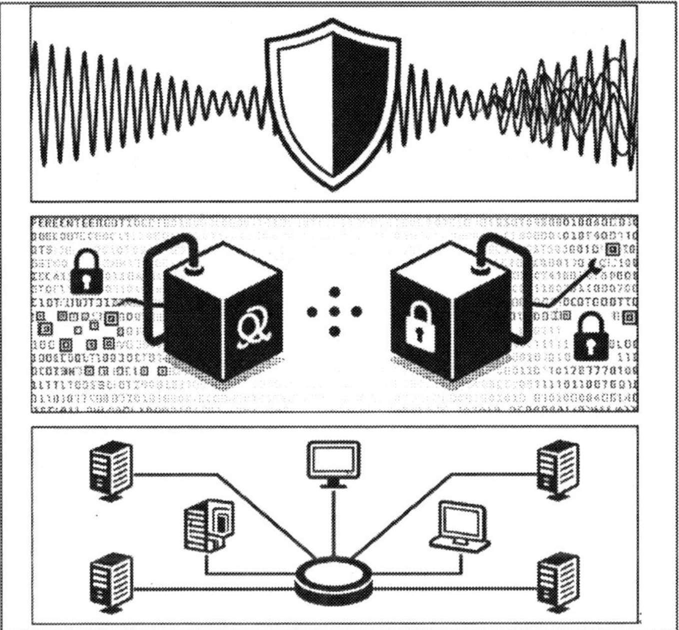

Fig. 1.5: Quantum Security: Jamming-Proof, Encrypted, Connected

- **Quantum Key Distribution (QKD):** Entanglement enables mathematically unbreakable encryption. Military units can share encryption keys using QKD, ensuring that any interception attempt is immediately detected. This is mission-critical in protecting targeting data, unit positions, and sensor feeds.

- **Interoperability:** In future multilateral naval exercises (like MALABAR or MILAN), entanglement could underpin quantum-secure blue force tracking and cross-domain data sharing, enabling real-time coordination across diverse platforms such as submarines, drones, P-8I aircraft, and shore-based command centres.

Quantum Coherence and Decoherence: The Silent Battle for Quantum Superiority

In the theatre of quantum-enabled warfare, quantum coherence is the unseen operational backbone. This is akin to the discipline of radio silence or stealth in a covert operation. It refers to the ability of a quantum system, such as a qubit, to retain its delicate quantum state (superposition or entanglement) over time. As long as coherence is preserved, quantum systems can perform extraordinary feats: ultra-sensitive detection, real-time intelligence fusion, encrypted communication, and rapid decision modelling.

But coherence is fragile intolerant to environmental "noise" such as temperature fluctuations, electromagnetic interference (EMI), vibration, and even cosmic particles. When coherence is lost, a phenomenon known as decoherence occurs, where the system reverts to classical behaviour, much like a compromised comms link during a strike operation. The result: corrupted data, failed computations, or blind sensors.

Consider a Garud Commando detachment deployed in a radar-saturated zone near the LoC, relying on absolute electromagnetic silence to avoid detection. One stray transmission or an unshielded emission could give away position and compromise the mission. In quantum systems, decoherence is the equivalent of that costly mistake. Even the smallest interference can collapse a quantum state, nullifying its tactical advantage.

The challenge is amplified when quantum assets are deployed in dynamic and contested military environments:

- A quantum sensor mounted on an Indian Navy P-8I aircraft must retain coherence despite constant vibration, EMI from onboard radar systems, and atmospheric noise.

- A quantum computer housed in a mobile battlefield HQ (e.g., during a combined India-USA Yudh Abhyas exercise) must maintain stable operation while surrounded by RF emitters, power fluctuations, and rugged field conditions.

Decoherence as a Tactical Threat

From a commander's perspective, decoherence is not just a lab problem, but an operational vulnerability. Just as jamming and EW threats can disable GPS or SATCOM, environmental noise can disable quantum systems. Hence, ensuring prolonged coherence is essential for:

- Quantum Radar Systems: Detecting stealth aircraft or submarines requires the sensor to remain coherent long enough to extract signal signatures.

- Quantum Navigation: In GPS-denied environments such as deep valleys in Arunachal or dense ocean zones, quantum gyroscopes must operate stably for prolonged durations.

- Quantum Communications: The longer entangled states remain coherent, the farther and more reliably quantum-encrypted messages can be sent without compromise.

Just as the Indian Armed Forces invest in electromagnetic shielding, hardened comms, and EMP protection, defence scientists are working to "quantum-harden" systems. This includes:

- Cryogenic enclosures to reduce thermal noise.
- Topological qubits and error correction protocols that self-repair disrupted states.
- Isolation chambers to shield from ambient radiation and vibrations in mobile deployments.

In the near future, quantum field kits may require the same planning and support as traditional C4ISR systems: power management, EMI shielding, temperature stabilisation, and mobility under rugged conditions.

Ultimately, quantum coherence is the operational heartbeat of next-generation war fighting systems just as discipline in radio silence, radar cross-section management, and cyber hygiene are vital today. Decoherence is the adversary's opening, a silent saboteur that erodes tactical advantage. The armed force that can best shield and stabilise its quantum assets will dominate:

- Information assurance,
- Autonomous sensing,
- Rapid, multi-domain decision-making.

Just as the military trains for EW scenarios and GPS-denied operations, planners will be required to articulate SOPs where coherence management is as critical as logistics, fuel, or ammunition.

The Quantum-Classic Divide: Working Together on the Battlefield

In the evolving battle space, the rise of quantum technologies does not signal the obsolescence of classical computing, but rather its augmentation. Quantum systems are not intended to replace traditional digital infrastructure, but to serve as precision assets deployed selectively for mission-critical operations that exceed the limits of classical systems. Much like how precision-guided munitions (PGMs) are reserved for high-value, high-impact targets rather than routine fire missions, quantum computing and sensing technologies are strategic enablers, brought to bear where conventional methods fall short. These include use cases such as optimizing complex military logistics in real time across multiple theatres, performing unbreakable quantum encryption, or deploying ultra-sensitive quantum sensors for submarine detection, GPS-denied navigation, or early missile launch warnings.

From an operational standpoint, military commanders must develop a hybrid mindset understanding when and where to deploy quantum capabilities within a broader mission system. Quantum computers can work in tandem with classical systems, handling specific functions such as cryptographic key generation or simulating battlefield scenarios with many unknown variables, while classical systems manage routine processing, control, and communication. This collaborative model, often referred to as quantum-classical hybrid architecture, mirrors modern combined arms doctrine where air, land, and sea forces operate in unison, leveraging each other's strengths.

For the military to fully capitalise on quantum potential, integration is key. Commanders, staff officers, and war fighters must not only trust these systems but also understand their limitations and optimal-use cases. Training programs and operational doctrines must evolve to ensure quantum tools are not misapplied or over-relied upon, but instead used where they offer a clear tactical or strategic edge. In sum, the future of warfare will not be purely quantum, but intelligently hybrid, where quantum precision supports and extends the capabilities of proven classical systems creating a force multiplier effect on the digital battlefield.

Further, in order to prepare for the quantum future, some foundational steps such as Digital Twinning are considered advisable. These use AI as against quantum computing, and would help run quick SWOT analyses and readiness checks for the transition. It is to be remembered that quantum computing is not a panacea, but offers a vastly improved way of fighting. It gives us an edge, and we must take it, given the stakes in an unstable neighbourhood, and the speed of battle that has the possibility to not only overwhelm our decision-making processes, but ultimately increase the cost of victory.

The Quantum Shield: Military significance of quantum technologies

Harnessing quantum-powered AI to enable faster, smarter decisions from tactical edge to strategic command centres

Breaking Classical Encryption: The End of Today's Security Assumptions

Modern military operations rely on sophisticated digital infrastructure spanning communications, intelligence, logistics, navigation, targeting, and command systems where encryption plays a central role in safeguarding the confidentiality, integrity, and authenticity of sensitive information across all domains. Today, the most widely used encryption protocols, such as RSA (Rivest–Shamir–Adleman) and ECC (Elliptic Curve Cryptography), derive their security from the computational difficulty of specific mathematical problems. These include factoring large prime numbers or solving discrete logarithms, tasks so complex that even the most advanced classical supercomputers would require impractical timeframes, often measured in thousands of years, to crack them.

Fig. 1.6: The Quantum Threat: Breaking Military Encryption in Hours

However, quantum computing introduces a paradigm shift that renders this foundational assumption obsolete. Quantum algorithms, most notably Shor's algorithm, are capable of solving these very problems exponentially faster than classical approaches. In principle, a sufficiently powerful quantum computer could break RSA and ECC encryption in a matter of hours or minutes, posing an existential threat to current digital security infrastructures across both civilian and military domains.

This is not a distant academic concern but an immediate operational vulnerability. Quantum decryption capabilities, if developed by adversarial states or actors, would enable them to covertly access critical military communications, decrypt classified intelligence transmissions, and potentially manipulate satellite-based command-and-control systems. From a defence standpoint, this scenario is tantamount to an adversary possessing an undetectable master key capable of unlocking secure bases, intercepting operational directives, and compromising joint-force coordination, all without triggering traditional security alarms. Encrypted communication links used in nuclear command-and-control architectures, strategic missile systems, naval fleet coordination, and airborne early warning platforms could be rendered transparent to hostile surveillance. Moreover, adversaries equipped with quantum decryption capabilities could retrospectively decrypt previously captured encrypted data scenario known as "harvest now, decrypt later" thereby exposing years of sensitive historical communications.

The risk extends beyond information theft. In future high-intensity conflicts where electromagnetic and cyber domains play a decisive role, compromised encryption could lead to false flag operations, spoofed commands, or disruption of C4ISR frameworks. The loss of secure communications in the fog of war could paralyse mission execution, disrupt joint operations, and erode decision-making superiority.

Given this context, the Indian Armed Forces face a strategic imperative that cannot be deferred: the rapid transition to post-quantum cryptography (PQC). PQC refers to cryptographic systems that are designed to be secure against both classical and quantum attacks. These algorithms, currently being standardised by global initiatives including those led by the US National Institute of Standards and Technology (NIST), offer viable alternatives that do not rely on number-theoretic assumptions.

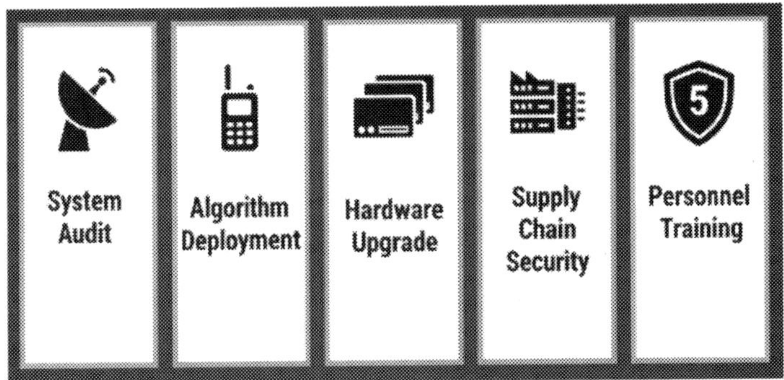

Fig. 1.7: Five Pillars of Quantum-Safe Defence Infrastructure

Implementing PQC in Indian defence infrastructure demands a holistic and layered approach:

- Inventory and audit of vulnerable systems, including satellite uplinks, battlefield management systems, and encrypted radio networks;
- Deployment of quantum-resistant algorithms in mission-critical communications and data exchange protocols;
- Upgrading cryptographic hardware to support the computational requirements of new algorithms;
- Securing the supply chain to ensure cryptographic tools are free from foreign tampering;
- Training of cyber and signals personnel to operate and manage hybrid classical-quantum security frameworks

This transition must also be synchronised across the services Army, Navy, and Air Force as well as with allied institutions such as the Defence Research and Development Organisation (DRDO), the National Technical Research Organisation (NTRO), and strategic cyber commands. Failing to act decisively risks ceding informational superiority and cyber dominance to adversaries who are already investing heavily in quantum research.

Quantum-Secure Communications: Beyond Eavesdropping and Jamming

In the evolving battle space of the 21st century, communication has become the strategic nervous system of modern warfare. Quantum-secure communication, particularly through Quantum Key Distribution (QKD), offers a transformational leap in the area coordinating between frontline units, synchronising joint operations, or commanding nuclear assets, with the assurance that messages are received securely, unaltered, and without compromise. Unlike classical encryption which relies on mathematical complexity, QKD draws its strength from the laws of quantum mechanics. Using phenomena such as entanglement and the no-cloning theorem, QKD allows two parties to generate and share encryption keys with the guarantee that any attempt to intercept or tamper with the transmission will be instantly detected. This is because any measurement or eavesdropping on a quantum state irreversibly disturbs it, revealing the intrusion.

To illustrate this in operational military terms: envision a scenario where strategic orders are delivered through an invisible, autonomous stealth drone programmed to initiate a full mission abort and alert command if it detects even the slightest attempt at interference. This metaphor encapsulates the unparalleled integrity and security offered by QKD. Unlike traditional encryption, QKD not only guarantees the confidentiality of information but also embeds intrinsic mechanisms that immediately signal any attempt at interception or tampering ensuring both the trustworthiness and survivability of critical communications under adversarial conditions.

The operational relevance for India's tri-services is profound. In naval warfare, QKD can be employed to establish ultra-secure links between submerged submarines and naval command centres, ensuring uninterrupted command and control even in high-interference zones. For the Army, quantum-secure links can be deployed in forward operating bases or mobile command posts, ensuring that strategic orders and battlefield intelligence are transmitted without vulnerability to electronic surveillance or spoofing. The Air Force can leverage QKD-enabled satellite communications to maintain resilient and tamper-proof links between airborne early warning systems, unmanned platforms, and ground control in contested airspace.

Moreover, as India prepares for future joint-force and network-centric operations under doctrines such as Theatre Commands and Integrated Battle Groups (IBGs), a secure quantum backbone will be essential to maintain coherence across services. In cyber-contested environments, where traditional networks may be jammed, spoofed, or corrupted, QKD-enabled systems provide the last line of trusted communications capability that remains reliable even when everything else is compromised.

For the Indian military, adoption of quantum-secure communication is a strategic necessity. As quantum technologies mature globally and near-peer adversaries continue to invest in them, India must accelerate its efforts to deploy and operationalise QKD at a scale beginning with space-based systems, critical command infrastructure, and defence research corridors.

The future of secure military communications lies not in stronger passwords or longer keys, but in the unchangeable laws of quantum physics. Embracing this shift will help ensure that the Indian Armed Forces maintain secure, survivable, and unbreakable lines of communication across every theatre of war.

Quantum-Enhanced Decision-Making: Simultaneous Futures, Better Choices

Modern military decision-making demands the rapid assessment of evolving tactical scenarios, real-time outcome simulation, and the swift execution of optimal responses across multi-domain operations. Traditional computational methods remain inherently sequential and increasingly inadequate, struggling with exponential complexity as variables and constraints multiply an acute limitation in high-pressure battlefield environments where time is scarce and information incomplete.

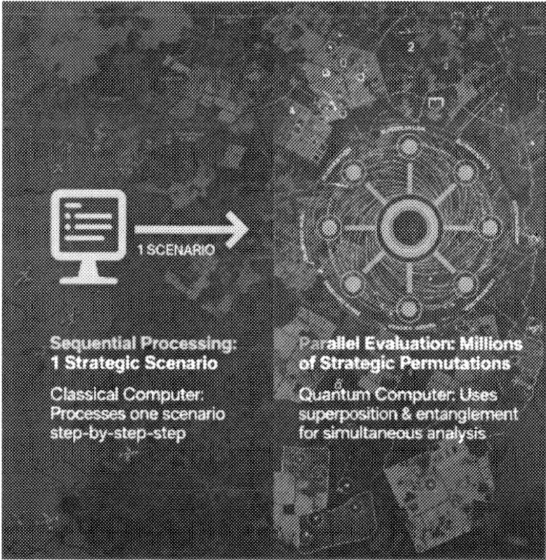

Fig. 1.8: Quantum Leap: From Sequential to Simultaneous Strategy

Quantum computing introduces a fundamentally different paradigm. By harnessing quantum superposition and entanglement, quantum processors can represent and process vast numbers of potential solutions simultaneously. Rather than executing simulations one at a time, quantum-enhanced systems evaluate thousands or even millions of strategic permutations in parallel offering an insight into entire landscapes of decision possibilities. This capability is not simply a matter of computational speed; it is a structural advantage in tackling combinatorial optimisation and probabilistic simulation core to military operations such as:

- Determining the optimal distribution of battalions across varying terrain to maximise defence and mobility;
- Identifying resilient supply routes under conditions of dynamic threat and contested airspace
- Simulating multiple enemy courses of action (COAs) across time horizons, and ranking them based on projected risk and strategic impact

In classical war gaming and simulation environments, commanders and analysts are constrained to explore only a limited subset of scenarios due to time and resource constraints. This can introduce selection bias and risks overlooking low-probability, high-impact events. Quantum algorithms such as quantum annealing, quantum Monte Carlo methods, and the Quantum Approximate Optimisation Algorithm (QAOA) can evaluate all plausible configurations at once, providing a broader and deeper assessment of operational environments.

For example, in a quantum-enhanced red-teaming scenario, thousands of adversarial options can be run concurrently to identify weaknesses in base layouts, logistics chains, or electromagnetic emission profiles. In logistics, a quantum-assisted command centre could instantly evaluate the most efficient, risk-mitigated placement of forward depots, mobile radar systems, or drone recharging nodes based on terrain, weather, enemy ISR coverage, and projected battlefield churn.

This is not merely a faster calculator; it is a transformational shift from reactive planning to proactive, systemic foresight. Quantum-enhanced decision systems bring a decisive edge by fusing data ingestion, simulation, and decision-making into a unified computational process. In doing so, they support a range of critical operational functions:

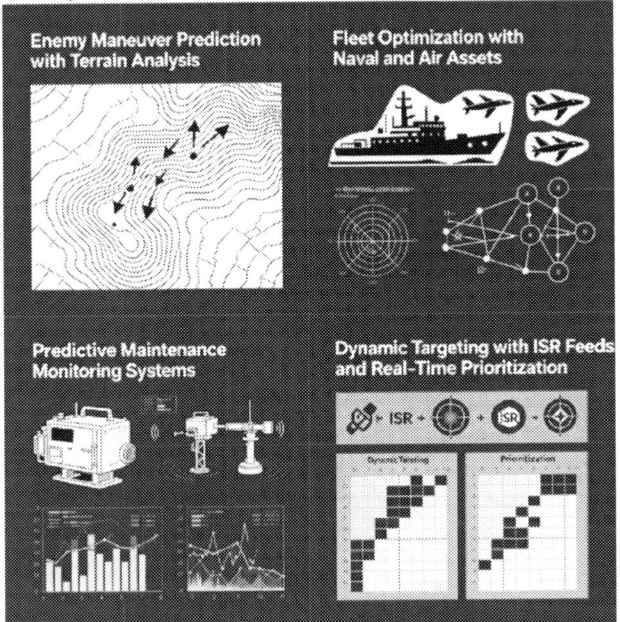

Fig. 1.9: Quantum-Powered Operations: Four Force Multipliers

- **Anticipating enemy manoeuvres:** Quantum models can simulate thousands of potential adversary movement patterns by integrating terrain data, past behaviour, doctrine repositories, and current surveillance feeds. Unlike classical systems which extrapolate from historical data, quantum systems can generate emergent threat pathways in real time, preserving initiative and situational awareness.

- **Fleet and asset optimization:** In naval and air operations, where assets must evade radar, account for fuel logistics, and maximise sensor overlap, quantum optimisation algorithms can determine the most effective trajectories and deployment configurations adjusting continuously for weather, risk zones, and satellite coverage.

- **Predictive maintenance and failure prevention:** Strategic platforms such as missile batteries, AWACS aircraft, or nuclear submarine sonar arrays are vulnerable to critical component failures under high stress. Quantum-enhanced analytics can process live telemetry and maintenance logs to detect early failure signatures, enabling condition-based maintenance and reducing mission risk.

- **Real-time dynamic targeting**: As ISR inputs flow from satellites, UAVs, radar arrays, and SIGINT intercepts, threat landscapes shift second-by-second. Quantum algorithms can reprioritize high-value targets dynamically, incorporating the most current sensor data without bottlenecks vital in missile defence, close air support, or counter-UAV engagements.

Globally, countries such as the USA, China, and France are actively pursuing military applications of quantum optimisation for mission planning, logistics, and autonomous swarms. These efforts are no longer confined to theoretical labs; prototypes and early-stage operational systems are already in development. If the utility of quantum computing is understood clearly, then a Decision Support System (DSS) in its true sense can become a reality.

1.2 QUANTUM: BATTLE READY

Quantum Fortress: Hardening Military Networks and Navigation for the Next War

Quantum-secure networks and next-gen navigation shielding mission-critical systems from cyber strikes and GPS denial in the wars of tomorrow

Threats and Counter-measures in Quantum Cyber Security: Emerging Threats, Adaptive Defences

As quantum computing matures, it introduces a disruptive vector into the cyber domain quantum-enabled cyber threats. Unlike traditional cyber-attacks that rely on exploiting software vulnerabilities or social engineering, quantum threats centre around brute-forcing cryptographic systems, intercepting secure communications, and even deploying quantum-enhanced hacking techniques. With the advent of Shor's algorithm, encryption standards like RSA and ECC that are cornerstones of military data security can be broken in minutes by a sufficiently powerful quantum computer. This makes national defence networks vulnerable to "crypto-breaking" attacks that were previously considered implausible. The threat landscape also expands to include quantum-enhanced signal interception, manipulation of quantum communication channels (like QKD), and sabotage through quantum sensors or entanglement-based interference.

In the evolving battle space of the 21st century, quantum-enabled adversaries can be conceptualized as a new generation of digital special operations forces; highly sophisticated, low-observable, and precision-focused actors capable of penetrating hardened digital infrastructures. These adversaries, equipped with quantum computing power, possess the ability to rapidly compromise public key cryptosystems (such as RSA and ECC) that currently protect military-grade communications, C4ISR platforms, and secure satellite uplinks. Their advantage lies in the ability to bypass traditional encryption barriers not through software exploits, but via mathematical supremacy, enabled by quantum algorithms like Shor's and Grover's. For instance, stored encrypted battlefield telemetry or drone control signals, once considered secure under AES or RSA, become exposed to future decryption if adversaries "harvest now, decrypt later" using future quantum resources. This necessitates a shift from traditional cyber hygiene to anticipatory quantum readiness.

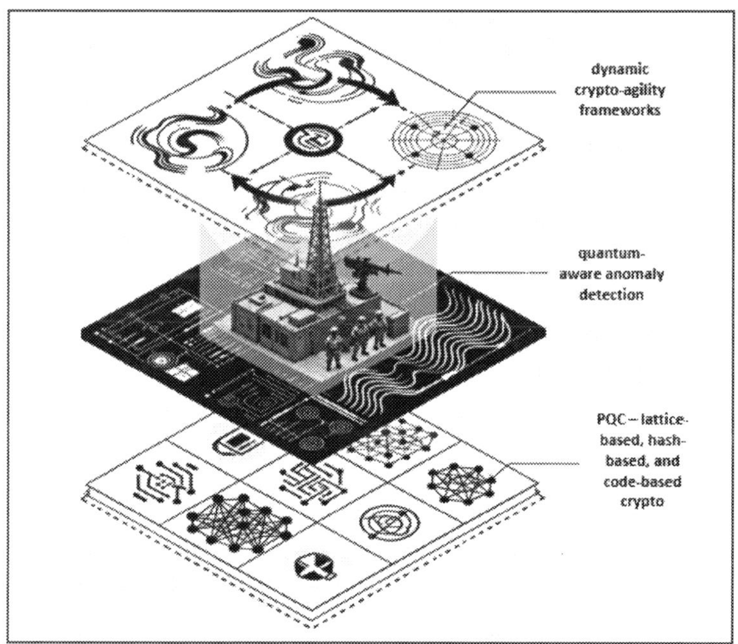

Fig. 1.10: Layered Defence: Building Quantum-Resilient Cyber Security

To counter such threats, defence cyber doctrine must evolve towards a multi-tiered, quantum-resilient cyber security architecture, mirroring concepts from layered kinetic defence. At the foundational layer, post-quantum cryptographic (PQC) algorithms such as lattice-based, hash-based, and code-based crypto systems must be adopted across command infrastructure. On the perimeter, quantum-aware anomaly detection systems utilizing AI/ML models trained to identify the signature patterns of quantum-enhanced cyber intrusion act as early warning radar systems. Furthermore, dynamic crypto-agility frameworks, allowing rapid switch-over to newer cryptographic standards, provide the electronic countermeasure (ECM) equivalent in cyberspace, ensuring adaptability under duress.

This transition demands doctrinal, technological, and organizational alignment. Static cyber defences must be replaced with active cyber resilience operations (ACRO), involving continuous vulnerability assessment, cyber red-teaming with quantum threat emulation, and simulation of zero-day exploits facilitated by quantum computing. These efforts should be embedded within national and joint force cyber exercises (e.g., Table-Top Exercises (TTXs), cyber range drills), integrating both conventional cyber and quantum-augmented threat scenarios. From a command readiness perspective, cyber units must be trained to recognise and respond to quantum-enabled threats just as EW officers are trained against radar jamming or GPS spoofing.

Inter-force interoperability and coalition defence strategies must include standardization of PQC adoption, information-sharing on quantum threat intelligence, and joint procurement of quantum-resilient technologies. NATO, QUAD, and other strategic alliances should include a quantum-resilient posture as a formal capability pillar within their cyber doctrines. In short, quantum cyber security is not an abstract future concern; it is a current operational planning requirement that will determine the survivability of national defence networks in a rapidly converging threat environment.

Operationally, defence forces must prioritize identifying and assessing quantum cyber vulnerabilities across classified networks, satellite uplinks, and battlefield IoT systems. This includes developing quantum-aware cyber security protocols that can detect anomalies unique to quantum attack signatures, crafting robust incident response strategies, and embedding quantum threat scenarios into cyber defence exercises and war games. Furthermore, given the global nature of this threat, there is a pressing need for multinational coordination sharing threat intelligence, standardizing post-quantum practices, and collaborating with industry partners who are pioneering quantum-secure technologies. In this new theatre of warfare, foresight and interoperability will define cyber survivability. Quantum cyber security is not just a technical challenge; it is a strategic imperative for command dominance in the coming decades.

Quantum Key Distribution (QKD): Unbreakable Links, Trusted Command

Quantum Key Distribution (QKD) represents one of the most practical and battlefield-relevant applications of quantum physics today. At its core, QKD is a method of securely distributing encryption keys using the principles of quantum mechanics, ensuring that any attempt to intercept the key is not only unsuccessful but immediately detectable. Unlike classical key exchange protocols vulnerable to interception and decryption especially with the impending threat of quantum computers.

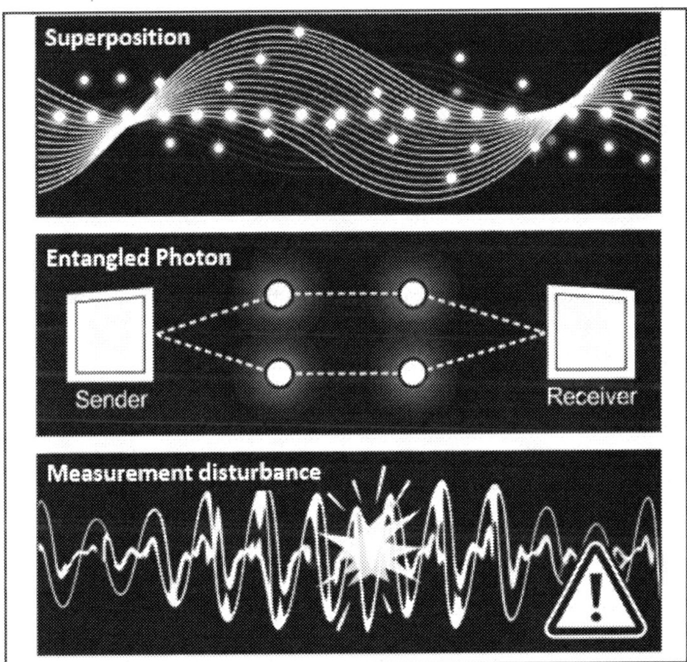

Fig. 1.11: Quantum Key Distribution: Unbreakable Communication Through Physics

QKD exploits quantum phenomena such as superposition, entanglement, and the no-cloning theorem to create an inherently secure communication channel. Superposition allows particles like photons to exist in multiple states simultaneously. This allows a photon used in key distribution to encode information in a probabilistic state, which only resolves into a definitive value upon measurement. Entanglement takes this security a step further. It links two particles such that their states remain perfectly correlated, regardless of the distance between them. When used in QKD, one entangled photon is sent to the recipient while the other remains with the sender. Any attempt to tamper with or observe the photon in transit disrupts the entangled state, immediately signalling a breach. The no-cloning theorem is a built-in failsafe of quantum systems that prevents an adversary from making an exact replica of a quantum message. This is the quantum equivalent of field orders written in disappearing ink that cannot be photocopied or transcribed without destroying the original. Traditional encryption can be cracked given enough computational power and time; quantum-encrypted keys, by contrast, *cannot* be cloned or copied under the laws of physics, making brute force or algorithmic attacks irrelevant.

Together, these quantum effects form a security architecture where any unauthorized attempt to intercept a key, be it by physical intrusion, signal interception, or cyber compromise thereby automatically triggers an alert due to the measurable disturbance of the quantum state. For the military commander, this translates to absolute assurance that mission-critical communications, targeting data, or nuclear command codes are transmitted through unbreakable links, and that any threat to communications integrity is detected and countered in real time. In high-stakes scenarios such as joint-force coordination, forward-deployed drone control, or nuclear command-and-control (NC2), this capability ensures that operational command remains trusted, secure, and uncompromised, even under hostile surveillance or cyber-attack. To translate this into a military context, consider QKD as a "quantum courier" a highly trusted agent delivering sealed orders in a briefcase that explodes if tampered with. Traditional encrypted radio communications are akin to padlocked messages: secure for now, but vulnerable if the enemy eventually finds the key. In contrast, QKD acts like a tamper-proof envelope that shreds its contents at the slightest hint of interference. This property is critical in high-stakes operational environments where adversaries actively seek to intercept or spoof communications.

Operationally, QKD holds immense potential in securing command and control networks, particularly in contested electromagnetic environments. It can safeguard communications between forward-deployed units and headquarters, ensure satellite uplinks and drone feeds are immune to hijacking or jamming, and enable real-time coordination across land, air, sea, space, and cyber domains with quantum-assured confidentiality. Several defence establishments have already begun experimenting with QKD-enabled infrastructure. China's *Micius* satellite has demonstrated quantum key exchange over thousands of kilometres via space-based links, and NATO's Science and Technology Organization is actively assessing QKD for future coalition networks. India, too, has begun laying the groundwork through DRDO and ISRO's QKD trials over terrestrial fibre networks. As the quantum arms race accelerates, QKD will serve as the cryptographic backbone for future multi-domain operations, ensuring that the right orders reach the right units securely, instantly, and without compromise.

Post-Quantum Cryptography: Future-Proof Codes, Defences that Endure

As quantum computing advances, the cryptographic foundations that currently secure military communications, weapon systems, and classified networks face an unprecedented threat. Post-Quantum Cryptography (PQC) refers to a new class of cryptographic algorithms specifically designed to resist attacks from quantum computers. These are machines capable of breaking widely-used encryption standards like RSA and Elliptic Curve Cryptography (ECC) in a fraction of the time it would take classical systems. Unlike QKD, which leverages quantum physics, PQC operates on classical hardware but is mathematically fortified to withstand quantum threats. Prominent PQC families include lattice-based cryptography, hash-based signatures, and code-based encryption. Each of these PQC families are designed to remain secure even against adversaries with large-scale quantum computing capabilities.

From a military perspective, adopting PQC is akin to upgrading from conventional padlocks to quantum-proof vaults anticipating that tomorrow's adversaries will wield siege weapons capable of penetrating today's best defences. It is like reinforcing hardened bunkers, not because the enemy is already at the gate, but because intelligence confirms they are building a weapon that will render existing bunkers obsolete. Similarly, military-grade encryption protocols, whether embedded in command-and-control systems, satellite links, ISR data transfers, or secure tactical radios must evolve now to protect data that remains sensitive for decades.

Operationally, the shift to PQC requires proactive modernization of military digital infrastructure. Communications and logistics networks must be updated, legacy encryption phased out, and future systems built with post-quantum resilience in mind. Transition planning must also address the threat of "harvest-now, decrypt-later" attacks, where adversaries store encrypted information today with the intent to decrypt it once quantum capability matures. For defence planners, this means not just replacing cryptographic algorithms but reshaping procurement, doctrine, and training to ensure long-term information dominance and cyber security continuity. PQC readiness must be treated as a strategic imperative on par with force modernization because in the age of quantum warfare, the side with secure communications controls the battle space.

Quantum Eyes on the Battlefield: Next-Generation Detection, Navigation, and Awareness

"Unveiling the Unseen for Tactical Precision and Strategic Superiority"

Quantum Navigation: GPS Independence, Precision hat Prevails

Modern military operations are highly dependent on GPS for precision navigation, targeting, and synchronization. However, this dependence introduces a critical vulnerability. Adversaries capable of jamming, spoofing, or disabling satellite signals can severely degrade combat effectiveness, especially in contested or denied environments. To overcome this, quantum navigation technologies are emerging as game-changers. Quantum accelerometers and gyroscopes, leveraging atom interferometry, measure motion and position using the quantum properties of ultra-cold atoms. These systems do not rely on external signals, making them immune to traditional electronic warfare (EW) tactics or satellite denial. Essentially, quantum navigation functions as a self-contained, highly accurate inertial navigation system (INS) capable of sustaining precision over long durations without GPS updates.

A relevant military parallel is the inertial navigation system employed by submarines, which must maintain navigational accuracy for extended durations while submerged and without reliance on satellite-based signals. Quantum navigation systems advance this concept significantly by employing quantum accelerometers and gyroscopes to deliver exceptionally high precision accurate to the scale of centimetres for terrestrial, aerial, maritime, and missile platforms. These systems function as self-contained, non-emitting navigation solutions that are inherently resistant to electronic warfare tactics such as jamming and spoofing. They provide assured positioning, navigation, and timing (PNT) capabilities in GPS-denied or degraded environments, including contested electromagnetic spectrum zones, dense urban terrain, and subterranean operational theatres.

Fig. 1.12: Quantum Navigation: Precision Positioning Without GPS

Globally, leading defence agencies are investing heavily in this capability. The UK's Defence Science and Technology Laboratory (DSTL), in partnership with the University of Birmingham and other research bodies, is pioneering a "quantum compass" for maritime and airborne platforms. This device uses quantum accelerometers to allow naval and aerial assets to navigate without GPS, enhancing autonomy and survivability. In the USA, DARPA's Quantum-Assisted Sensing and Readout (QuASAR) and Quantum Inertial Navigation Systems (QINS) programs are advancing deployable quantum INS systems for aircraft and long-range missile platforms, focused on enhancing accuracy and miniaturization for field deployment. These initiatives indicate a shift from lab demonstrations to battlefield readiness within the next operational cycle.

Operationally, the implications are profound. Quantum navigation allows for precision missile strikes in GPS-denied zones, enables autonomous unmanned systems to operate behind enemy lines, and enhances command-and-control mobility in EW-heavy environments. It also reduces reliance on vulnerable satellite constellations, offering strategic resilience in space-contested scenarios. As defence forces modernize, quantum INS must be integrated into future navigation architectures, joint operational doctrines, and platform upgrade paths. Just as night vision once redefined night combat, quantum navigation will redefine manoeuvre warfare in electromagnetic and denied-access theatres, ensuring mission assurance regardless of signal availability.

Quantum Radar: Stealth Broken, Detection Reinvented

Modern battle spaces are increasingly becoming dominated by low-observable platforms such as stealth aircraft, autonomous drones, and hypersonic glide vehicles. With the conventional radar systems facing growing limitations in detection and tracking, quantum radar emerges as a shift in technology that exploits the principles of quantum entanglement and quantum illumination to overcome these limitations. Unlike classical radar, which relies on the transmission and reception of microwave pulses reflected off a target, quantum radar utilizes pairs of entangled photons specifically signal and idler photons generated via spontaneous parametric down-conversion (SPDC). The signal photon is transmitted toward the target while the idler photon is retained in a secure reference channel. Even when the signal photon is scattered or absorbed, any reflected component retains a measurable quantum correlation with its idler counterpart.

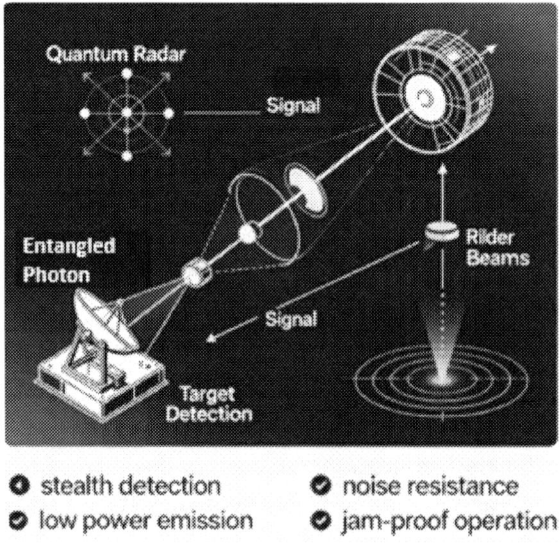

stealth detection   noise resistance
low power emission   jam-proof operation

Fig. 1.13: Quantum Radar: How It Works and Why It Wins

This correlation allows the system to distinguish true signal returns from thermal noise and electronic countermeasures with significantly enhanced sensitivity and noise rejection. Traditional radar systems are vulnerable to jamming, clutter, and stealth technologies that reduce radar cross-section (RCS). In contrast, quantum radar exploits the fact that quantum correlations decay in a predictable manner when disturbed by noise, allowing for superior discrimination of legitimate targets even those designed to minimize electromagnetic signatures. This makes it especially potent in electronic warfare (EW) environments, where adversaries actively attempt to blind or spoof conventional sensors.

Quantum radar does not rely on high-energy emissions and echo detection alone but instead "interrogates" the space with quantum-linked photons, enabling passive and covert target acquisition. Moreover, it is conceptually akin to sonar in underwater operations, detecting targets through signal correlation even when reflection strength is weak or suppressed. However, unlike sonar, which suffers from resolution loss in complex acoustic environments, quantum radar benefits from entanglement-enhanced resolution and reduced false-alarm rates due to its quantum statistical filtering mechanisms.

Ongoing research by defence institutions such as U.S. DARPA and other agencies launched programs in quantum radar are exploring ways for quantum technology to improve detection of radio-frequency signals. The Chinese CETC (China Electronics Technology Group) is reportedly building an experimental quantum radar. These efforts underscore the strategic race to operationalise quantum radar. Experimental demonstrations such as those conducted by the University of Waterloo's Institute for Quantum Computing and China's claimed prototype test over 100 km. This highlights the technology's potential viability in near-peer conflict scenarios. While current systems may be limited in range and require cryogenic conditions for photon management, advances in photonic integration and quantum transceiver miniaturization are rapidly closing the gap towards field-deployable systems.

Operationally, the deployment of quantum radar has profound implications for air defence and situational awareness. In a contested battle space where adversaries employ electronic warfare (EW), jamming, or radar-evading platforms, quantum radar offers a resilient alternative that can maintain tracking fidelity. It can provide early warning against low-RCS (radar cross-section) threats, bolster naval and airborne surveillance coverage, and enhance missile defence accuracy. Integrating quantum radar into layered defence networks can bridge current detection gaps, particularly against hypersonic glide vehicles and autonomous aerial systems operating at low altitude or high speed. As such, quantum radar is not merely a technological upgrade it is a transformational capability that redefines the future of sensor superiority and threat detection in modern warfare.

Quantum Sensors: Instant Awareness, Tactical Dominance

Quantum sensors offer a transformative leap in battlefield sensing capabilities by leveraging fundamental quantum mechanical properties such as superposition, entanglement, and spin state sensitivity. These sensors can detect minute variations in gravitational, magnetic, or electric fields with levels of precision orders of magnitude beyond conventional technologies. For instance, quantum gravimeters can identify the presence of underground voids or tunnels by sensing minuscule shifts in the local gravitational field, while quantum magnetometers can detect the magnetic signatures of concealed vehicles or weapon systems. This capability enables not only detection but real-time tracking of hidden threats, even when visual, thermal, or acoustic cues are absent or masked.

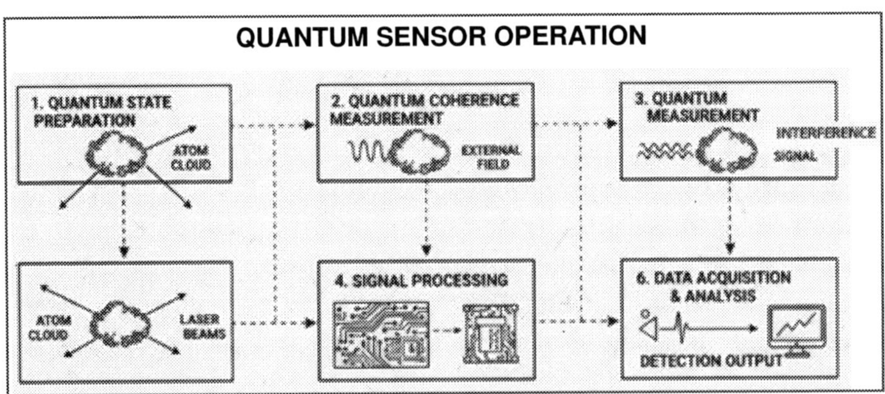

Fig. 1.14: Quantum Sensors: Detecting the Invisible Through Atomic Precision

A practical military analogy is that of deploying "quantum tripwires" sensor systems so sensitive that even the smallest enemy movement, disruption, or subterranean excavation triggers an alert. It is as if every commander and operator were equipped with a sixth sense a non-line-of-sight capability to perceive battlefield anomalies without direct exposure. Just as early radar systems gave navies the edge in maritime surveillance during World War II, quantum sensors could give ground and special operations forces the equivalent edge in 21st-century conflicts particularly in environments where conventional ISR assets are blinded or limited by terrain, cover, or electromagnetic interference.

From an operational standpoint, the implications are significant. Quantum sensors enable early detection of underground facilities, tunnel networks used for asymmetric operations, or the emplacement of improvised explosive devices (IEDs) without the need for intrusive ground-penetrating operations. In intelligence, surveillance, and reconnaissance (ISR) roles, these sensors offer passive, hard-to-detect alternatives to active radar or acoustic probes crucial in stealth-sensitive missions. Moreover, quantum sensing can enhance perimeter security, monitor troop or vehicle movements in contested areas, and provide real-time threat mapping in dense urban settings or dense foliage where GPS, thermal imaging, or radio frequency-based systems are constrained.

With military research initiatives from the U.S. Army Research Laboratory, UK Quantum Technology Hubs, and DARPA pushing the boundaries of deployable quantum gravimeters, magnetometers, and field-sensing arrays, quantum sensors are poised to become integral components of next-generation ISR architectures. Their integration into tactical units, autonomous platforms, or stationary surveillance systems could redefine the tempo and effectiveness of both offensive and defensive operations across all domains of warfare.

Quantum Warfare: Strategic Implications and Ethical Considerations

Shaping the Future of Conflict with Precision, Power, and Ethical Intelligence

Quantum Disruption in Strategy: New Physics, New Doctrines

Quantum technologies represent not merely an advancement in capabilities but a fundamental shift in the character of warfare. Unlike conventional technologies that deliver incremental improvements, quantum systems disrupt foundational assumptions that have underpinned strategic planning for decades. They offer capabilities that were previously considered either infeasible or firmly within the domain of science fiction such as unbreakable quantum communications, undetectable submarine tracking, non-jammable inertial navigation, and ultra-precise sensing in contested environments. These technologies compress the decision-making window by providing near-instantaneous, highly secure information exchange, while simultaneously enabling novel operational reach into denied, stealth-shielded, or electromagnetically silent domains. The resulting effect is the emergence of strategic "black box" capabilities that may be invisible to adversaries until they are decisively applied.

From a doctrinal perspective, quantum technology is as transformational as the introduction of nuclear weapons or satellite-based navigation. It is not simply a new tool in the war fighter's kit. It offers a tectonic shift that alters the rules of engagement and the structure of deterrence. A historical analogy would be the deployment of radar during World War II or reconnaissance satellites in the Cold War technologies that fundamentally redefined who controlled the air and space domains. In the quantum era, control of these new technologies will determine who sets the tempo of war fighting across land, sea, air, cyber, and space. Quantum supremacy in defence is poised to redraw the balance of power in much the same way GPS transformed precision warfare and nuclear weapons redefined strategic deterrence.

The operational implications of quantum disruption are both immediate and profound. Armed forces must begin rethinking force structure, command architectures, and strategic doctrines to reflect the capabilities and vulnerabilities introduced by quantum systems. Traditional concepts such as Mutually Assured Destruction (MAD), first-strike deterrence, and escalation ladders must be re-evaluated in the light of quantum-enhanced sensing, communications denial, and counter-stealth operations. Moreover, the risk of strategic surprise looms large: an adversary achieving first deployment of operational quantum capabilities such as untraceable underwater navigation, detection of stealth assets, or the interception of classical encryption could temporarily blind, misdirect, or paralyze traditional forces before kinetic engagement even begins. To mitigate such risks, militaries must invest not only in quantum R&D and capability acquisition but also in the equally critical domain of quantum-aware doctrine development and war gaming.

Quantum Autonomy: Machines that Choose, Faster than Thought

Quantum autonomy marks a paradigm shift in military decision-making, leveraging the unique capabilities of quantum computing to dramatically accelerate and enhance autonomous systems. At its core, quantum computing operates using qubits units of quantum information that exist in multiple states simultaneously due to the principle of superposition. This enables quantum processors to perform quantum parallelism, evaluating a vast number of potential outcomes at once. When applied to autonomous military platforms, this allows machines to process complex, high-dimensional battlefield data and generate optimal decisions in real time, even under conditions of uncertainty, degraded communication, or adversarial deception. Unlike classical AI systems limited by sequential logic and finite states, quantum-enhanced autonomy brings exponential gains in decision-making speed, adaptability, and foresight.

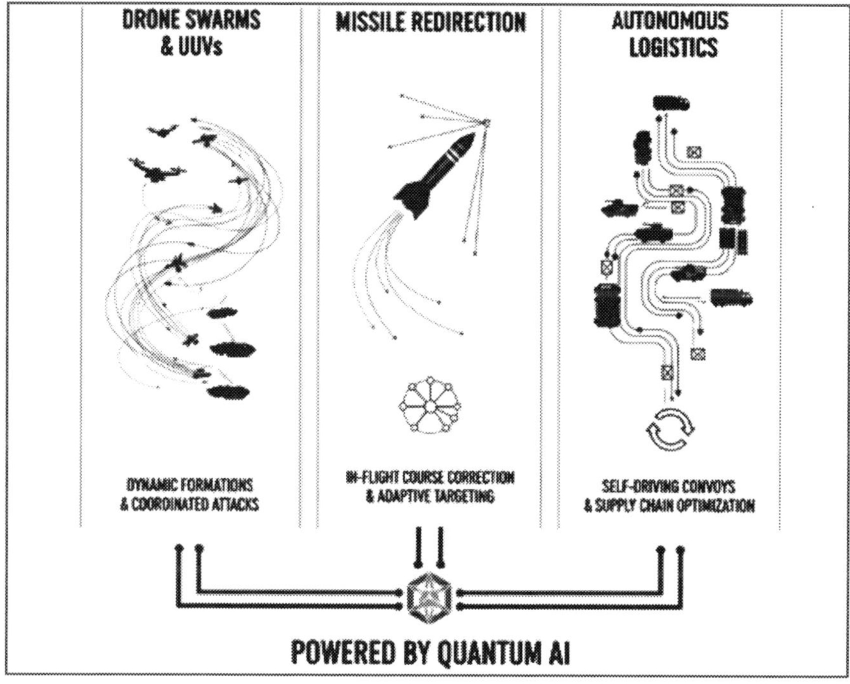

Fig. 1.15: Quantum Autonomy: Intelligent Swarms, Smart Missiles, Adaptive Logistics

This capability is not just theoretical. The U.S. Defence Advanced Research Projects Agency (DARPA) is advancing this vision through programs like ONISQ (optimisation with Noisy Intermediate-Scale Quantum Devices) and Quantum Benchmarking, both designed to explore real-world use cases of quantum computing in mission planning, logistics, and autonomous manoeuvring. Similarly, the UK Defence Science and Technology Laboratory

(DSTL), in partnership with companies like Riverlane and Oxford Quantum Circuits, is investing in quantum-enhanced AI for defence decision loops, demonstrating the growing international push to operationalise quantum autonomy. Quantum autonomy thus blends the foresight of strategic planning with the agility of frontline decision-making, embedded in machines that function at a computational speed far beyond that of traditional processors or human cognition. Most critically, quantum autonomy enables the compression of the OODA loop (Observe–Orient–Decide–Act).

Operationally, the implications span all war fighting domains:

- Air and Naval Combat: Swarms of autonomous drones and unmanned underwater vehicles (UUVs) could dynamically reconfigure formations, evade threats, and engage targets without needing centralized control, all powered by quantum-accelerated decision engines. Programs like DARPA's OFFSET and the U.S. Navy's Ghost Fleet Overlord point toward this future.

- Missile Systems: Quantum-enhanced control logic could enable missiles to reassess and redirect mid-flight based on evolving target profiles or intercept threats using intelligent evasion algorithms.

- Logistics and Sustainment: Platforms such as autonomous ground convoys or quantum-informed logistics routing algorithms (as envisioned in ONISQ) could navigate supply chains under fire, optimise medevac routes, and allocate resources during real-time combat scenarios ensuring sustained operational tempo in contested environments.

Quantum Ethics: Invisible Lines, Global Risks

The advent of quantum technologies in military operations introduces a new class of ethical and strategic dilemmas that transcend traditional doctrines. Quantum-enabled systems, whether in sensing, communication, navigation, or decision-making, pose unprecedented challenges to accountability, transparency, and proportionality in warfare. In particular, the integration of quantum computing into autonomous platforms risks delegating life-and-death decisions to systems whose processes are inherently opaque and non-intuitive. The same quantum characteristics that provide operational advantage such as non-determinism, superposition, and entanglement also render it difficult to trace causal chains, evaluate intent, or conduct post-action review, thereby straining the principles of lawful warfare and established rules of engagement.

One pressing concern is the erosion of the "human-in-the-loop" paradigm. Traditional military ethics and legal doctrines emphasize meaningful human control in targeting decisions. However, quantum-augmented autonomy may outpace human ability to monitor or intervene in real time, especially in fast-evolving operational environments. This could lead to delegated authority without clear oversight, challenging not only tactical accountability but also strategic command responsibility. Furthermore, the emergence of quantum supremacy at the point at which quantum systems decisively outperform classical ones could introduce strategic instability by undermining verification, deterrence, and global arms control frameworks.

The operational relevance of quantum ethics is both immediate and strategic. First, there is an urgent need for the development of ethical frameworks and legal doctrines governing the deployment of quantum-augmented autonomous systems. particularly those with kinetic capabilities. This includes clarity on proportionality, non-combatant immunity, and decision accountability. Second, global military stakeholders must engage in diplomatic efforts to define and enforce norms around destabilizing uses of quantum technology such as autonomous kill chains, untraceable communications, or undetectable surveillance systems. Just as treaties exist for nuclear testing and chemical weapons, emerging discussions around "quantum arms control" including potential frameworks for QKD verification, algorithmic transparency, and intergovernmental notification protocols must be advanced before adversaries weaponize ambiguity.

Finally, the burden lies on technologically advanced militaries to lead by example. Building coalitions for responsible quantum military development involving allies, defence labs, and international bodies can help shape norms, reduce strategic miscalculation, and prevent a quantum-fuelled arms race. Without proactive governance, the very attributes that make quantum technologies powerful speed, opacity, and asymmetry could undermine the moral and legal foundations of 21st-century warfare.

1.3   QUANTUM DEFENCE DOCTRINE SHAPING THE FUTURE BATTLEFIELD FOR INDIA

Quantum Information Operations: Digital Battlefield in the Post-Classical Era

Outpacing cyber adversaries Mastering the Art of Cyber Warfare with Unbreakable Intelligence

Quantum Information Warfare: Decode Faster, Deceive Smarter

Quantum computing is poised to radically enhance the effectiveness and speed of military information operations. In the context of information warfare, this translates into accelerated data analytics, cryptanalysis, natural language processing (NLP), and pattern recognition capabilities. Military intelligence units could use such capabilities to rapidly decrypt enemy communications, analyse intercepted traffic, or detect fabricated media including AI-generated misinformation and deep fakes.

Beyond passive defence, quantum-enhanced tools can power cognitive countermeasures that neutralize adversarial information campaigns. These include detecting sentiment manipulation across digital platforms, exposing synthetic content, and filtering adversarial propaganda embedded in social media or open-source information. Offensively, quantum tools may one day enable the design of advanced influence operations that exploit algorithmic vulnerabilities in adversary information ecosystems constituting a new tier of cognitive warfare.

A powerful military analogy for quantum-enhanced information warfare is the deployment of a superhuman code-breaker on the battlefield one capable of instantly decrypting encrypted enemy signals, uncovering hidden narratives, and delivering precisely calibrated counter-information before the adversary's message gains traction. This represents a quantum leap from the traditional capabilities of electronic warfare and cyber-intelligence units.

For Indian armed forces quantum-enhanced information warfare presents both a necessity and an opportunity. Firstly, quantum-accelerated SIGINT platforms can significantly reduce the time required to decrypt, analyse, and exploit enemy communications, particularly in high-stakes border standoffs or during transitory peace-time operations along the Line of Actual Control (LAC) or Line of Control (LoC). Secondly, India's adversaries have already been investing in AI-driven influence operations, disinformation campaigns, and cyber intrusions. Quantum-enabled AI tools could form the backbone of a national defence framework against deep fakes, psychological operations, and cognitive disruption. Such systems would allow real-time identification of narrative manipulation, especially during times of internal unrest, border provocations, or electoral interference.

Operationally, this capability fits seamlessly within India's evolving doctrine on Information Dominance and Net-Centric Warfare. It strengthens tri-service fusion centres and supports joint cyber operations commands that require speed, precision, and predictive situational awareness. In hybrid warfare environments, particularly in contested maritime zones in the Indian Ocean Region (IOR) or sensitive border areas, quantum-enabled ISR (Intelligence, Surveillance, and Reconnaissance) systems can filter genuine intelligence from noise ensuring commanders act on truth, not deception.

Quantum Arms Race: Strategy, Readiness, and Ethics in the New Era of Military Power

Racing to redefine military superiority

Quantum Arms Race: Deterrence Rewritten, Alliances Reshaped

The quantum arms race represents a transformative shift in global military dynamics, reshaping deterrence, operational doctrines, and alliance structures. Unlike previous technological evolutions, quantum technologies redefine strategic advantage without triggering kinetic escalation. The ability to decrypt enemy communications instantaneously, neutralize stealth technologies, or ensure impenetrable command-and-control (C2) networks creates a new dimension of warfare where superiority in the "invisible layer" of the battle space determines success in the visible one. This competition is no longer confined to laboratories it is a strategic priority for major powers. The USA, China, the European Union, India, and others have launched national quantum missions, investing billions to secure quantum dominance. For India, programs such as the National Mission on Quantum Technologies and Applications (NM-QTA) aim to ensure that the Indian military is not left behind in this epochal shift. Quantum supremacy will re-order strategic equations. In this new race, the ability to "own" the quantum layer undetectable sensing, unbreakable communications, and superior computational power means controlling the tempo, the narrative, and the outcomes of future conflicts. It's akin to commanding the electromagnetic spectrum or satellite space: whoever dominates this invisible layer holds decisive battlefield leverage.

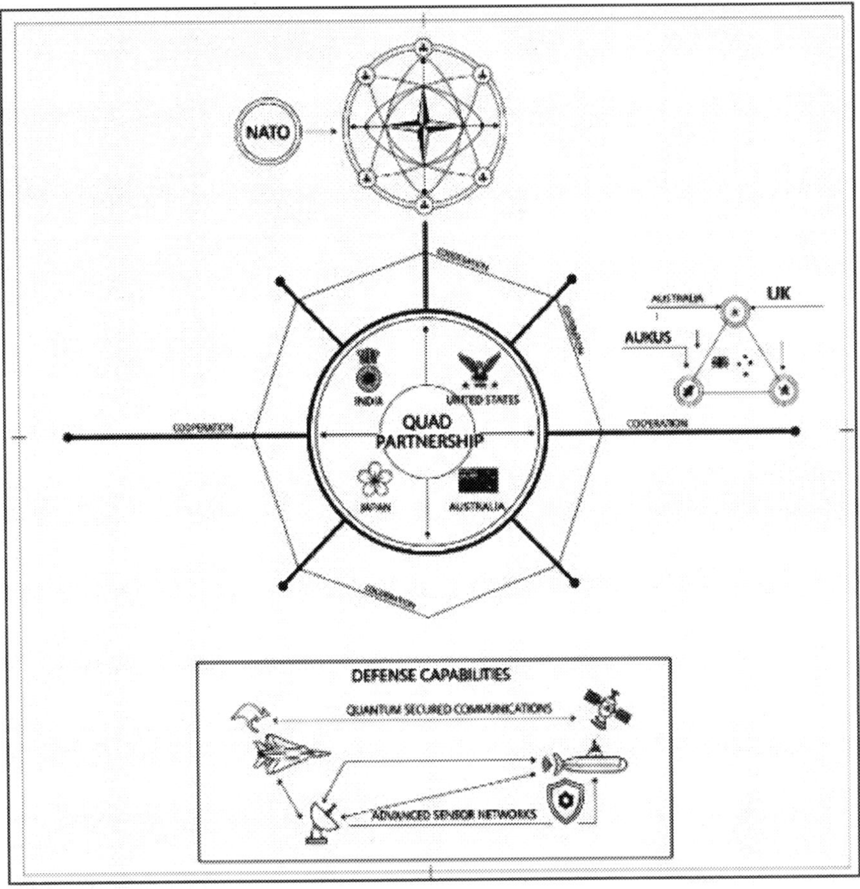

Fig. 1.16: Strategic Quantum Partnerships: Securing India's Quantum Edge

Quantum arms race demands doctrinal shifts and renewed strategic planning. Early warning systems, force projection models, and cyber defences must all adapt to account for quantum-enabled capabilities. In alliance frameworks such as NATO, the QUAD, or AUKUS, quantum cooperation will become a litmus test for technological trust and interoperability. For India, strategic collaboration in quantum research, particularly with QUAD partners like the USA., Japan, and Australia will be critical to balancing regional threats and technological asymmetries.

Quantum Readiness: Train the Force, Adapt the Mission

The accelerating integration of quantum technologies into military operations demands a fundamental shift in how forces are trained, organized, and led. Quantum readiness is not simply about acquiring new hardware; it is about cultivating a mindset and doctrine that can harness the strategic implications of quantum-enhanced capabilities. At its core, quantum readiness involves ensuring that commanders and staff at all levels possess basic quantum literacy. This encompasses an understanding of how quantum sensors, communication systems, navigation tools, and computational engines work and how adversaries may use them. As systems like C4ISR (Command, Control, Communications, Computers, Intelligence, Surveillance, and Reconnaissance) and ISR platforms embed quantum capabilities, military forces will need to evolve operational planning, simulation environments, and capability development to keep pace with this disruptive shift.

Operational relevance for Indian military modernization is significant. Quantum-secure communications will need to be embedded in tri-service networks, especially along sensitive border regions. Quantum sensors could revolutionize early warning systems in the Andaman & Nicobar Command or detect subterranean threats in border areas like Jammu & Kashmir or Arunachal Pradesh. To prepare, all Indian Professional Military Education (PME) institutions must introduce quantum technologies in their syllabi not just as technical curiosities but as mission-critical capabilities. Training simulations and war games should model quantum-enabled adversaries to stress-test current doctrines, while red-teaming exercises can help forces anticipate strategic surprise. Without quantum readiness, future forces risk being technologically outpaced and strategically blindsided. With it, they stand to define the new war fighting doctrines in a quantum-dominated battle space. At this juncture, Gen Alpha is growing up with AI and will see quantum technologies being adopted. This means that for these young people, learning about quantum tech is easy, since it is pedagogical. However, for those serving in the military at the time of writing, getting a radically new idea in involves andragogy, or adult education. This is not without challenge.

Quantum Rules of War: Responsibility in the Shadow Domain

As quantum technologies advance into the military domain, they introduce complex ethical and legal challenges that existing frameworks are ill-equipped to address. Quantum warfare blurs traditional lines between aggression and espionage, offence and defence, and critically between human accountability and machine autonomy. Quantum-enabled systems, particularly those involving autonomous decision-making, can execute operations without direct human oversight, raising serious concerns about proportionality, attribution, and the legitimacy of targeting. Furthermore, the inherently covert nature of quantum technologies such as undetectable communication or passive sensing complicates transparency and heightens the risk of unintended escalation. Existing legal frameworks, including the Geneva Convention and the Laws of Armed Conflict (LOAC), were crafted in an era when the concept of machines deciding acts of war, or weapon systems operating within imperceptible domains, was purely theoretical. Quantum now forces these theories into reality. Quantum warfare is following a comparable path; that of the development of anti-satellite (ASAT) weapons and stealth aircraft which introduced game-changing capabilities that were not explicitly illegal but fundamentally altered the diplomatic and strategic landscape. The mere existence of Quantum technology will transform the rules of competition and conflict, creating a "shadow domain" where the absence of attribution or traceability challenges traditional doctrines of deterrence and retaliation.

Operationally, this places a new burden on military officers. Beyond tactical expertise, commanders and planners must be familiar with emerging norms governing the ethical use of quantum-enhanced capabilities. Training in LOAC and international humanitarian law must evolve to include quantum-specific scenarios such as non-kinetic disruption of critical infrastructure, quantum-deceptive manoeuvres, or autonomous kill-chain decisions powered by quantum AI. Rules of engagement (ROE) must anticipate operations conducted without physical violence but with potentially strategic impact such as blinding ISR satellites via quantum jamming or interfering with command networks using quantum algorithms. India, in particular, as a rising military power with a clear stake in responsible tech use, must actively participate in international dialogues on quantum ethics and arms control. Coordination between the armed forces, legal advisors, diplomats, and AI ethics experts will be essential to develop national guidelines that balance strategic advantage with adherence to global norms before adversaries set their own rules in the quantum shadow.

Chapter Two

Quantum Computing Revolutionising Warfare

"Quantum Computing: Revolutionising Warfare," provides the reader with a clear, mission focused understanding of how quantum algorithms will reshape operations, intelligence, and decision superiority. It opens with a plain spoken orientation to foundational algorithms—Shor's for cryptanalysis and Grover's for accelerated search—explaining what they do, why they matter, and how they translate into concrete military advantages. Grover's Algorithm is presented as a force multiplier for target discovery, pattern matching, and intelligence triage across vast, noisy datasets. The Quantum Approximate Optimization Algorithm (QAOA) is introduced as a commander's tool for complex optimization under constraints—routing, tasking, scheduling, and strike asset pairing—supporting both deliberate planning and time sensitive, real time decisions at the tactical edge and in joint operations centers.

The chapter then addresses cryptanalysis and intelligence with operational candor. It explains how quantum computing threatens classical public key systems that underpin secure communications, logistics, and weapon system updates, and why a transition to quantum resistant approaches is essential. Readers are guided through the implications for signals security, data at rest, and over the air updates, while also seeing the upside: quantum accelerated data processing to fuse multi INT feeds, surface weak signals in massive collections, and elevate threat detection, indications and warnings, and targeting fidelity.

A dedicated section introduces quantum simulation as a new instrument of operational art. Officers learn how quantum models can represent complex, interacting systems—contested air maritime environments, anti access/area denial architectures, electromagnetic congestion, and attrition dynamics—supporting campaign design and risk analysis. Practical vignettes demonstrate value for naval resource allocation and routing under threat, red blue course of action comparison, and predictive analytics that anticipate adversary moves, logistics choke points, and escalation pathways. As hardware and software mature, these capabilities promise sharper estimates, faster planning cycles, and more agile re tasking under fog and friction.

The chapter concludes with applications to war gaming and strategic forecasting, translating theory into commander ready tools. Quantum driven simulations increase realism, stress test concepts of operation, and provide decision support for naval and joint command structures through scenario based optimization of force packages, sensor shooter pairing, and sustainment. The end state for the reader is practical: an appreciation of where quantum delivers advantage today, a roadmap for near term adoption alongside classical high performance computing and AI, and concrete implications for Indian force development, training, and interoperability—so that readiness and strategic dominance are preserved as the battlespace evolves.

2.1 INTRODUCTION TO QUANTUM ALGORITHMS FOR MILITARY APPLICATIONS

From digital trenches to quantum command redefining the rules of military computation

Quantum computing is not just a leap in raw processing power; it brings in a paradigm shift in how problems are approached and solved. At the core of this revolution are quantum algorithms. These algorithms are purpose-built sets of instructions that have been designed to harness the unique capabilities of qubits, such as superposition and entanglement, to solve problems that are computationally challenging enough for classical machines. Understanding these algorithms is akin to learning a new form of battlefield doctrine, one that operates in probabilistic logic rather than deterministic order.

In contrast to traditional battlefield mapping, which requires sequential assessment of scenarios, quantum algorithms operate within a probabilistic computational space enabling the simultaneous evaluation of multiple possibilities. While classical algorithms solve problems step by step, akin to a reconnaissance unit methodically clearing buildings one at a time, quantum algorithms function more like a highly coordinated special operations team capable of scrutinizing all targets concurrently. This capability significantly reduces the time required to identify high-value objectives.

Differences Between Classical and Quantum Computing in Military Contexts

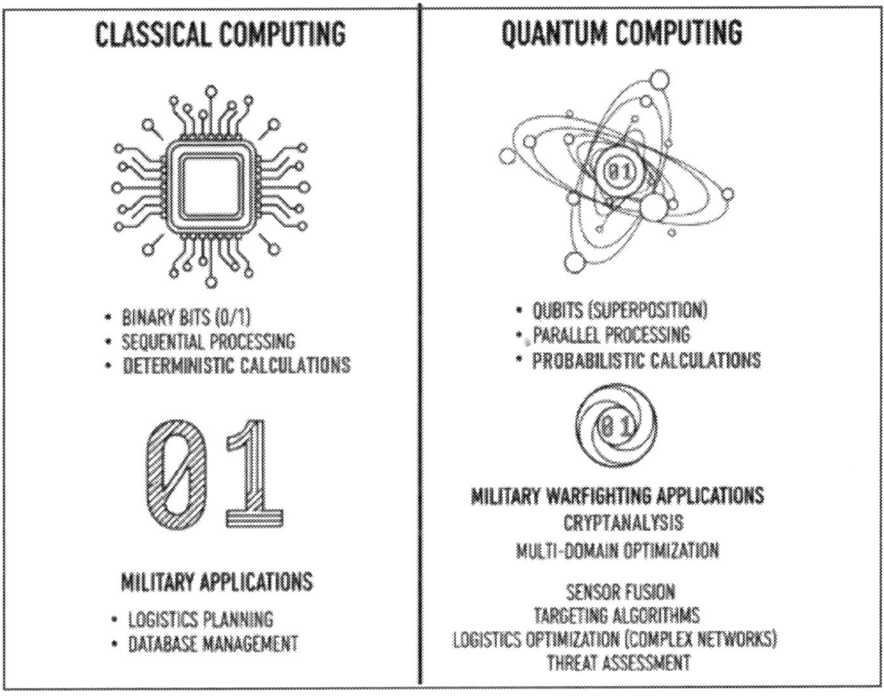

Fig. 2.1: Beyond Binary: Quantum Computing's Warfighting Edge

Classical vs. Quantum Computing

Aspect	Classical Computing	Quantum Computing
Basic Unit of Data	Bit (0 or 1)	Qubit (0, 1, or both at once through superposition)
Processing Model	Deterministic (step-by-step)	Probabilistic (parallel state evolution)
Military Analogy	Infantry clearing a city block-by-block	Special forces simultaneously surveying all buildings via advanced drone swarm
Speed for Complex Problems	Exponential time for large-scale simulations or cryptographic tasks	Polynomial or near-instantaneous time for certain tasks (e.g., factoring, searching)

Aspect	Classical Computing	Quantum Computing
Use Cases	Routine data processing, legacy systems, logistics, tactical-level C2	High-dimensional war gaming, rapid ISR analysis, advanced logistics, quantum cryptography
Encryption/Cyber Impact	Secures data using RSA, ECC vulnerable to future quantum attacks	Breaks existing encryption (via Shor's Algorithm); enables unbreakable quantum communication (via QKD)
Simulation Capability	Limited in modelling quantum physics, complex chemical/nuclear systems	Simulates chemical reactions, weapons effects, and sensor behaviour with atomic-scale precision
Scalability for Planning Ops	Struggles with many-variable planning (e.g., multi-domain, multi-theatre operations)	Handles exponential complexity with ease ideal for scenario planning and logistics optimization
Communication and Sensing	Classical encryption and RF-based sensor data fusion	Entanglement-based communication; quantum radar and sensing for stealth/anti-stealth
Technology Maturity	Mature and widely deployed	Emerging but accelerating; strategic advantage for early adopters

Quantum computing represents a paradigm shift in how information is processed, with profound implications for both military operations and national defence. By processing information in parallel and at scale, these systems can accelerate situational awareness and enable more informed, and on time decision-making in dynamic threat environments.

Quantum computing offers transformative potential in solving computationally intensive problems that classical systems find increasingly challenging. Tasks such as route optimization for logistics in contested zones, or deciphering encrypted enemy communications, grow exponentially more complex as operational variables increase. Quantum algorithms explore massive solution spaces simultaneously, to handle these complex problems. In real-world terms, this capability enables rapid and precise optimization of resources such as fuel, munitions, and personnel deployments across fluid operational theatres providing a clear tactical edge in fast-paced, multi-domain environments.

Quantum computing also challenges the foundations of current military cyber security. Classical encryption protocols like RSA and ECC depend on the computational difficulty of factoring large integers. Quantum algorithms, most notably Shor's algorithm, can solve these problems exponentially faster, rendering many current encryption methods vulnerable. This presents a strategic threat often termed the "crypto-apocalypse," where adversaries equipped with quantum capabilities could compromise secure military communications. To safeguard operational integrity, defence institutions must proactively adopt quantum-resilient cryptographic frameworks, including post-quantum cryptography and quantum key distribution (QKD), which offer forward-looking, robust protection against emerging cyber threats.

Another critical area where quantum computing excels is in simulation. Classical machines struggle to model the intricacies of quantum systems, but quantum computers naturally simulate phenomena such as nuclear reactions, electromagnetic interference, and material behaviours at atomic precision. These capabilities offer a decisive advantage by enabling the virtual testing of stealth materials, evaluating the effects of electromagnetic pulse (EMP) attacks, and modelling the dispersion of biochemical agents, all without resorting to costly or hazardous physical trials. As a result, quantum simulations can significantly accelerate the R&D cycle and improve the efficacy and safety of emerging defence technologies.

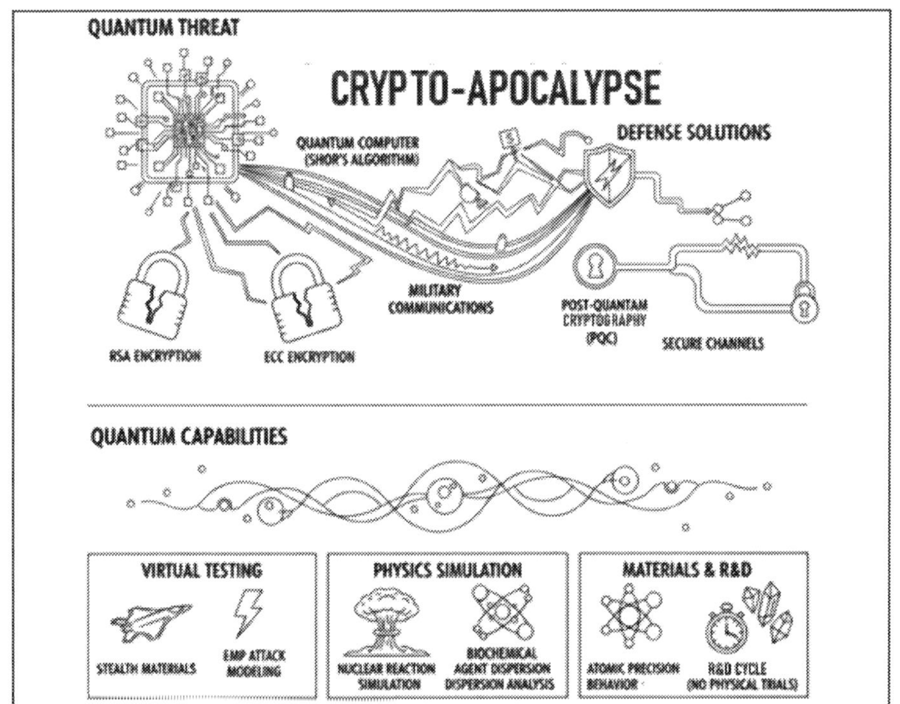

Fig. 2.2: Quantum's Double Edge: Breaking Codes, Accelerating Innovation

Complementing quantum computing, quantum sensors offer another frontier of capability. Unlike traditional sensors, quantum sensors can detect signals at extremely low energy thresholds, allowing them to bypass conventional stealth and jamming counter-measures. When integrated with quantum processors, these sensors enhance threat detection by delivering faster, more precise assessments across diverse operational environments. Their military relevance is substantial: technologies such as quantum radar and entanglement-based sensing hold the potential to detect previously undetectable threats, including stealth aircraft and submerged submarines. This capability could fundamentally reshape the doctrine of Anti-Access/Area Denial (A2/AD) operations and elevate situational awareness across air, sea, and land domains.

2.2   QUANTUM ALGORITHMS FOR WARFARE

Quantum Algorithms: The Strategic Arsenal of the Next Digital Battlefield

"From cracking codes to tracking threats – quantum algorithms unlock new dimensions of strategic superiority"

Quantum algorithms lie at the heart of quantum computing's disruptive power. Unlike conventional algorithms that process data linearly, quantum algorithms exploit the unique properties of qubits such as superposition and entanglement to solve certain classes of problems exponentially faster than classical computers. In a military context, this translates into the ability to break encryption once thought invulnerable, accelerate target acquisition cycles, and optimize decision-making in high-stakes, time-sensitive operations. Two quantum algorithms Shor's and Grover's are particularly significant due to their direct implications for military cyber security and intelligence, surveillance, and reconnaissance (ISR).

Shor's Algorithm: Cracking the Uncrackable

Shor's Algorithm, created by Peter Shor in 1994, is a major breakthrough in quantum computing because it can break modern encryption methods. Most online security systems rely on very large numbers that are created by multiplying two prime numbers together. Finding these two prime numbers is extremely difficult for normal computers, which is why encryption is secure.

But Shor's Algorithm makes this process much faster using quantum computing, meaning it could break encryption that protects sensitive data worldwide. This is why governments and cyber security experts are racing to develop new, quantum-resistant encryption methods before quantum computers become powerful enough to crack current security systems.

Conventional public-key cryptographic protocols, most notably RSA (Rivest–Shamir–Adleman) and Elliptic Curve Cryptography (ECC), depend on the computational difficulty of factoring large numbers or solving discrete logarithm problems. These problems are considered "computationally hard" for classical computers, meaning that no efficient algorithm is known to solve them in a reasonable amount of time as the key size increases. For example, factoring a 2048-bit RSA key using the most advanced classical methods would take thousands of years using today's supercomputers, thereby ensuring its practical security.

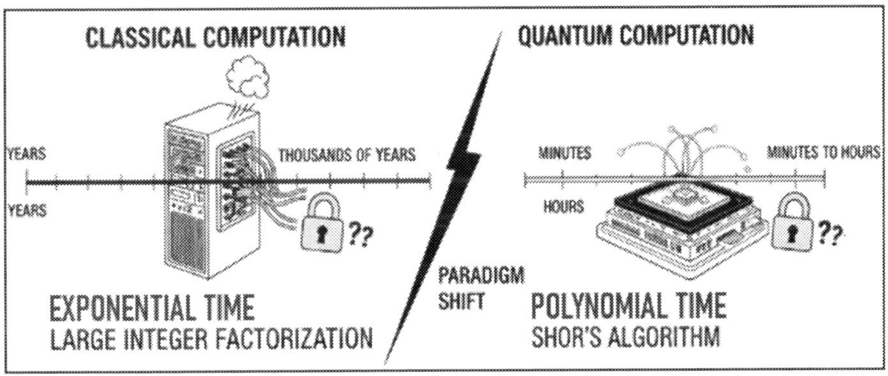

Fig. 2.3: Millennia to Minutes: Shor's Algorithm Shatters Encryption Time

However, Shor's Algorithm changes this paradigm. It operates in polynomial time on a quantum computer. It can solve these problems exponentially faster than classical methods. In concrete terms, what would take millennia for a classical system could be achieved in minutes or hours on a sufficiently powerful quantum computer. This is not merely a theoretical concern. With rapid advancements in quantum hardware, the day when Shor's Algorithm becomes practically executable on real quantum machines is drawing closer.

The military implications of this are profound:

- RSA and ECC-based security protocols, which are commonly used to protect secure communications, battlefield information systems, satellite command links, authentication keys, and even nuclear deterrent infrastructure, would no longer be secure. Once a hostile actor possesses a quantum computer capable of executing Shor's Algorithm at scale, they could decrypt any intercepted communications or files protected under these schemes.

- "Harvest now, decrypt later" strategies become viable. Adversaries can collect encrypted military traffic today, store it indefinitely, and decrypt it retroactively once quantum resources become available. This poses a long-term strategic intelligence threat.

- Cryptographic obsolescence forces a fundamental transition towards quantum-resilient security. This includes implementing post-quantum cryptographic (PQC) algorithms currently being standardized (e.g., lattice-based, hash-based cryptography) and exploring Quantum Key Distribution (QKD) systems, which leverage the laws of quantum mechanics themselves for secure key exchange.

Shor's Algorithm poses a direct and existential threat to classical cryptographic systems that underpin the confidentiality and integrity of secure military operations. Encryption protocols such as RSA and ECC are extensively used across defence communication channels, logistical data networks, command-and-control systems, and even nuclear safeguard mechanisms. Once a quantum computer capable of executing Shor's Algorithm at scale becomes operational, these encryption systems could be compromised almost instantaneously, jeopardizing the security of critical information infrastructures.

The algorithm also introduces a new vector for strategic espionage. Adversaries can engage in what is known as a "harvest now, decrypt later" approach intercepting and storing encrypted military communications today with the intention of decrypting them retroactively once quantum computing capabilities mature. This delayed compromise of sensitive information can have far-reaching implications, including exposure of historical operational plans, intelligence sources, strategic capabilities, and alliance communications.

To mitigate this threat, militaries must initiate an urgent transition towards quantum-resilient cryptographic standards. This includes adopting post-quantum cryptography (PQC) methods such as lattice-based or code-based algorithms that remain secure even in a quantum threat environment. Additionally, emerging technologies like Quantum Key Distribution (QKD) offer the potential for theoretically unbreakable encryption based on the fundamental laws of quantum physics. A proactive shift to these next-generation cryptographic solutions is essential to maintain strategic and information superiority in an era of accelerating quantum advancement.

Grover's Algorithm: Accelerating Search and Decision in the Battle Space

Grover's Algorithm developed by Lov Grover in 1996 was designed to search through an unsorted dataset significantly faster than any classical method. In classical computing, searching an unsorted list of *N* items such as identifying a single target from a massive intelligence database requires, on average, *N/2* steps in the worst case. In contrast, Grover's Algorithm requires only about "N steps to find the desired item. This represents a quadratic speed-up.

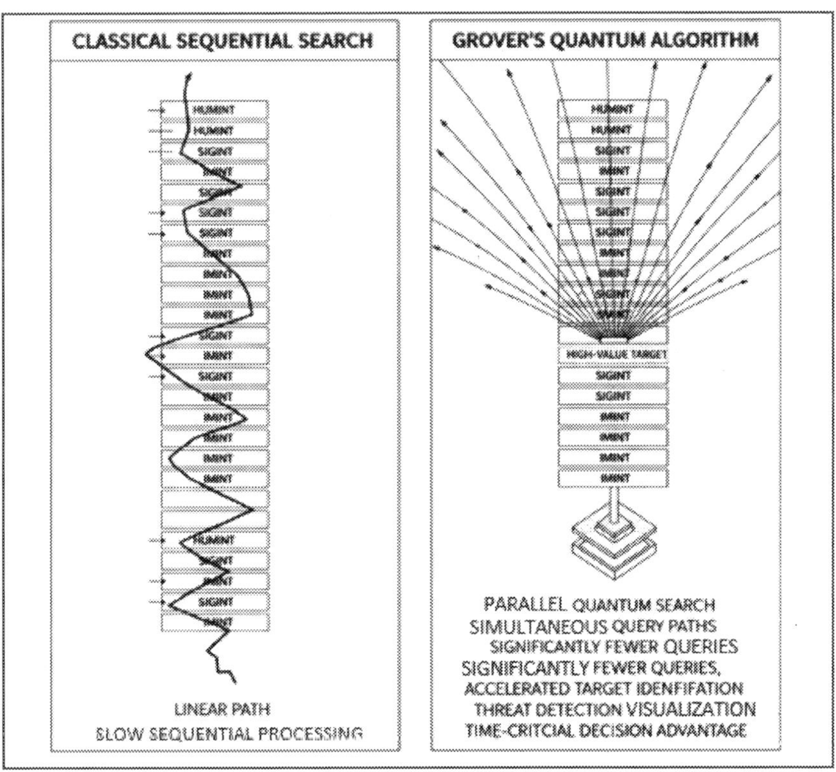

Fig. 2.4: Parallel Intelligence: Grover's Quantum Search Advantage

To understand the significance, *Indicators* are a good example. Imagine a situation where intelligence staff must search through vast streams of data from HUMINT to IMINT in real time. Classical systems must sift through each entry sequentially or rely on probabilistic heuristics, which may miss critical indicators. Grover's Algorithm, however, allows a quantum system to search the entire dataset in parallel and identify the matching entry with significantly fewer queries. This enables in faster identification of high-value targets, quicker threat detection, and more agile operational decision-making in time-critical scenarios. Imagine identifying the Enemy Course of Action (ECOA) 'in play' at a much earlier stage than is possible today. The deployment, redeployment, turnaround decisions and patterns would all change, whilst simultaneously reducing the time between Int refresh cycles.

This is especially useful for maritime planners, given the vast ocean spaces of the Indo Pacific and the staggering number of vessels plying the AO at any given time. While AI can partially achieve similar outcomes, its limitations in exhaustive search and pattern detection could be complemented by quantum approaches such as Grover's Algorithm.

Grover's Algorithm acts like a highly efficient special reconnaissance unit that can sweep through a massive, unstructured battle space and pinpoint specific threats or objectives – be it a hidden IED signature in sensor data, a rogue signal in electronic warfare channels, or the presence of anomalous activity in cyber security logs.

Operationally, the advantages are multifold:

- **Improved Intelligence, Surveillance, and Reconnaissance (ISR):** Enables rapid processing of satellite and UAV imagery to locate targets or changes in terrain.

- **Faster Cyber Threat Detection:** Assists in scanning large volumes of network traffic to identify malicious code or intrusion signatures.

- **Cryptographic Implications:** Although Grover's Algorithm does not break symmetric cryptography outright, it reduces the effective security level by half. For instance, a 128-bit symmetric key (e.g., AES) would offer only 64 bits of security against a quantum adversary using Grover's method necessitating longer key lengths in future-proof systems.

Grover's Algorithm: Accelerating Search and Intelligence Analysis

"Search Power Re-imagined: Square Root Speed-Up in Target Identification"

Speed as a Strategic Weapon in Information Dominance

- **Enhances Identification of Enemy Units or Assets in Satellite/Drone Imagery** – Modern military reconnaissance collects vast volumes of aerial and orbital imagery through high-resolution satellites, drones, and surveillance aircraft. Parsing through this data to locate enemy troop formations, hidden missile systems, or mobile command posts is a labour- and compute-intensive task. Traditionally, image recognition systems scan these feeds sequentially or use heuristics that may miss obscured or camouflaged targets. Grover's Algorithm transforms this challenge by enabling a quantum system to perform a pattern-matching search across the entire imagery database at a significantly reduced computational cost about $\sqrt{N}$ steps for N potential matches. In military terms, this means faster identification of hostile assets hidden within complex terrain or urban cover, increasing the speed and accuracy of target acquisition. This is particularly effective in environments where adversaries use deception, decoys, or rapid mobility to evade detection.

- **Rapid Processing of ISR (Intelligence, Surveillance, Reconnaissance) Data for Time-Sensitive Targeting** – Time-sensitive operations such as decapitation strikes, hostage rescue missions, or neutralization of mobile launchers depend on ISR data being processed in near real-time. However, ISR data includes not just imagery but also signals intelligence (SIGINT), electronic intelligence (ELINT), human intelligence (HUMINT), and open-source intelligence (OSINT), all of which must be searched, validated, and correlated under extreme time pressure. Grover's Algorithm can act as a quantum-accelerated ISR engine, rapidly querying these diverse data streams to locate specific threats, keywords, or indicators (e.g., a unique RF signature, heat profile, or command phrase). By reducing the time to sift through petabytes of data, commanders gain faster sensor-to-shooter linkage, allowing for precise, lawful engagement before the target moves, escapes, or escalates a conflict.

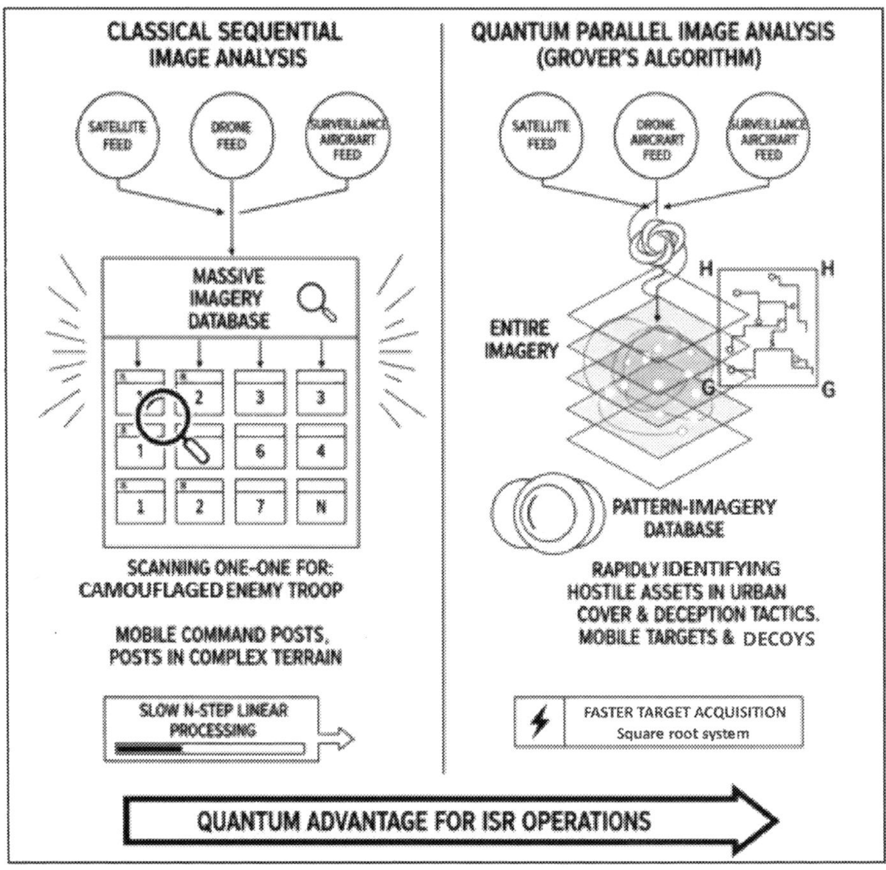

Fig. 2.5: Seeing Through Deception: Quantum-Accelerated Target Recognition

- **Ideal for "Find the Needle" Scenarios in Digital Battlefields** – Digital warfare introduces vast "signal noise" millions of data points across networks, sensors, and cyberspace. A cyber adversary might hide a single line of malicious code in a network, or an enemy submarine might emit one irregular sonar ping in a sea of acoustic clutter.

  Grover's Algorithm is uniquely suited to these "needle in a haystack" problems. By reducing the number of required queries to identify a match, quantum systems leveraging Grover's Algorithm can extract critical anomalies or indicators that signal infiltration, tampering, or covert movement. In EW (Electronic Warfare), for example, identifying a unique radar signature amidst background emissions could mean the difference between ambush and anticipation.

Counter-Insurgency and Threat Pattern Detection

Modern militaries must sift through massive, unstructured datasets to detect the faint signals of insurgent activity, sleeper cell communications, and asymmetric threats. Grover's Algorithm, redefines how quickly intelligence can be transformed into mission-ready action.

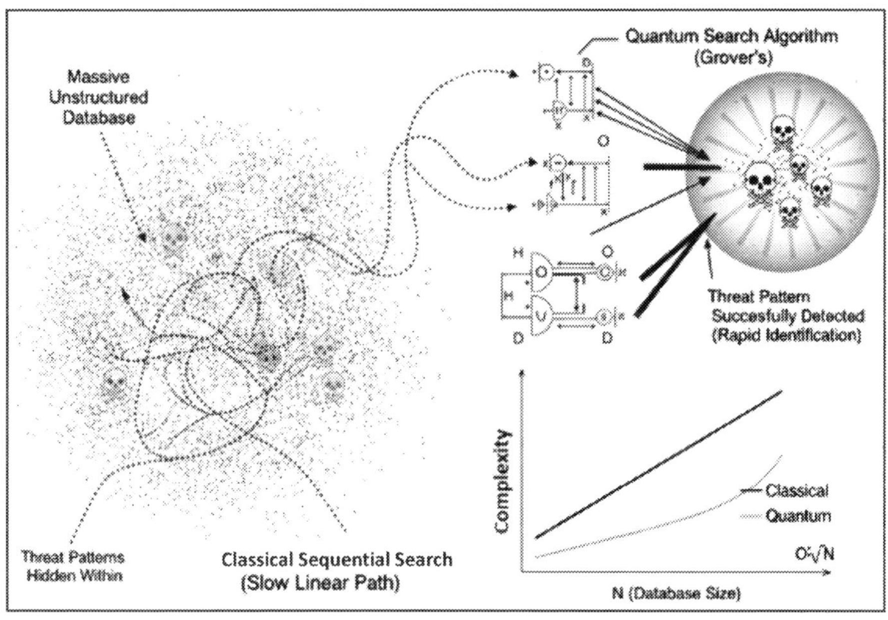

Fig. 2.6: Finding Needles Faster: Quantum Search for Hidden Threats

In the realm of counter-terrorism and asymmetric threat detection, identifying a single hostile communication such as the timing and location of an improvised explosive device (IED) attack within a sea of intercepted phone calls, chat logs, or financial transactions is a challenge. Using classical computing, this task is akin to deploying a team to interrogate every potential source sequentially, an approach that is both time-intensive and operationally limiting. In contrast, Grover's Algorithm empowers a quantum-enabled system to rapidly sift through this unstructured data, isolating the high-risk communication with remarkable efficiency. This quantum search capability enables military intelligence and special operations forces to identify and neutralize threats before the attack is executed enhancing the speed, precision, and effectiveness of interdiction operations in real-time.

This capability enables quantum-enhanced pattern recognition across vast HUMINT and SIGINT datasets, rapidly surfacing communication anomalies, behavioural irregularities, or metadata patterns that may indicate hostile intent. By compressing the timeline from signal to action, Grover's Algorithm supports a new tempo for counter-terror and counter-insurgency operations, especially in contested zones or urban environments saturated with civilian traffic and noise.

Operationally, the value of such speed and precision is most visible in scenarios like identifying IED emplacement cycles, mapping smuggling corridors used by non-state actors, or tracking activity on the dark web where procurement of weapons, explosives, or chemical agents may occur. These data are often unstructured, noisy, and incomplete; ideal terrain for Grover's quantum advantage to thrive.

In this new battle space of information dominance, the edge belongs to those who can detect weak signals faster than the adversary can act. Grover's Algorithm, when integrated into future military quantum platforms, will act not just as a computational tool but as a force multiplier turning what would have been a blind search into a precision-targeted intelligence strike. For modern forces, it is the difference between reactive security and proactive dominance.

Cyber Defence Acceleration: Quantum-Assisted Anomaly Detection

In the cyber domain, speed and precision are paramount. Grover's Algorithm offers a critical edge in scanning vast, unstructured datasets such as system log files, network traffic, and software codebases.

Grover's Algorithm identifies anomalous elements with high precision, even when cloaked within legitimate system activity. This capability is particularly critical in detecting advanced cyber threats such as polymorphic malware or zero-day exploits, which are designed to evade conventional defences by constantly altering their signatures. By surfacing hidden threats in real time, Grover's Algorithm significantly enhances the resilience of mission-critical digital infrastructure against sophisticated and stealthy cyber intrusions.

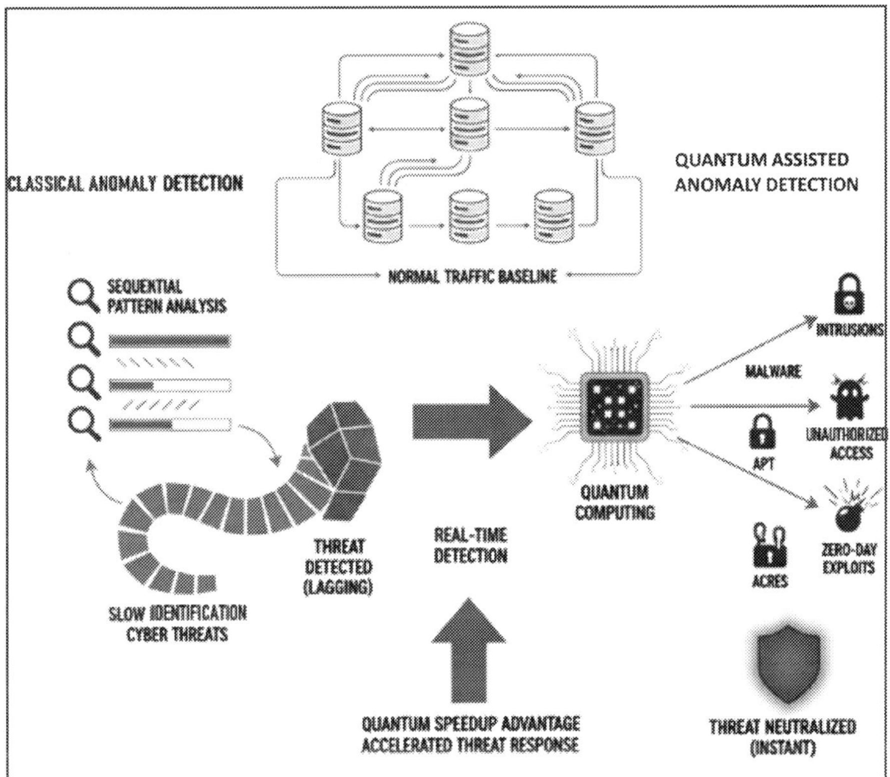

Fig. 2.7: Real-Time Shield: Quantum-Accelerated Threat Detection

Operationally, Grover's Algorithm supports faster incident response times during live cyber operations by accelerating the detection and isolation of rogue processes, unauthorized code injections, or data exfiltration attempts. In contested environments such as during EW or hybrid warfare campaigns, speed can mean the difference between containment and catastrophic breach. Also, the algorithm helps to strengthen the integrity of command networks operating under continuous threat. Quantum-enhanced detection tools, powered by Grover's Algorithm, can monitor system health and flag minute behavioural anomalies that precede full-scale attacks, offering commanders pre-emptive defence opportunities. In the thick of cyber conflict, Grover's Algorithm enables near-real-time digital forensics, allowing military analysts to reconstruct attack chains, trace adversary tactics, and implement countermeasures with minimal latency.

Quantum Logistics: Getting the Right Resource, Right Now

In high-stakes military operations, the ability to deliver the right resource at the right time can be the decisive factor between mission success and operational failure. Grover's Algorithm speeds-up searching unsorted data offering a transformative edge in military logistics by dramatically reducing the time needed to identify optimal supply solutions within distributed and often fragmented inventories. By enabling quantum systems to locate specific items in "N time, Grover's Algorithm empowers logistics planners to rapidly sift through vast, unstructured datasets such as supply manifests, real-time battlefield demands, transport availability, and storage constraints. This enhanced capability proves mission-critical in dynamic operational environments where supply routes are contested, demand patterns shift quickly, and redundancy is minimal ensuring that essential materiel reaches the point of need with speed and precision.

- **Improves Decision-Making Under Resource Scarcity** – In scenarios where critical supplies such as fuel, ammunition, medical kits, or spare parts are limited, commanders must make rapid decisions on how best to allocate them across competing operational priorities. Classical computers struggle when faced with large, unstructured datasets and multiple interdependent variables. Quantum computers, through algorithms like Grover's and quantum approximate optimization algorithms (QAOA), can evaluate vast combinations of allocation scenarios in parallel, identifying the most effective distribution strategies much faster than traditional systems.

- **Supports Dynamic Reallocation in Contested or Remote Environments** – Today's battlefields are characterized by rapidly evolving variables enemy movements, weather shifts, infrastructure degradation, cyber intrusions, and sensor data streams all of which demand swift interpretation and response. Quantum computing, excels in such environments. Through quantum parallelism, it can evaluate multiple evolving scenarios simultaneously. When integrated into command-and-control (C2) systems, quantum processors can ingest real-time ISR (Intelligence, Surveillance, and Reconnaissance) data, logistics reports, and threat assessments, and then dynamically re-optimize mission plans or asset deployments in fractions of the time required by classical algorithms.

At the heart of this capability lie quantum algorithms such as the Quantum Approximate Optimization Algorithm (QAOA) and Variational Quantum Algorithms (VQAs). These are especially adept at solving combinatorial optimization problems ideal for combat scenarios where resources must be reallocated across multiple objectives, each with changing constraints.

For instance, consider an amphibious assault where a shift in weather threatens planned landing zones, while satellite data reveals unexpected enemy reinforcements. A quantum-enhanced mission planner can weigh alternate beachheads, recalculate naval asset positioning, adjust UAV tasking, and re-prioritize logistics while minimizing risk exposure and maintaining mission objectives.

Furthermore, quantum-enabled fusion of data from multi-domain sensors (space, air, land, cyber, and sea) can lead to better predictive modelling. These systems can forecast enemy actions, anticipate logistical bottlenecks, or flag weaknesses in the command structure as scenarios unfold. By doing so, they empower commanders with real-time, actionable intelligence not just after-the-fact assessments.

Quantum Approximate Optimization Algorithm (QAOA): Strategic Advantage in High-Pressure Decision Spaces

"From battlefield confusion to quantum-enabled clarity – fighting smarter, not harder."

The Quantum Approximate Optimization Algorithm (QAOA) holds revolutionary promise in modern warfare where commanders face a barrage of high-stakes decisions often under extreme time constraints and amid a chaotic battle space with incomplete or conflicting data in situations such as choosing optimal troop movements and resupply routes to allocating scarce ISR assets and assessing emergent threats. The operational tempo demands decision-making systems that are both faster and more adaptive than current technologies allow.

QAOA is a quantum algorithm designed to tackle combinatorial optimization problems where the objective is to find the best combination of choices from a vast set of possibilities. Unlike classical algorithms that must explore a large number of permutations sequentially (often growing exponentially with complexity), QAOA uses the principles of quantum superposition and interference to explore multiple options simultaneously. The solution discovered converges on near-optimal solutions with significantly fewer steps.

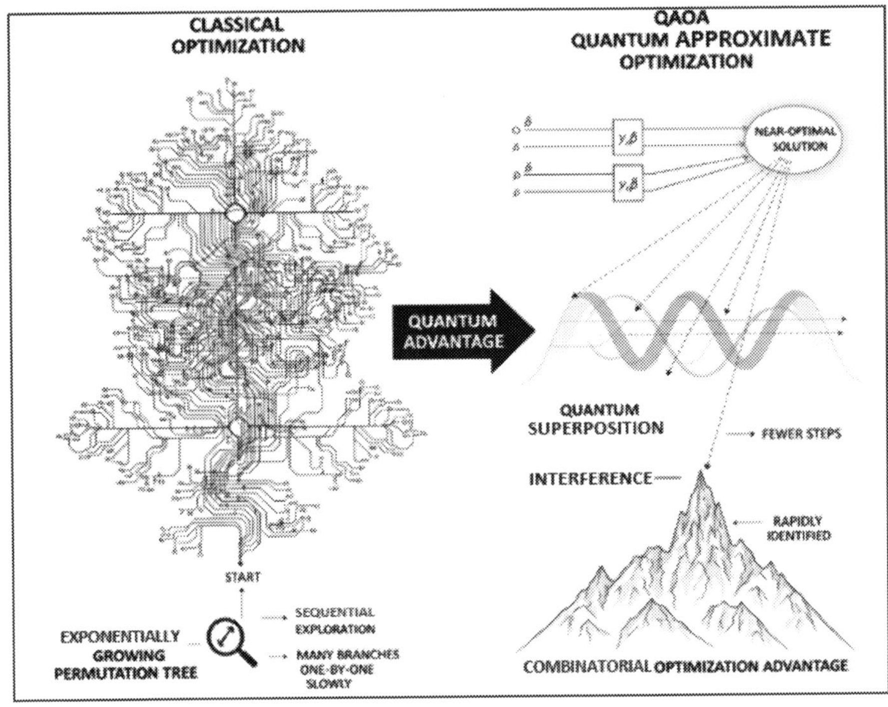

Fig. 2.8: Optimal Faster: QAOA's Quantum Superposition Advantage

- **Enhances command decisions involving many variables** – QAOA operates by translating complex military optimization problems into a mathematical structure known as a cost function. For instance, if a commander needs to determine the safest and most efficient route for a supply convoy under threat of ambush, the problem is encoded such that suboptimal choices such as routes passing through hostile zones or those with time delays incur penalties. The quantum system is then initialized in a superposition state, simultaneously representing all possible configurations such as routes, timings, unit allocations, and logistical priorities. The algorithm then proceeds through repeated application of two key components. First, the Problem Hamiltonian which embeds the mission's constraints and objectives, such as minimizing exposure time, avoiding known threats, or prioritizing the evacuation of wounded personnel. Second, the Mixer Hamiltonian which then introduces a probabilistic element, allowing the algorithm to explore and refine alternative configurations by shifting through the solution space in a manner akin to adaptive war gaming. After several rounds called *QAOA layers* the system converges with high probability on the optimal or near-optimal solution set, ready for actionable execution.

- **Supports near-real-time tactical planning under time pressure** – In modern high-tempo combat environments traditional decision-support systems, operating on classical algorithms, often struggle to optimize multifactor challenge to swiftly meet battlefield exigencies. In these environments, the time-to-decision is as critical as the decision itself, making speed, accuracy, and adaptability essential attributes of any computational framework supporting the war fighter. QAOA, a hybrid quantum-classical algorithm, is engineered to tackle combinatorial optimization problems by searching for the best solution among numerous possible combinations far more efficiently than conventional algorithms. Its utility lies in translating tactical military scenarios into mathematically structured problems as cost functions. These functions can represent objectives such as minimizing the risk to convoys, maximizing speed of delivery, or optimizing asset deployment across a battle space while obeying operational constraints like supply limitations or enemy activity.

Once the problem is encoded, QAOA executes a quantum search using alternating layers of quantum gates: phase operators, which encode the cost function (the "what needs to be optimized"), and mixing operators, which help the system explore different possible solutions. This configuration allows the quantum system to home in on high-quality, low-cost solutions by amplifying the probability of optimal states. The process is not purely quantum solution. It is enhanced through classical feedback, where the algorithm iteratively refines parameters based on the performance of previous rounds. Despite relying on near-term noisy intermediate-scale quantum (NISQ) hardware, QAOA can already provide substantial performance improvements over classical heuristics in time-critical decision spaces.

The strategic edge QAOA offers lies in its ability to deliver speed with complexity. Unlike traditional tools that struggle under the weight of rapidly evolving parameters and unpredictable adversarial moves, QAOA thrives in uncertainty by rapidly converging on actionable outcomes. For defence planners, the implications of integrating QAOA into command-and-control (C2) systems are profound. It can power faster scenario-based war gaming, support smarter real-time mission adjustments, and offer commanders higher confidence in decisions made with partial or evolving data.

- **Deployment Optimization in Multi-Theatre Operations** – Modern warfare is no more unilateral. Joint operations demand synchronized planning where tasks, roles, and resources must be distributed optimally. QAOA enables this through its ability to handle heterogeneous inputs and constraints. For instance, in a multinational naval operation in contested waters, QAOA can model diverse force structures and mission requirements, and then output optimal task allocation, e.g., which ship should patrol which sector, who provides air cover, and how shared refuelling should be scheduled. This drastically reduces the risk of redundancy, overlap, or strategic blind spots in coalition missions, helping planners coordinate allied strengths for a more effective collective posture.

  Over-committing elite units or depleting critical supplies early in a campaign can lead to mission failure. Traditional planning tools struggle to account for the downstream impact of such decisions. QAOA addresses this by evaluating trade-offs across time, geography, and sustainability. It identifies solutions that meet mission goals without exceeding logistical or operational thresholds. For example, it might recommend using a mix of lower-intensity units with support drones in a secondary theatre, rather than depleting special forces who may be needed later. It might also suggest delaying a strike until resupply cycles can be realigned to sustain post-strike operations.

- **Enables Adaptive Decision-Making in Autonomous Systems** – In GPS-denied environments, electronic warfare zones, or unmapped/ hostile terrain, autonomous military platforms such as Unmanned Ground Vehicles (UGVs), Unmanned Aerial Vehicles (UAVs), or Autonomous Underwater Vehicles (AUVs) face rapidly changing conditions with degraded external guidance. Traditional algorithms that rely on pre-defined paths or central command inputs fail to respond efficiently in such scenarios. QAOA helps solve combinatorial optimization problems of the form:

Minimize cost function $C(x)$ over a set of feasible configurations x,

where the cost function could encode path risks, fuel constraints, line-of-sight visibility, terrain features, and threat likelihoods.

At the core of the Quantum Approximate Optimization Algorithm (QAOA) lies a quantum-enhanced decision-making process built upon parameterized quantum circuits. These circuits systematically alternate between two critical components: a phase separation operator, which encodes the cost function representing the operational constraints (such as risk exposure, fuel limitations, or tactical priorities), and a mixing operator, which allows the system to explore a wide range of alternative configurations within the feasible solution space. This dual-stage approach enables QAOA to outperform classical methods in solving combinatorial optimization problems, particularly when the solution space is exponentially large in conditions such as evaluating all possible paths through hostile or unpredictable terrain. Its advantage becomes even more pronounced in missions where multiple, often conflicting, objectives must be balanced in real time, such as choosing between routes that offer low visibility, minimal fuel consumption, and high survivability.

This capability is especially vital for autonomous military systems operating in dynamic threat environments. For example, a UAV conducting reconnaissance in contested airspace may need to continuously adjust its flight path to evade newly identified anti-aircraft systems, while simultaneously conserving fuel and maintaining sensor coverage. In such a scenario, QAOA allows the UAV's onboard quantum or hybrid edge processor to dynamically evaluate numerous possible trajectories, each assessed against a constantly evolving cost function. The algorithm penalizes paths that pass through high-risk zones, rewards those that enhance stealth or efficiency, and adapts to changes in the threat landscape as they occur. This real-time, quantum-enhanced adaptability enables the UAV to rapidly select and follow the optimal path thus, maximizing mission success while minimizing exposure and resource depletion. As such, QAOA represents a breakthrough in embedding resilient, adaptive intelligence into next-generation military platforms.

- **Enhances Collaborative Human-Machine Operations** – In multi-domain operations involving swarms or heterogeneous assets, unmanned systems need to act in synchrony with human intent, not merely execute scripts. Traditional autonomy struggles to interpret dynamic mission constraints and the evolving priorities of field commanders. The Quantum Approximate Optimization Algorithm (QAOA) empowers distributed autonomous systems such as unmanned aerial swarms or ground-based robotic units to perform complex, multi-variable constrained optimization tasks in real time. These systems operate in fast-changing combat environments where mission objectives, threat contours, and coordination demands are in constant flux. Unlike traditional algorithms that falter under the weight of NP-hard problems such as multi-agent path planning with obstacle avoidance or dynamic task allocation under bandwidth and power constraints QAOA excels by delivering near-optimal solutions within drastically reduced computational timeframes. This is achieved through a hybrid quantum-classical approach, enabling tactical machines to make locally optimal decisions while remaining in synchrony with a broader strategic directive.

  Individual drones in a surveillance swarm each potentially equipped with a lightweight quantum processor can independently determine their next best action (e.g., selecting the next high-priority zone to monitor) based on localized input. Simultaneously, these autonomous agents remain coordinated with a global optimization function that governs swarm-wide objectives, such as maximizing ISR coverage, minimizing redundant paths, and balancing risk exposure across the unit. QAOA processes dynamic mission updates and commander-imposed constraints in real time, allowing these platforms to instantly reconfigure their roles without requiring centralized re-tasking. This interaction fosters what can be termed a "shared decision loop," where human commanders provide strategic intent and evolving mission parameters, and quantum-enhanced machines autonomously execute optimally aligned tactical behaviours. In doing so, QAOA bridges the critical gap between human initiative and machine autonomy unlocking unprecedented agility, responsiveness, and coordination in multi-domain operations.

2.3 BREAKING CLASSICAL ENCRYPTION USING QUANTUM ALGORITHMS

Shattering the unbreakable – how quantum weapons aim straight at the heart of secure communication

Shor's Algorithm: The End of RSA and ECC

The Missile That Bypasses the Bunker – Breaking the Codes that Protect Command and Control

Modern military-grade encryption rests on the foundational belief that factoring very large numbers is computationally infeasible for classical computers. This assumption is the bedrock of cryptographic systems like RSA and Elliptic Curve Cryptography (ECC), which protect the confidentiality of strategic communications, nuclear command and control channels, and defence logistics platforms. For example, breaking a 2048-bit RSA encryption key would require classical computers millions of years rendering such systems effectively unbreakable by today's standards.

However, this cryptographic approach is rendered vulnerable by Shor's Algorithm, a quantum algorithm capable of factoring large integers exponentially faster than any known classical method. Leveraging the unique strengths of quantum mechanics, Shor's Algorithm reduces the time needed for the mathematically complex task from centuries to mere hours or minutes, provided a sufficiently powerful quantum computer is available. What was once computationally impossible becomes a solvable challenge in polynomial time a dramatic shift with profound implications for national security. At its core, Shor's Algorithm transforms the factoring challenge into a period-finding problem, which quantum systems can resolve with high efficiency using the Quantum Fourier Transform. The algorithm begins by selecting a random number (not a factor of the target composite number), then identifies the periodicity of a function rooted in modular arithmetic. With that period, it can extract a non-trivial factor of the original number. Classical computers are ill-equipped to navigate this path efficiently but quantum machines excel due to their inherent ability to explore vast computational landscapes in parallel.

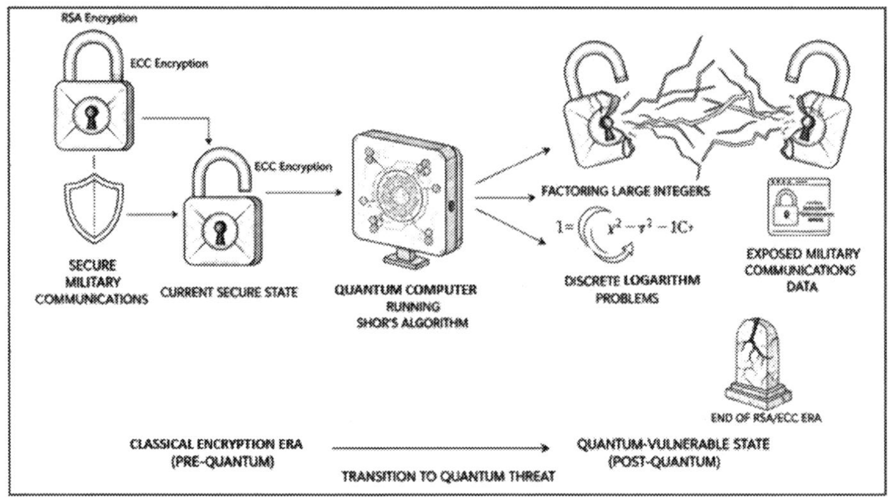

Fig. 2.9: The Encryption Endgame: Shor's Algorithm Breaks RSA/ECC

Shor's Algorithm poses an existential threat to current encryption standards that underpin secure command communications, intelligence-sharing networks, weapons systems authentication, and encrypted data archives. The danger is not only in future quantum capabilities but in the "harvest now, decrypt later" risk, where adversaries can store intercepted encrypted transmissions today and decrypt them retroactively when quantum technology matures.

Military planners and cyber security strategists must begin transitioning to Post-Quantum Cryptographic (PQC) algorithms classical techniques believed to be quantum-resistant and explore Quantum Key Distribution (QKD) as a next-generation security mechanism. These technologies must be integrated into the design of future C4ISR systems, military satellites, encrypted battlefield communication networks, and nuclear command infrastructures.

Military Domain	Vulnerable Asset	Risk Enabled by Shor's Algorithm
Satellite Comms	UHF, X-band, Ka-band terminals	Decryption of real-time imagery, spoofed control
NC3/NC2 Systems	ICBM commands, sub communications	Forged launch messages, command spoofing
Tactical Networks	Secure SDRs, MANETs	Key compromise, traffic decryption
ISR Assets	Secure drone telemetry	Data exfiltration, counter-drone attacks
Cyber Defence	VPNs, IDS encryption logs	Retrospective traffic decryption, command injection

The consequences for military operations are:

- **Compromise of Secure Communications:** In modern warfare, from encrypted satellite links that relay orders across global combat zones to tactical data uplinks used by forward-operating units, confidentiality and authenticity of information flows are critical to maintaining operational superiority. However, the advent of quantum computing and use of Shor's Algorithm threatens to dismantle the cryptographic foundations upon which these secure communications rest.

  Most military-grade communication protocols rely on asymmetric encryption schemes, such as:

  - ❑ **RSA (Rivest-Shamir-Adleman):** Based on the difficulty of factoring the product of large primes.

  - ❑ **ECC (Elliptic Curve Cryptography):** Based on the elliptic curve discrete logarithm problem.

  Once these foundational elements of public key infrastructure are broken, any encrypted communication relying on them becomes transparent either in real-time or retroactively. Three high-risk domains where the impact of Shor's Algorithm would be particularly consequential are:

  - ❑ **Encrypted SATCOM Links** – Military satellite communications (SATCOM) are the enabling of real-time operational control, and to secure voice, data, and telemetry exchanges between command elements and deployed units. Systems like the U.S. Advanced Extremely High Frequency (AEHF), India's GSAT series, or NATO's SATCOM Post-2000 use RSA- or ECC-based public key encryption maintain confidentiality and authentication. However, Shor's Algorithm breaks these cryptographic foundations by exponentially accelerating the factoring of large prime numbers, the core mathematical challenge that classical encryption relies upon. If adversaries gain access to scalable quantum systems, they could dismantle decision superiority, degrade force synchronization, and expose vulnerabilities across the electromagnetic spectrum:

  - ❑ Decrypt archived SATCOM transmissions, exposing years of sensitive operational planning, ISR data, and joint-force coordination.

- ❑ Forge cryptographic handshakes, allowing hostile actors to masquerade as legitimate users within the communication network potentially injecting false commands or misinformation.

- ❑ Jam or spoof command-and-control links, disrupting UAV guidance, missile targeting, or satellite payloads critical to air-ground integration.

- ❑ **Blue Force Tracking (BFT)** – Blue Force Tracking systems used across NATO and allied militaries provide near real-time positional data of friendly units across terrestrial, aerial, and maritime domains. These feeds rely on encrypted GPS and data uplinks to prevent enemy interception. Should adversaries use Shor's Algorithm to decrypt or mimic these encrypted data channels, they could:

- ❑ Monitor live troop deployments, compromising manoeuvre elements, logistics routes, and combat service support chains.

- ❑ Target high-value units such as command posts, artillery batteries, or special operations forces with precision munitions.

- ❑ Conduct psychological operations by manipulating or spoofing unit positions creating confusion, triggering friendly fire incidents, or forcing premature retreats.

- ❑ **Nuclear Command, Control, and Communications (NC3)** – Perhaps the gravest implication of quantum-enabled cryptographic defeat lies in the realm of nuclear command and control. NC3 architectures which facilitate the secure transmission of launch authority to delivery platforms such as ICBMs, SSBNs, and strategic bombers rely on tamper-proof encryption protocols to ensure that only authorized commands are accepted and that communications cannot be replayed, forged, or intercepted. A quantum-enabled adversary could, in theory:

- ❑ Decrypt or replay past nuclear command signals, threatening the integrity of escalation protocols and potentially enabling unauthorized launch sequences.

- ❑ Spoof command traffic, injecting false launch or stand-down orders, thereby destabilizing mutual assured destruction (MAD) paradigms.

❏ Nullify second-strike capability, if retaliatory communications or targeting instructions are compromised, thus eroding strategic deterrence

• **"Harvest Now, Decrypt Later" Threat:** At the heart of the "Harvest Now, Decrypt Later" (HNDL) threat lies the future-breaking capability of Shor's Algorithm, a quantum algorithm designed to factor large composite numbers in polynomial time, exponentially faster than the best-known classical methods. Modern encryption schemes such as RSA, Diffie–Hellman, and Elliptic Curve Cryptography (ECC) derive their strength from the mathematical intractability of problems like integer factorization and discrete logarithms tasks that classical computers cannot solve efficiently at scale.

However, Shor's Algorithm breaks this security premise. Once a sufficiently powerful, fault-tolerant quantum computer is operational, any stored cipher-text encrypted using RSA or ECC can be retroactively decrypted. This is because the public keys used in these systems are widely distributed. Only the private key (derived via factorization or solving the discrete logarithm) needs to be recovered for decryption. Shor's Algorithm allows for this key recovery exponentially faster than brute-force or number-theoretic attacks using classical systems.

The military threat environment amplifies the dangers of HNDL due to the longevity and sensitivity of the data transmitted through secure military channels. Stored encrypted communications about operational plans, troop movements, or war-gaming exercises could be decrypted years later giving adversaries a deep forensic view of command patterns and doctrines. Authentication signals, targeting coordinates, and silo readiness reports, even if no longer in use, could reveal procedural vulnerabilities or system architectures.

System	Risk Enabled by Shor's Algorithm	Consequence
SATCOM (UHF/X-band/ Ka-band)	Decryption of command uplinks/downlinks	Loss of C2 integrity
Tactical Radios & SDRs	Breaking session key exchanges	Traffic decryption and spoofing
Secure Messaging (JSIN, STRATNET)	Retrospective message compromise	Exposed nuclear or covert operations
Coalition Interoperability Links	Intercepted authentication between allies	Breakdown in trust and shared situational awareness

To mitigate the HNDL threat enabled by Shor's Algorithm, militaries must:

- **Adopt Post-Quantum Cryptography (PQC):** Transition to quantum-resistant algorithms such as lattice-based, code-based, or multivariate polynomial cryptosystems as standardized by NIST's Post-Quantum Cryptography initiative.

- **Deploy Quantum Key Distribution (QKD):** Use QKD for forward secrecy, as any attempt to measure or intercept the key during distribution alters the quantum state triggering immediate detection.

- **Apply Crypto-Agility to Legacy Systems:** Enable rapid protocol replacement and over-the-air cryptographic patching in vulnerable systems like tactical radios, BFT platforms, and secure routers.

- **Risk-Aware Data Retention Policies:** Classify and restrict long-term storage of data encrypted using classical methods that may be vulnerable to quantum decryption.

Grover's Algorithm: Lockpicking at Quantum Speed

Imagine a secure facility protected by a lock with a million possible combinations. A traditional intruder tries each combination one by one, taking days or even months to succeed. Grover's Algorithm, by contrast, is akin to a digital skeleton key, a tool that can intelligently test only a fraction of those combinations, finding the right one in hours instead of days. In military terms, this is equivalent to the adversary rapidly bypassing encryption barriers that were previously considered secure for the duration of a mission or operation cycle.

Grover's Algorithm, developed in the mid-1990s, provides a quadratic speedup for searching unsorted data spaces, a feature particularly relevant to cryptographic brute-force attacks. Classical brute-force techniques for symmetric key systems (like AES) require checking all possible key combinations, making the time complexity $O(N)$, where N is the number of possible keys. Grover's Algorithm reduces this to $O(\sqrt{N})$, effectively halving the exponent in brute-force time. For example, a 128-bit key, considered secure under classical computing, can be attacked with the same effort as a 64-bit key under Grover's quantum search. This imposes a direct threat to many existing symmetric cryptographic protocols currently deployed in military communication and battlefield systems.

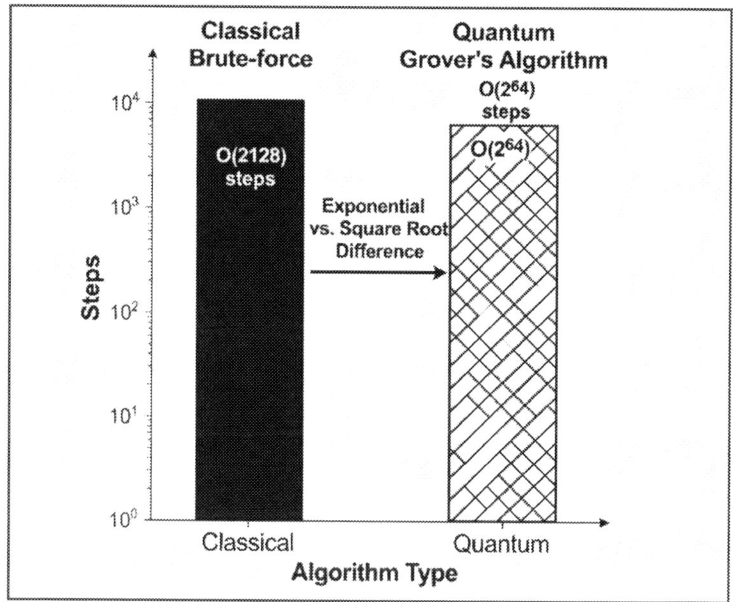

Fig. 2.10: Quantum vs Classical: Breaking AES-128 with Grover's Algorithm

Symmetric encryption algorithms like AES (Advanced Encryption Standard), 3DES, and Blowfish protect data using a shared secret key. The security of these systems relies on the computational infeasibility of trying every possible key commonly known as a brute-force attack. For example, AES-128 has 2^{128} possible keys while classical brute-force attack requires $O(2^n)$ operations (i.e., 2^{128} for AES-128), which is considered computationally infeasible even for powerful supercomputers. If a symmetric cipher uses an n-bit key (e.g., 128 bits), the classical search requires $O(2^n)$ steps. Grover reduces this to $O(2^{n/2})$ steps using quantum superposition and amplitude amplification. For AES-128 Classical search requires 2^{128} operations while Quantum search using Grover requires 2^{64} quantum operations.

This has immediate implications for military systems:

- **Compromised Symmetric Encryption:** Grover's Algorithm presents a substantial threat to symmetric encryption making it vulnerable to feasible quantum attacks. This reduced strength not only compromises current operational security but also shortens the cryptographic longevity of systems designed to safeguard classified data over extended periods. Lightweight and edge devices, including tactical radios, IoT-based surveillance nodes, and unmanned platforms, are particularly at risk, as they often use shorter keys or computationally efficient ciphers due to hardware constraints. The resulting vulnerability creates an attractive target for adversaries seeking entry points into larger command networks. In high-tempo combat environments, Grover-enabled quantum attacks could disrupt encrypted GPS links, blue force tracking, and communications uplinks, undermining battlefield coordination. These implications collectively demand an urgent and comprehensive cryptographic upgrade across all military systems, transitioning to AES-256 or higher and incorporating crypto-agile architectures capable of adapting to evolving post-quantum threats.

Impact Area	*Grover's Implication*
Key Security	Halves brute-force resistance (e.g., AES-128 becomes 64-bit secure)
System Lifetime	Shortens effective cryptographic shelf-life
Edge Device Risk	Compromises lightweight/tactical comms
Operational Security	Enables faster decryption of intercepted data
Required Action	Mandates migration to AES-256 and crypto-agility

- ❏ **Halved Effective Security Strength:** Since Grover's Algorithm provides a quadratic speed-up in brute-force key search, it cuts the effective security of symmetric encryption in half. Systems currently relying on AES-128 (e.g., tactical radios, UAV telemetry, secure field communications) will become vulnerable once quantum computing scales, exposing them to faster compromise than originally projected.

- ❏ **Reduced Encryption Longevity:** Encryption systems are often designed to remain secure for decades. Grover's Algorithm compresses this timeline, reducing the effective lifespan of cryptographic systems unless key sizes are increased. Sensitive communications stored today (e.g., satellite imagery, nuclear launch codes, ISR logs) could be decrypted retroactively if intercepted now and cracked later using a "harvest now, decrypt later" approach. This is especially critical for long-lived assets like strategic deterrent systems and black-site intelligence repositories.

❏ **Increased Vulnerability of Lightweight and Edge Devices:** Increased Vulnerability of Lightweight and Edge Devices. These systems are the first points of weakness. A quantum-enabled adversary could exploit Grover's Algorithm to breach tactical networks through these devices, leading to potential reconnaissance compromise, supply route interference, or false data injection into command chains.

❏ **Tactical System Disruption:** Grover's Algorithm reduces the time needed to brute-force encryption on time-sensitive or mission-critical systems. These systems are the first points of weakness. A quantum-enabled adversary could exploit Grover's Algorithm to breach tactical networks through these devices, leading to potential reconnaissance compromise, supply route interference, or false data injection into command chains.

❏ **Forced Cryptographic Upgrades:** To counter Grover's Algorithm, symmetric algorithms like AES must double key lengths (e.g., from AES-128 to AES-256) to regain equivalent classical security levels. Armed forces must initiate crypto-modernization programs, especially for strategic communications systems (SATCOM, NC3), secure data-at-rest and data-in-transit platforms and mission-critical embedded systems.

• **Vulnerability of Lightweight Tactical Systems:** Grover's Algorithm introduces a techno-functional threat to lightweight tactical systems by significantly reducing the effective security of symmetric cryptographic protocols, particularly AES-128, which is widely deployed in edge devices due to its low computational overhead. This creates a vulnerability window in platforms such as UAVs, unmanned sensors, soldier-wearable devices, and IoT-enabled communication tools. These systems, often positioned at the tactical edge, become exploitable entry points for adversaries using quantum-accelerated brute-force attacks to compromise encrypted data streams or inject malicious commands. This technical compromise directly impacts operational functionality, degrading tactical autonomy, ISR reliability, and real-time decision loops in disconnected or contested environments. Additionally, many of these platforms lack the computational resources or design flexibility to adopt stronger encryption natively, making cryptographic agility and lightweight post-quantum cryptographic protocols an immediate functional necessity. Ensuring end-to-end resilience requires not just upgrading encryption strength (e.g., AES-256), but embedding future-proof, mission-aware cryptographic design into tactical hardware and communication stacks.

Vulnerability Area	Grover's Algorithm Implication
Encryption Weakness	Halves strength of AES-128; unsuitable for long-term security
Perimeter Breach Risk	Enables adversaries to exploit tactical nodes as access points
ISR and Autonomy Impact	Disrupts data trust and autonomous mission flow
Upgrade Constraints	Legacy and constrained devices face difficulty in adopting stronger crypto
Required Action	Mandates AES-256, crypto agility, and lightweight post-quantum cryptography

❑ **Weakening of Lightweight Cryptographic Protocols:** Lightweight tactical systems such as unmanned aerial vehicles (UAVs), soldier-wearable devices, edge sensors, field radios, and IoT-based surveillance tools rely on symmetric encryption with shorter key lengths (e.g., AES-128) due to power, memory, and processing constraints. Grover's Algorithm, which provides a quadratic speed-up in brute-force attacks, effectively halves the key strength of these systems. For example, AES-128 offers only 64-bit security in a quantum context, leaving such devices significantly more susceptible to quantum-enabled adversaries.

❑ **Exploitation as Entry Points:** As these lightweight systems often act as perimeter nodes or first-line interfaces in broader command-and-control (C2) networks, a compromise at this level via quantum-assisted decryption using Grover's Algorithm can enable an adversary to intercept real-time tactical data, inject false data or spoof device identities, pivot into more secure backend systems, escalating the breach from tactical to operational or even strategic level.

❑ **Degradation of Tactical Autonomy and ISR:** Autonomous platforms and forward ISR nodes often operate in disconnected, hostile, or GPS-denied environments. Their reliance on encrypted communication and onboard decision-making means that any degradation in encryption reliability due to Grover's impact directly affects their survivability, data integrity and operational effectiveness.

Enhancing Intelligence Gathering through Quantum Data Processing

Quantum Intel: From Data Swamps to Tactical Gold – Speed, depth, and clarity in the fog of information warfare

Quantum Speed in Processing Signals Intelligence (SIGINT)

Quantum computing introduces a fundamental shift in how encrypted communications and signal-rich environments are processed, particularly in the domain of Signal Intelligence (SIGINT). Unlike classical computing systems that operate using binary logic and linear processing methods, quantum computers utilize quantum bits (qubits), which can exist in superpositions of states. This allows quantum systems to evaluate multiple possibilities simultaneously, making them ideally suited for high-dimensional data analysis, such as real-time signal interception and interpretation in the electromagnetic (EM) spectrum.

In classical SIGINT workflows, intercepted signals are analysed through a series of deterministic processes signals that are demodulated, filtered, decoded, and then analysed for patterns or threats. This workflow is sequential and constrained by clock speed and memory bandwidth, especially when dealing with massive, noisy, or encrypted datasets. Furthermore, traditional systems often rely on pre-programmed heuristics, which may fail to detect novel or obfuscated patterns that adversaries deliberately introduce. Quantum systems, particularly when paired with quantum machine learning (QML) and quantum pattern recognition algorithms, offer the ability to process entangled states representing complex correlations across multiple signals and parameters. For example, a quantum computer running a variational quantum classifier (VQC) can be trained to distinguish hostile versus benign communication patterns, even when these are obscured by encryption or camouflaged within civilian traffic. Quantum amplitude amplification methods like those used in Grover's Algorithm further increase the probability of quickly identifying rare or anomalous patterns, such as a brief insurgent broadcast hidden among civilian chatter.

In practical battlefield scenarios, this capability becomes a force multiplier. Consider a theatre with overlapping signals from hundreds of sources radio, satellite, cellular, and tactical mesh networks. Classical systems must isolate one frequency at a time and often require signal de-confliction to avoid misinterpretation. Quantum systems, however, operate as if examining all frequencies, all at once, thanks to quantum parallelism. This enables simultaneous multi-spectral signal scanning, drastically reducing latency between signal interception and threat detection.

- **Accelerated Decryption and Analysis:** Unlike classical systems, which sequentially process intercepted signals through pre-programmed filters and algorithms, quantum computers exploit quantum parallelism which evaluates vast and entangled signal spaces simultaneously. When paired with quantum algorithms such as Shor's Algorithm for decryption and Quantum Machine Learning (QML) models for pattern recognition, these systems enable rapid, intelligent sifting through high-volume, high-noise electromagnetic environments. Imagine a digital battlefield teeming with intercepted communications radio chatter, satellite relays, drone telemetry, and encrypted messages. A classical SIGINT system is like a squad of analysts, each scanning one channel at a time. A quantum SIGINT platform, by contrast, operates like a single elite analyst with the cognitive capacity to listen to all channels at once, flag only threat-relevant communications, and instantly translate enemy intent into action.

Operational Relevance:

- ❑ **Accelerated Decryption:** Quantum systems can exploit algorithms like Shor's to rapidly factor encryption keys protecting enemy SATCOM, command links, or field-level communications turning what was previously computationally infeasible into a tactical reality.

- ❑ **Real-Time Pattern Analysis:** QML models running on quantum processors can extract actionable signals such as insurgent planning, IED coordination, or adversarial manoeuvre discussions from vast volumes of intercepted voice, text, or telemetry data, even when encrypted or obfuscated.

❑ **Shortened Intelligence Cycle:** Traditional SIGINT processes may require hours or days to detect, decrypt, and verify threat-relevant signals. Quantum-enabled systems can compress this to minutes or seconds, enabling pre-emptive strikes, electronic warfare jamming, or force reallocation in near-real time.

❑ **Enhanced Survivability:** By rapidly identifying electronic emissions or signals associated with hostile actors, these platforms support electronic counter-measure activation, mission abort recommendations, or stealth re-routing of vulnerable platforms.

- **Shortened Intelligence Cycle:** A traditional intelligence cycle is often constrained by sequential decryption processes, heuristic-based filtering, and post-event analysis. In high-tempo combat environments, this lag can render intelligence obsolete before it reaches decision-makers. Unlike classical systems that search for patterns one frequency or protocol at a time, quantum computers can process entire high-dimensional data spaces in parallel. Quantum algorithms such as Shor's (for breaking classical encryption), Grover's (for rapid search in large signal logs), and quantum-enhanced machine learning models enable instant identification of threat-relevant communications even amidst a sea of civilian or deceptive transmissions. Conventional SIGINT deploys a team of analysts to manually tune into one radio frequency at a time. Quantum-enabled SIGINT, in contrast, is capable of processing all frequencies, decoding all signals, and alerting commanders only when adversarial chatter is detected, all in real time.

- **Enhanced Real-Time Operations:** Traditional SIGINT platforms (e.g., ELINT, COMINT, FISINT) process signals using classical algorithms that must scan frequencies, decrypt, demodulate, and analyse signal metadata sequentially or in limited parallel threads. In contrast, quantum computers operate on qubits that can represent multiple states simultaneously via superposition. Quantum systems can analyse multiple frequency bands and protocols concurrently. This will enable in signals modulated in unknown or dynamic formats (e.g., spread-spectrum, frequency-hopping, or burst transmissions) to be detected and characterized more efficiently. This capability will be vital for real-time surveillance in dense and contested electromagnetic environments, such as urban warfare zones or anti-access/area-denial (A2/AD) theatres. Quantum algorithms like the Quantum Principal Component Analysis (qPCA) or Quantum Boltzmann Machines can learn patterns embedded in noisy or low-SNR signals where classical signal processing techniques fail. These models excel at identifying anomalous signal structures (e.g., covert burst comms, steganographic channels), extracting underlying features (e.g., source location, modulation type, encryption structure). Such capabilities are critical in electronic counter-countermeasures (ECCM) where the adversary attempts to mask or mimic friendly transmissions.

For intercepted encrypted traffic (e.g., from VHF/UHF military radios, SATCOM, or data links), Shor's Algorithm poses a future capability to break public-key encryption (e.g., RSA, ECC). In real-time, SIGINT archived traffic (e.g., past SATCOM bursts) can be decrypted as soon as quantum capacity permits. Tactical communications using elliptic curve-based key exchange can become vulnerable in the near future if not transitioned to post-quantum cryptography (PQC).

Pattern Recognition across Disparate Intelligence Sources

In modern combat and counter-terrorism operations, intelligence is scattered across heterogeneous sources voice intercepts, electro-optical/infrared (EO/IR) satellite imagery, SIGINT metadata, HUMINT field reports, cyber telemetry, and geo-location pings. Classical systems typically analyse these streams in isolation or sequential pipelines, using rules-based filters or statistical models, which limits their ability to uncover non-obvious correlations or fuse incomplete datasets. Quantum-enhanced machine learning (QML), in contrast, applies quantum computing principles, especially superposition, entanglement, and amplitude amplification to process and correlate massive, noisy, and multi-modal data in parallel, navigating exponentially large solution spaces that are computationally intractable for classical systems.

For example, in QML, quantum kernel methods can classify non-linear patterns across different intelligence modalities by embedding the data into a high-dimensional Hilbert space, where patterns that are inseparable in classical feature space become distinguishable. Similarly, variational quantum classifiers or quantum Boltzmann machines can be trained on hybrid datasets (e.g., thermal IR signatures, movement logs, and signal strength fluctuations) to detect and predict anomalous behaviour such as insurgent rendezvous patterns or arms cache placements. Quantum systems can infer latent variables connecting incomplete or seemingly unrelated clues, thereby transforming fragmented raw data into actionable threat assessments.

This capacity is particularly valuable in battlefield scenarios where time and ambiguity dominate. Imagine a quantum-enhanced ISR node at a forward operating base tasked with detecting sniper activity. Classical methods might scan for direct visual cues or sound signatures. A QML-based system, however, could correlate satellite imagery of environmental shadows, gaps in local radio traffic, subtle motion trajectories from drone video, and fragmented HUMINT chatter to flag a likely sniper team. This is not mere acceleration; it is quantum-native fusion of intelligence cues, with the potential to expose weak signals deeply embedded in noise.

Operationally, this empowers joint task forces to conduct rapid link analysis, track insurgent financial flows across dark web forums, triangulate terror network command nodes, or even correlate biometric data with location-based access logs. In strategic terms, this minimizes analyst workload, reduces cognitive overload during crisis response, and cuts down intelligence fusion time lines from hours to real time. The result is faster kill chains, pre-emptive threat neutralization, and enhanced situational awareness in denied, degraded, or contested operational environments (DDIL).

Predictive Analysis of Adversarial Behaviour

Modern military decision-making is fundamentally a complex optimization problem involving incomplete information, rapidly changing variables, and adversaries that are themselves adaptive. Classical computational models, even when supported by AI, often falter in such non-deterministic environments due to their reliance on sequential processing, limitations in simulating uncertainty, and exponential time required for evaluating multi-variable interactions. This is where quantum computing provides a paradigm shift. At the core of this transformation lie Quantum Machine Learning (QML) and Quantum Game Theory, both of which allow modelling and forecasting adversarial behaviour with unprecedented depth and speed. These methods leverage the unique capabilities of quantum systems. Superposition enables the quantum processor to evaluate multiple potential future states (e.g., enemy actions, counter-measures, response chains) simultaneously, rather than sequentially while entanglement allows for encoding correlations between strategic variables, for instance, how adversarial troop movement patterns relate to electronic warfare activities or satellite overpass timings. Amplitude amplification in quantum algorithms like the Quantum Approximate Optimization Algorithm (QAOA) or Variational Quantum Classifiers (VQC) help identify optimal or high-risk adversarial paths, even in stochastic or non-linear scenarios.

A Quantum-Adversary simulation pipeline for example may include as input layer ISR data feeds, historical movement patterns, adversarial military doctrine, social media sentiment (for insurgency prediction), and geospatial constraints. Encoding will be carried out by quantum feature maps that convert classical data into high-dimensional Hilbert space representations that can be entangled and processed. Quantum circuit-based simulation will then generate strategic future combinations of red-force tactics, deception manoeuvres, cyber actions, and timing. As a part of decision layer, quantum-enhanced classifiers or optimizers will rank and return the most probable or most dangerous enemy options, enabling pre-emptive action.

Operationally, this means:

- Red teaming in seconds: Instead of relying on days-long simulations, quantum processors can generate and evaluate thousands of red-force scenarios in minutes.

- Grey zone detection: Quantum algorithms detect emergent behaviours or anomalous troop patterns that classical systems would dismiss as noise.

- Doctrine mirroring: By feeding in open-source doctrine and operational playbooks of adversarial nations, quantum models can reverse-engineer their likely courses of action.

- Temporal war gaming: Simulate not just one battle outcome, but the cascading second- and third-order effects on logistics, morale, public opinion, and strategic posture.

2.4 QUANTUM SIMULATION FOR WAR STRATEGY AND SCENARIO PLANNING

Quantum War Games: Infinite Scenarios, Superior Strategies – Training for the future before it arrives

Anticipating Adversary Movements and Strategic Shifts

In an era defined by multi-domain threats and rapidly shifting geopolitical realities, anticipating adversarial movements before they occur is no longer a strategic luxury; it is an operational imperative. Quantum computing provides a breakthrough capability in this domain by harnessing quantum algorithms to process and correlate massive volumes of historical, tactical, and real-time ISR (Intelligence, Surveillance, Reconnaissance) data. Using advanced quantum machine learning and probabilistic modelling techniques, these systems can simulate adversary behaviour and forecast shifts in posture, doctrine, or deployment patterns.

Capability	*Description*	*Strategic Impact*
Real-Time Posture Forecasting	Anticipate troop, naval, and ISR asset shifts	Enables pre-emptive deployments and ISR prioritization
Doctrinal Shift Detection	Identify changes in rules of engagement or targeting philosophy	Refines targeting, rules of engagement, and escalation ladders
Adversary War-gaming Against New Tech	Predict adversary counters to novel systems like hypersonic glide vehicles or space-based ISR	Informs resilient technology rollout and operational integration
Political and Policy Sensitivity Analysis	Quantum models simulate political pressure points, election cycles, sanctions impact	Strengthens deterrence signalling and geopolitical alignment

At the heart of quantum-enabled adversary anticipation lies quantum simulation the ability to model and analyse highly complex, interdependent systems where classical computing struggles. Unlike traditional war-gaming or Monte Carlo simulations that explore a limited number of scenarios sequentially, quantum simulators evaluate millions of permutations simultaneously, making them ideal for forecasting how adversaries might adapt their posture, doctrine, or alliances in response to changing conditions. Key enablers include quantum-enhanced game theory, quantum ML and high-dimensional modelling.

- **Quantum-enhanced game theory** allowing simulation of iterative, adaptive decision-making among state and non-state actors. Quantum-enhanced game theory utilizes quantum superposition and entanglement to process multiple strategy permutations simultaneously allowing war fighters to simulate all possible adversarial responses in parallel.

  - ❑ **Quantum Superposition:** Instead of modelling one strategic move at a time, quantum systems can represent and process all possible moves and counter-moves simultaneously. This enables near-real-time evaluation of "if-we-do-this, they-might-do-that" chains.

  - ❑ **Quantum Entanglement:** Links between different adversarial decisions and domains (e.g., cyber operations linked to kinetic retaliation) can be modelled with entangled quantum states capturing non-obvious dependencies and second-order effects that classical models often miss.

  - ❑ **Quantum Payoff Optimization:** Using quantum optimization techniques (e.g., QAOA or quantum Nash equilibrium solvers), planners can determine the most stable or risky configurations under various adversarial doctrines.

Application Area	Quantum Game Theory Advantage	Military Impact
Doctrine Adaptation Modelling	Simulate how adversaries might shift posture/doctrine in response to new systems (e.g., hypersonic, AI ISR)	Enhances procurement decisions and force structuring
Escalation Ladder Forecasting	Model strategic stability scenarios involving tactical nukes, cyber-attacks, space denial	Improves deterrence calibration and crisis management
Red-Teaming Emerging Tactics	Predict adversary countermeasures to swarming drones, maritime hybrid warfare, or psyops	Supports proactive doctrine evolution
War-gaming Alliance Dynamics	Explore multilateral game models across allies and adversaries	Refines coalition interoperability planning and response synchronization

- **Quantum Machine Learning (QML):** From salami-slice tactics in the South China Sea to grey-zone cyber disruptions, the challenge lies in forecasting not just what the enemy might do, but why, when, and how they will adapt. Quantum Machine Learning (QML) provides this anticipatory edge by merging the adaptive power of AI with the computational superiority of quantum algorithms, which identifies latent behavioural patterns and strategic inflection points across historical and real-time data.

❑ **High-Dimensional Pattern Detection:** QML algorithms can embed massive intelligence datasets (such as troop movements, supply chain anomalies, propaganda narratives, and cyber footprint patterns) into a quantum Hilbert space. This allows for nonlinear correlations to be discovered across data that appears unrelated to classical systems like matching a sudden spike in encrypted radio traffic with satellite heat signatures and social unrest near a border.

❑ **Temporal and Strategic Forecasting:** Quantum recurrent neural networks (QRNNs) and quantum-enhanced time series analysis allow for trajectory prediction forecasting how an adversary might evolve their posture over time based on doctrine, logistics, past manoeuvres, and political pressures.

❑ **Learning from Sparse and Noisy Data:** Military intelligence often comes from incomplete or compromised data (e.g., low-resolution HUMINT, partial SIGINT). QML can learn effectively even with sparse signals, outperforming classical models in de-noising and inference.

Operational Relevance

Capability	Quantum ML Application	Tactical/Strategic Impact
Adversary Posture Prediction	QML identifies subtle cues in movement, force readiness, and information ops	Enables forward positioning and pre-emptive defensive measures
Behavioural Pattern Analysis	Learns adversary responses to past engagements and political triggers	Informs red teaming and deception strategies
Multi-Domain Threat Fusion	Correlates ISR, HUMINT, cyber, and open-source intelligence	Reduces fog-of-war and enhances situational awareness
Early Warning Systems	Detects low-probability, high-impact threat scenarios (e.g., surprise attacks)	Supports escalation control and crisis readiness

- **High-dimensional modelling** captures the interaction between political, military, economic, cyber, and social domains with unprecedented granularity. High-dimensional modelling, especially when enabled by quantum computing, embeds these complex scenarios into multi-axis mathematical spaces (Hilbert spaces) where advanced algorithms can detect emergent adversarial behaviours, hidden dependencies, and nonlinear threat patterns.

  ❑ **Multi-Variable Intelligence Fusion:** Classical systems might analyse troop movements, satellite imagery, and diplomatic statements independently. High-dimensional quantum models fuse all these inputs into a unified, correlated data structure where interdependencies emerge automatically.

  ❑ **Quantum Feature Space Embedding:** Quantum systems embed data points into a vast Hilbert space, allowing them to compute relationships across hundreds of features doctrinal preferences, supply chain anomalies, cyber probes, terrain factors, and psychological operations in parallel.

  ❑ **Trajectory Inference:** Using quantum-enhanced regression and clustering techniques, systems can extrapolate adversary behaviour trends, even when the observable indicators are sparse or masked.

  ❑ **Scenario Differentiation at Scale:** High-dimensional models can differentiate between hundreds of plausible adversarial futures, helping commanders prioritize the most likely courses of action with probabilistic confidence scores.

Operational Relevance

Capability	Enabled by High-Dimensional Modelling	Strategic Advantage
Anticipatory Posture Adjustment	Detects early signals of adversarial build-up or doctrinal pivot	Supports re-positioning of assets and proactive defence
Behavioural Intent Mapping	Models' enemy decision-making under different pressures	Enhances red-teaming and psychological operations
Doctrine-Aware War-gaming	Integrates strategic norms, play-books, and historical patterns	Improves war-game fidelity and doctrinal counterplay
Cross-Domain Awareness	Links movements in cyber, space, air, and land into unified threat models	Supports coordinated multi-domain response

Forecasting Geopolitical Instability and Conflict Zones

Quantum-enhanced forecasting offers a transformative capability in understanding and anticipating geopolitical volatility. At its core, this capability leverages quantum computing's ability to process high-dimensional, interdependent datasets including political signals, economic metrics, climate anomalies, military deployments, and historical unrest patterns. Classical forecasting systems are fundamentally limited by their sequential processing and reliance on linear models. These models often reduce global volatility to simplified cause-effect chains, ignoring nonlinear interdependencies, for example, how a sudden currency devaluation in one region could influence civil unrest in another through trade, energy flows, or military alliances.

Quantum algorithms such as quantum Boltzmann machines and variational quantum circuits can simulate complex, multi-domain futures in which a minor trade disruption, combined with drought-induced migration and a military build-up, could rapidly evolve into regional instability. The result is a multi-scenario heat map of geopolitical risks ranked by likelihood, severity, and temporal proximity generated far faster and more accurately than classical methods allow. Quantum systems overcome this through quantum parallelism, where multiple probable futures are simulated simultaneously rather than iteratively. Algorithms such as Quantum Boltzmann Machines (QBMs), useful for learning and sampling from complex probability distributions, and Variational Quantum Circuits (VQCs), ideal for optimizing across large, uncertain datasets are able to assess how political, economic, environmental, and military variables co-evolve. For instance, a QBM can be trained on historical data from past regional conflicts factoring in climate triggers (drought, floods), energy dependencies (pipelines, resource disputes), population pressure, foreign troop presence, and information warfare. The result is a geopolitical stress index that doesn't just react to existing conditions; it anticipates their progression and identifies weak signals of escalation before they materialize.

Scenario Heat Maps and Strategic Forecasting

Through quantum-enhanced scenario planning, military analysts can generate multi-scenario heat maps that visualize future conflict zones:

Trigger	*Pathway Simulated via QML*	*Predicted Outcome*
Energy Export Ban	Trade contraction → Inflation → Urban Protests	Instability in resource-dependent regions
Arctic Ice Melt	Territorial claim overlap → Naval posturing → Skirmish	Maritime escalation forecasted
Water Scarcity in Sahel	Migration → Ethnic tension → Cross-border militia activity	Sub-Saharan crisis risk signal
Coup in Strategic Ally	Reduced intel sharing → Security vacuum → Regional insurgency	Deterrence failure modelled

These are not mere data visualizations, they are quantum-evaluated futures ranked by probability, impact, and emergence timeframes, helping planners decide where to posture, where to watch, and where to invest resources.

❑ **Quantum Boltzmann Machines (QBMs):** Quantum Boltzmann Machines (QBMs) are a powerful class of quantum machine learning models designed to capture and model complex, high-dimensional probability distributions. In the context of forecasting geopolitical instability and conflict zones, QBMs enable defence and intelligence communities to move beyond classical linear prediction models and simulate how multiple interdependent geopolitical factors evolve simultaneously to trigger future conflicts. A Boltzmann Machine is a type of stochastic recurrent neural network used for modelling probability distributions over complex datasets. A Quantum Boltzmann Machine is its quantum analogue leveraging quantum superposition and entanglement to evaluate

Strategic Application for Defence and Intelligence

Application Area	*QBM-Enabled Benefit*
Early Warning Systems	Detects weak signals before escalation (e.g., minor unrest leading to strategic conflict)
Scenario-based War-gaming	Simulates adversary behaviour under various geopolitical shocks
Intelligence Allocation	Helps allocate ISR assets based on predicted regional instability
Strategic Basing and Prepositioning	Identifies optimal regions for temporary or permanent force deployments
Crisis Response Modelling	Anticipates ripple effects of interventions, sanctions, or natural disasters

many possible states of a system simultaneously, rather than sequentially. QBMs excel in particular in the areas of modelling nonlinear interactions between variables, finding hidden correlations in sparse or noisy data and generating probabilistic forecasts across vast, multidimensional spaces.

❏ **Variational Quantum Circuits (VQCs):** Variational Quantum Circuits (VQCs) ensure forecasting geopolitical instability and conflict zones by enabling adaptive, high-dimensional, and data-efficient modelling of complex and dynamic environments where classical methods struggle to scale. VQCs form the foundation of many hybrid quantum-classical machine learning algorithms, combining the representational power of quantum systems with optimization strategies from classical computation. This fusion makes VQCs ideal for tackling real-world military forecasting tasks that are nonlinear, uncertain, and data-scarce, like anticipating geopolitical conflict flashpoints.

A Variational Quantum Circuit is a parameterized quantum circuit that can be trained to solve optimization and machine learning problems. It works by preparing quantum states using quantum gates parameterized by tuneable angles, measuring outputs (expectation values) and feeding them into a classical optimizer, iteratively tuning parameters to minimize or maximize a cost function. This hybrid structure allows the circuit to learn patterns, simulate stochastic systems, and approximate probability distributions.

Classical vs. VQC for Geopolitical Simulation

Feature	Classical Simulation	VQC-Driven Forecasting
Variable Interdependence Modelling	Limited or manual	Naturally encoded via entanglement
Data Requirement	High	Lower, robust to sparsity
Scenario Generation	Sequential, time-intensive	Parallel, exponential coverage
Adaptability to Real-Time Updates	Slow re-training	Dynamic reconfiguration with fewer iterations
Interpretability of Patterns optimization	Requires post hoc analysis	Captures latent dynamics during

Geopolitical instability results from a complex interplay of variables such as military deployments, climate-induced migration, trade embargoes, proxy interventions, political unrest and more such variable conditions. Classical models like regression or time series forecasting fail to model these variables interactively and simultaneously in nonlinear systems. VQCs, however, leverage quantum entanglement to represent interdependencies between factors, and superposition to explore many possible input combinations at once. VQCs can simulate multiple evolving futures in parallel. This makes them suitable for anticipating conflict chain reactions, forecasting responses to shocks and more. They evaluate thousands of potential conflict pathways simultaneously, updating predictions dynamically as new data streams in from ISR, HUMINT, and OSINT. In many theatres, especially unstable or authoritarian regions, real-time data is scarce or manipulated. VQCs are well-suited for this because they require fewer data samples than deep classical models to generalize. This is critical for military attachés and defence planners who must generate foresight despite intelligence gaps.

Operational Relevance of VQCs for Conflict Forecasting

Use Case	VQC-Enabled Advantage
Strategic Basing Decisions	Simulate long-term instability potential across regions to select optimal staging areas.
Crisis Monitoring	Forecast escalation likelihood after events like elections, coups, or military exercises.
Intelligence Allocation	Prioritize ISR and HUMINT resources in regions with predicted high volatility.
Joint Planning with Allies	Run multi-scenario simulations across allied theatres for coordinated early action.
Conflict Prevention	Provide inputs for diplomacy or deterrence measures before conflict hardens.

Optimizing Force Posture and Readiness across Theatres

In the age of multi-theatre, multi-domain conflict, achieving real-time force readiness and optimal deployment is no longer a matter of periodic strategic reviews. It demands constant recalibration in response to evolving threats, terrain dynamics, and logistic constraints. Quantum computing introduces a capability to military readiness through its ability to simultaneously evaluate and optimize countless deployment configurations across geographically dispersed theatres. By processing a set of interdependent variables such as fuel availability, transport times, and weather forecasts to enemy capabilities, cyber threat levels, and terrain complexity, quantum systems can deliver near-instantaneous recommendations on optimal force posture.

Quantum algorithms such as the Quantum Approximate Optimization Algorithm (QAOA) and Variational Quantum Eigen solvers (VQE) are particularly suited to military logistics and force optimization tasks, where classical computing falters due to the combinatorial explosion of possibilities. These systems provide commanders with predictive readiness models that adapt not only to current battlefield data but also simulate near-future evolutions, allowing for highly responsive adjustments in posture.

- **Quantum Approximate Optimization Algorithm (QAOA):** QAOA being variational quantum-classical hybrid algorithm is designed to solve complex combinatorial optimization problems. This means optimizing over enormous and constraint-bound search spaces such as how to deploy forces across multiple theatres with limited transport capacity, fuel availability, readiness states, and evolving threats. QAOA works by encoding the optimization problem into a cost Hamiltonian (an energy function representing the objective), and iteratively applying parameterized quantum operations to find the minimum energy configuration, i.e., the best solution. Military theatre readiness and force posture decisions typically involve NP-hard problems, such as "Which battalions should be pre-positioned where?", "How should reinforcements be routed across contested terrain?", and "What configuration of joint forces across land, sea, and air achieves the highest deterrence value with lowest vulnerability?" Classical solvers face exponential growth in compute time as variables (nodes, constraints, adversarial responses) increase. QAOA, by leveraging quantum parallelism and superposition, explores thousands (or millions) of force permutations simultaneously, evaluating each against mission objectives and constraint functions.

Technical Workflow of QAOA in Force Posture Optimization

Step	Description	Military Contextualization
Problem Encoding	The force deployment challenge is converted into a binary optimization problem. Constraints and objectives are represented in a cost function (Hamiltonian).	Example: Each variable represents a decision (e.g., deploy 17 Para SF to Eastern Ladakh or not); constraints include supply chain capacity, readiness window, and airlift availability.
Quantum Circuit Initialization	A quantum circuit is prepared in a superposition of all possible states (deployment permutations).	Simulates all feasible joint-force configurations across land, sea, air, and cyber simultaneously.
Alternating Operators (QAOA Layers)	The algorithm alternates between two quantum operators: the cost operator (to encode the objective function) and the mixer operator (to explore the solution space). These are parameterized by angles (γ, β) optimized classically.	Simulates potential adjustments like reallocating artillery from Arunachal to central command if adversary airlift is detected.
Measurement and Sampling	After running the quantum circuit multiple times, the most frequent output state(s) represent the optimal or near-optimal solution.	Suggests deployment decisions with highest operational value, lowest logistics risk, and best deterrent impact.

There can be multiple application scenarios in theatre scale military planning based on QAOA:

- **Force Mobility and Placement Optimization:** Multiple variables such as terrain constraints, enemy ISR coverage, fuel and munitions supply, and weather windows can be solved with the objective to maximize force presence and readiness across strategic chokepoints with minimal vulnerability

- **Allied Coordination and Joint Readiness:** QAOA can simulate interoperability efficiency: What deployment schema of US-India-Japan-Australia forces ensures fastest QUAD activation in the IOR?

- **Dynamic Contingency Planning:** QAOA allows real-time re-computation of posture when adversarial manoeuvres are detected (e.g., PLA brigade repositioning, cyber-attack on logistics ERP)

- **Decoy and Misdirection Simulation:** QAOA can also simulate probabilistic adversary reactions selecting force postures that not only optimize defence, but also strategically mislead enemy sensors or intelligence estimates

- **Variational Quantum Eigen solvers (VQE):** VQE is a hybrid quantum-classical algorithm designed to find the lowest eigenvalue (ground state energy) of a given Hamiltonian, essentially the optimal configuration of a complex quantum system. In practical applications, this is used for solving non-linear optimization problems, which are prevalent in force deployment planning, resource allocation, and multi-domain military logistics. Unlike purely quantum algorithms like QAOA, VQE divides the computational load. The quantum processor prepares a quantum state and estimates the expectation value of the system Hamiltonian. A classical optimizer adjusts parameters to iteratively minimize the energy function (i.e., optimize the deployment solution).

With military posture planning across multiple theatres involving complex interdependencies (e.g., fuel availability affecting airlift; cyber threats degrading command mobility), non-convex objectives (e.g., maximizing deterrence while minimizing cost or risk), high-dimensional configuration spaces that classical solvers struggle with (the "combinatorial explosion" problem) VQE is powerful because it can find globally optimal solutions in these rugged, high-dimensional landscapes especially when traditional gradient methods fail.

Technical Workflow of VQE in Force Posture Optimization

Step	*Process*	*Military Contextualization*
Problem Mapping	The military deployment scenario is translated into a Hamiltonian (energy function).	Units, terrain, threat, and logistics are mapped as variables; constraints encoded into penalty terms.
Ansatz State Preparation	A parameterized quantum circuit (ansatz) prepares trial quantum states.	Think of each ansatz state as a specific deployment plan (e.g., air-defence in Taiwan + amphibious prepositioning).
Expectation Value Measurement	Quantum processor measures energy of each state.	Each measurement reflects the "cost" or "efficiency" of that deployment configuration.
Classical Optimization	A classical algorithm (e.g., gradient descent, Nelder-Mead) tunes the ansatz parameters.	The algorithm iteratively improves the solution like a digital CO planning force disposition.
Convergence	The process repeats until minimal energy (optimal configuration) is found.	Optimal posture is selected based on joint objectives: deterrence, mobility, readiness, logistics.

Countering Surprise through Pattern Recognition

In today's high-tempo, hybrid battle space, surprise is a weapon. Adversaries exploit asymmetric tactics, camouflage within civilian environments, and manipulate disinformation to conceal their intent. Quantum computing offers a game-changing capability in this realm by enabling advanced pattern recognition within massive, unstructured, and noisy datasets. Quantum Machine Learning (QML) combines quantum computing with statistical learning methods to enable powerful pattern recognition in extremely high-dimensional and noisy datasets. While classical ML algorithms depend on sequential data parsing and require substantial training data, QML models such as Quantum Support Vector Machines (QSVMs), Quantum Neural Networks (QNNs), and Quantum Kernel Estimators can explore complex correlations across data modalities simultaneously using principles of superposition and entanglement.

Technical Advantages over Classical ML

Capability	Classical ML	Quantum ML
Feature Space Handling	Linear or kernel-based; high-dimensionality is slow	Superposed, exponentially large feature spaces
Training Data Requirements	Large, structured datasets needed	Smaller, high-entropy datasets sufficient
Pattern Recognition	Sequential, siloed	Parallel, fused, multi-domain
Adaptability to New Threats	Needs retraining	Continual learning via quantum variational circuits
Speed	Limited by model size and CPU/GPU capacity	Massive parallelism in QPU enables near-real-time inference

This allows QML systems to fuse multimodal datasets (SIGINT, IMINT, HUMINT, cyber telemetry, satellite weather data) to uncover latent threat patterns (including low-observable or camouflaged behaviours) in order to predict the intent behind subtle and composite changes in adversarial posture. QML models can detect subtle deviations from historical baselines such as changes in vehicle movement, communication patterns, or even the lack of expected activity. These changes are often precursors to surprise attacks or deception operations. For instance, in an urban insurgency scenario, QML has the potential to identify an absence of market activity and altered mobile usage in a sector with no declared enemy presence flagging it as a potential staging ground for an IED ambush. QML can build composite pattern forming correlation across domains. While traditional models treat intelligence data in silos. QML, by contrast, integrates data across time, geography, and modality, helping uncover weak signals that classical AI would miss. A faint spike in encrypted RF traffic, a simultaneous drone flyover, and troop withdrawal in a separate AO analysed together may indicate an imminent cyber-kinetic operation. QML can flag this with high confidence due to non-linear correlation learning. QML algorithms trained using variational quantum circuits can continuously update their threat models as new data arrives, without re-training from scratch. This allows them to remain responsive in fluid battle spaces such as hybrid zones, where doctrines shift mid-conflict.

Operational Relevance

- **Supports Counter-Terrorism, Irregular Warfare, and Hybrid Threat Detection:** In operational theatres where enemy combatants blend with civilians, or cyber and psychological operations precede kinetic engagements, quantum systems enhance early-warning mechanisms by uncovering irregular behavioural signatures. A sudden shift in encrypted messaging patterns, the anomalous movement of vehicles in a contested zone, or the fusion of economic and weather disruptions can all serve as indicators of a coordinated strike or insurgent operation.

- **Identifies Shifts in Adversary Doctrine or Deception Strategies:** Quantum models can compare long-range behavioural baselines with real-time operational deviations, helping military intelligence identify when an adversary deviates from their standard playbook. Whether it's reduced electronic chatter before an electronic warfare assault, or logistical movements suggesting forward basing, such subtle changes can be flagged and assessed before they escalate.

- **Enhances Intelligence Fusion and Mission Assurance:** Quantum pattern recognition isn't siloed; it fuses signals across domains. For example, a drop in radio communications, weather anomalies, and rerouted supply chains may collectively point to a decoy operation. This fused insight enables commanders to adapt mission profiles in real time, re-route forces, reinforce vulnerable sectors, and pre-empt enemy surprise.

Fleet Movement Optimization in Multi-Theatre Scenarios

Modern naval operations involve a multidimensional battle space where force projection, deterrence, and responsiveness hinge upon effective fleet positioning. Traditional tools based on classical heuristics or linear programming struggle with the combinatorial explosion of variables inherent in planning across multiple theatres such as the Indo-Pacific, North Atlantic, and Arctic simultaneously. Quantum computing, particularly through algorithms like QAOA and VQE, addresses this challenge by encoding vast logistical and strategic variables into quantum states and solving optimization problems in exponentially large configuration spaces.

QAOA and VQE evaluate multiple variables simultaneously ranging from adversary fleet movement patterns, weather disruptions, fuel and resupply constraints, to undersea terrain and ISR asset availability. This enables naval strategists to simulate and select optimal fleet distributions that pre-empt emerging threats and maintain persistent presence across global areas of responsibility (AORs).

- **QAOA for Maritime Deployment Planning:** QAOA is well-suited for solving discrete optimization problems, such as where to place naval assets to maximize area coverage, minimize time-to-intervention, or optimize proximity to contested zones without escalating tensions. QAOA operates by encoding the fleet positioning problem into a cost Hamiltonian, where the "cost" represents suboptimal outcomes, e.g., delayed response, high fuel usage, or vulnerable placement. It uses a quantum-classical hybrid loop to iteratively find near-optimal solutions. Quantum processors explore the massive configuration space of potential deployments in superposition, while classical optimizers refine parameter settings. It enables multi-objective trade-off balancing such as between covering strategic chokepoints (e.g., Strait of Hormuz), minimizing re-supply delays, and ensuring redundancy in communications relay nodes (e.g., amphibious command ships). Optimizing the placement of three carrier strike groups such that at least one is within 36 hours of a projected flashpoint in each theatre, without overburdening any single logistics chain.

- **VQE for Resource Allocation under Uncertainty:** While VQE is often used in chemistry, it also solves continuous optimization problems, especially where the system is governed by uncertain variables, e.g., fuel consumption affected by sea state, or detection ranges impacted by weather. VQE minimizes the expected energy (or cost function) of a parametrized quantum circuit to find ground states in complex systems. In a military context, these "ground states" represent optimal configurations with minimal resource strain. It is particularly valuable in adaptive logistics: computing minimum-risk routes for resupply vessels under dynamic threat conditions, or adjusting planned routes in real time when intelligence indicates shifting submarine activity. VQE is robust in stochastic environments, such as simulating the risk-weighted likelihood of adversary fleet convergence in the South China Sea under escalating maritime exercises, determining the optimal fuel distribution between multiple support ships during a contested Pacific operation, while factoring in the probability of rerouting due to hostile surveillance or emerging typhoon paths.

- **Multi-Theatre Synchronization and ISR Integration:** Quantum systems does not merely optimize individual fleet moves; they integrate real-time ISR feeds (from space, cyber, sonar, and airborne assets) and model multi-theatre dependencies. Quantum simulations use high-dimensional entanglement of variables, allowing decision-makers to understand strategic ripple effects such as how moving a destroyer squadron from Guam to the Red Sea affects U.S. Indo-Pacific deterrence.

Operational Relevance: Dominance through Agility and Anticipation

- **Enhances Fleet Responsiveness to Crisis or Aggression:** In time-sensitive flashpoints such as a sudden escalation in the South China Sea or Arctic maritime disputes, quantum-enhanced fleet models allow rapid recalibration of naval posture. Ships can be rerouted with optimized risk profiles, ensuring they arrive at key positions before adversarial escalation peaks.

- **Reduces Logistical Strain During Force Projection:** By minimizing overlapping deployments, redundant re-supply chains, and inefficient routing, quantum optimization cuts down on fuel expenditure, wear-and-tear, and logistic tail vulnerabilities. This is particularly vital for blue-water navies operating across great distances with finite replenishment capacity.

- **Supports Deterrence through Agile Naval Posturing:** Strategic presence is not just about mass it's about precision timing and positioning. Quantum systems empower naval commanders to deploy force packages that appear unpredictable but with operational logic, signalling readiness without provocation. This bolsters strategic ambiguity and deterrence in gray-zone operations and contested sea lanes.

Logistics and Fuel Efficiency Across Naval Supply Chains

Naval operations rely on one unyielding truth: no fleet moves without fuel, and no campaign sustains without re-supply. Quantum computing introduces a leap in logistics optimization through high-fidelity modelling of fuel consumption, replenishment cycles, and re-supply routing. Unlike traditional linear optimization tools, quantum algorithms such as the QAOA excel at solving combinatorially complex problems that arise in naval logistics, where countless variables interact in real time. These include vessel fuel burn rates, maritime weather conditions, regional chokepoints, replenishment ship positioning, depot status, and threat proximity.

Naval logistics optimization problems often fall into the category of NP-hard combinatorial optimization problems. These include vehicle routing problem (VRP) with time windows, fuel limits, and risk weights, multi-depot logistics optimization, where multiple supply bases support multiple fleets with constrained resources, and dynamic re-supply planning, where changing

Technical Flow Example: Fuel-Efficient Convoy Planning Using

Step	Description
1. Model QUBO	Define binary variables: X_{ij} = 1 if tanker i services fleet j. Constraints: fuel capacity, distance, exposure level.
2. Hamiltonian Construction	Encode objective as cost: $H = \Sigma$ (distance * fuel usage * threat factor)
3. Quantum Circuit	Create QAOA circuit with p layers. Each layer applies a mixing and phase-separation operation.
4. Classical Optimization	Use gradient-based methods (e.g., Adam, Nelder-Mead) to optimize angles of quantum gates for minimal energy (optimal routing).
5. Measurement and Interpretation	Collapse quantum state to extract best routing configuration. Convert output bitstring into actionable fleet orders.

threat levels or weather conditions require real-time re-planning. Classical solvers suffer from exponential growth in computation time as the number of variables and constraints increases, especially when adapting to real-time battlefield updates. This is where quantum algorithms can bring transformational change.

- **Quantum Approximate Optimization Algorithm (QAOA):** QAOA is a variational hybrid algorithm designed to find approximate solutions to combinatorial optimization problems. It operates on problems expressible in Ising model or Quadratic Unconstrained Binary Optimization (QUBO) form. To minimize the total "energy" (i.e., cost) of visiting all required nodes within constraints the resupply scheduling is encoded as a QUBO matrix. Then, quantum circuits are designed to evolve towards the lowest-cost configuration post this tuning of variational parameters using classical optimization (e.g., COBYLA or SPSA) is applied. Parallelism inherent to quantum superposition allows QAOA to evaluate multiple supply chain configurations simultaneously, reducing time to optimality.

- **Variational Quantum Eigen solver (VQE):** VQE is used to find the minimum eigenvalue of a Hamiltonian (energy landscape), ideal for solving constrained optimization problems in logistics. With the objective to minimize fuel consumption across all fleet movements under sea-state conditions and enemy proximity and to evaluate lowest-risk depot placement for at-sea replenishment by minimizing exposure energy, a cost function (Hamiltonian) is defined that captures all logistic constraints and priorities. A parameterized quantum circuit that represents possible configurations is created. The quantum circuit is then run to evaluate expected energy (cost), and iterate until minimum energy (optimal plan) is found.

- **Quantum Annealing (via D-Wave-like architectures):** It is best suited for near-term quantum annealers. This method finds global optima of hard QUBO problems via a physical energy minimization process. This approach is effective for real-time reconfiguration of resupply convoys when disruptions (like enemy interdiction) occur and are robust in non-deterministic environments like maritime patrols where ISR data continuously changes.

Capability	Quantum Benefit
Real-time convoy re-routing	QAOA explores all alternatives at once under constraints
Energy-efficient fuel planning	VQE finds minimum-cost energy configuration
Response to disrupted sea lanes	Quantum annealing re-optimizes based on new ISR inputs
Coordinated multi-theatre support	QAOA can jointly optimize all support routes across oceans

Operational Relevance

- **Sustained Operations in Contested Waters:** Quantum-driven logistics models enable long-range operational planning even in theatres devoid of secure bases such as the South China Sea or the Arctic. By simulating thousands of replenishments and refuelling combinations, these systems can recommend optimal schedules and forward-deployed fuel stockpiles to support persistent naval presence.

- **Reduced Vulnerability of Supply Convoys:** In adversarial environments, supply lines are soft targets. Quantum logistics simulations help identify re-supply routes with the lowest exposure to enemy ISR and interdiction, optimizing time, distance, and security. This enhances survivability and mission assurance for underway replenishment operations.

- **Adaptive Support to Joint Task Forces:** Naval logistics must now integrate with joint and coalition forces across multi-domain operations. Quantum logistics planning accounts for variable demands from air wings, amphibious task forces, and even cyber assets needing afloat data relays synchronizing support chains dynamically to match evolving joint force requirements.

Port Security and Threat Anticipation

In the era of grey-zone conflict, transnational terrorism, and strategic competition over maritime access, ports have evolved from commercial gateways into critical military and security assets. The ability to anticipate threats before they materialize, be it from smuggled weapons of mass destruction, cyber-induced port disruptions, or disguised infiltration vessels demands analytical capabilities beyond traditional surveillance or human intelligence (HUMINT) alone. Quantum-enhanced models offer this edge by leveraging quantum machine learning (QML), variational algorithms, and high-dimensional pattern recognition to detect anomalies in maritime traffic, container manifests, and behavioural shipping data in real time.

Operational Translation: Real-Time, Predictive Naval Intelligence

Quantum Technique	*Security Function*	*Operational Impact*
Quantum Machine Learning (QML)	Classify vessel behaviour, flag anomalies	Early detection of deception, predictive threat alerts
Variational Quantum Algorithms	Optimize port clearance decisions under uncertainty	Reduce false positives, prioritize inspection queues
Quantum Boltzmann Machines	Detect hidden correlations in high-risk cargo movement	Identify smuggling, insider threats, or covert infiltration
Quantum Clustering & QPCA	Group and visualize shipping trends	Detect new smuggling routes or unusual geopolitical shifts

- **Quantum Machine Learning (QML):** Quantum Machine Learning merges classical data structures with quantum computational power to reveal patterns, correlations, and anomalies faster and more deeply than classical ML systems. Quantum kernel methods, quantum support vector machines (QSVM), and parameterized quantum circuits enable classification of complex shipping behaviour by encoding high-dimensional features (e.g., vessel flag, port of origin, deviation from expected course) into quantum Hilbert spaces.

  A QML model can learn to detect subtle anomalies like ghost ships (vessels that spoof or intermittently deactivate their AIS) or identify behavioural drift in trusted shipping lanes, e.g., a cargo vessel historically aligned with civilian movement now showing navigation paths that mirror military logistics routes. QML algorithms exploit quantum entanglement and interference to evaluate thousands of overlapping behavioural signals in parallel, reducing false positives and enabling real-time threat escalation scoring.

- **Variational Quantum Algorithms (VQAs):** VQAs, especially Variational Quantum Classifiers (VQC) and Variational Quantum Eigen solvers (VQE), are hybrid quantum-classical models optimized for near-term noisy intermediate-scale quantum (NISQ) devices. They are well-suited for maritime-domain use due to their ability to adaptively optimize under uncertain, noisy input data. A variational circuit is trained to minimize a cost function (e.g., probability of a vessel being a threat) based on feature encodings such as vessel tonnage, fuel consumption anomalies, last port of call, etc. VQAs can learn temporal patterns of deception, such as containers arriving just before port shutdowns or correlations between flagged cargo and time-stamped diplomatic movements in port-hosting nations. Unlike classical algorithms, VQAs do not require exhaustive training data and can operate efficiently under sparse, incomplete, or corrupted datasets common in adversarial intelligence environments.

- **Quantum Boltzmann Machines (QBM):** A Quantum Boltzmann Machine is a probabilistic model that encodes correlations between variables using quantum amplitudes, enabling more nuanced anomaly detection in nonlinear maritime logistics networks. QBMs learn probability distributions of normal vs. adversarial activity across interconnected data points such as cargo manifests, crew biometrics, time of arrival, and port proximity to known terrorist hubs. For e.g., detecting low-frequency, high-impact threats, such as a legitimate cargo vessel covertly carrying materials for chemical weapons production where each data point alone is benign but jointly forms a threat signature. QBMs use quantum annealing to reach optimal detection states faster than classical Boltzmann networks, which often get stuck in local minima due to the rugged nature of real-world data landscapes.

- **Quantum k-Means Clustering and Quantum Principal Component Analysis (QPCA):** These unsupervised quantum algorithms enable dynamic segmentation and trend analysis of complex, multi-source port traffic data. The algorithms can automatically group ships by behavioural similarity (e.g., traffic patterns, time-window entries, flag hopping), identifying outliers that might evade traditional filters. Quantum clustering reduces computational time complexity from exponential to polynomial (or better) in certain cases, enabling scalable maritime surveillance across hundreds of global ports simultaneously.

Operational Relevance

- **Enhanced Maritime Domain Awareness (MDA):** Quantum models fuse satellite data, AIS (Automatic Identification System) signals, cargo records, and historical maritime behaviour to deliver predictive insights on vessel intent and anomalies. This fusion enables forward naval commands and maritime operations centres to proactively monitor not just the sea lanes, but the patterns behind them.

- **Pre-emptive Interdiction and Threat Disruption:** By simulating countless potential shipping and cargo scenarios, quantum systems can detect signatures of covert operations such as irregular trans-shipment patterns, forged cargo data, or stealthy approach vectors of hostile actors long before they reach the littoral. This enhances the military's ability to interdict potential threats like nuclear smuggling, bio-weapon deployment, or asymmetric sabotage campaigns at the maritime doorstep.

- **Port Resilience and Continuity of Operations:** Ports in contested zones whether in the Indo-Pacific, Gulf, or Arctic are vulnerable not just to kinetic strikes but to cyber-attacks, AI-driven deception, and supply chain disruption. Quantum-enhanced simulations allow for proactive resilience planning, helping naval and coast guard commands assess how blockades, infrastructure failures, or intelligence leaks could impact port throughput and force projection. These systems enable dynamic reallocation of patrols, re-supply plans, and port defences in real time, ensuring that critical chokepoints remain secure and functional.

Submarine Patrol Route Planning with Stealth Maximization

In undersea warfare, every movement a submarine makes is a trade-off between mission effectiveness and survivability. The core computational challenge lies in optimizing route planning under multi-variable constraints; some static (like bathymetry or seabed maps) and others dynamic (like ASW patrol activity or thermal gradients). Classical systems are hamstrung by the combinatorial explosion of possible routes, especially when accounting for noise profiles, threat probabilities, sonar conditions, and satellite visibility windows. This is where quantum algorithms, especially QAOA and VQE, deliver a game-changing edge. The algorithms solve complex optimization problems in real time, identifying optimal paths that minimize acoustic signature exposure, exploit thermal layers for sound masking, and avoid satellite detection swaths, all while maintaining operational objectives such as ISR or launch readiness.

Comparative Advantage over Classical Methods

Challenge Area	Classical Algorithms	Quantum Algorithms (QAOA/VQE)
Route Permutation Exploration	Exponential time	Polynomial time with higher solution fidelity
Environmental Variable Integration	Static/simplified models	Dynamic, multi-variable quantum simulations
Acoustic Signature Minimization	Post-hoc estimation	Integrated into optimization cost function
Real-time Threat Re-planning	High latency or infeasible	Feasible via quantum circuit adaptation
Multi-Objective Optimization	Linear trade-offs	Simultaneous probabilistic trade-off balancing

- **Quantum Approximate Optimization Algorithm (QAOA):** QAOA is designed to solve combinatorial optimization problems, particularly those expressible in the form of a cost function over binary variables, a perfect fit for patrol pathing problems where discrete route decisions must be made (e.g., waypoint A or B at time t).

  For submarine routing, QAOA encodes each potential path segment as a binary variable in a graph. The cost function integrates multiple threat parameters: acoustic signature probabilities, sonar density maps, terrain occlusion, and predicted enemy movement vectors. The QAOA iteratively evolves a quantum state that concentrates probability mass on the most "stealth-optimal" routes, while adapting to time-variant constraints, e.g., if satellite overpasses change or if sonar buoys are deployed mid-patrol.

- **Variational Quantum Eigen solver (VQE):** While originally designed for quantum chemistry problems, VQE is effective in continuous-variable optimization tasks, particularly when environmental uncertainty is involved.

  In submarine route planning, VQE models the undersea environment as a quantum potential field, where terrain elevation, water salinity, and thermal layers affect sonar propagation. This solves for the "lowest energy" paths that represent minimum detectability trajectory paths where the acoustic profile of the submarine is naturally masked by the environment. This can incorporate continuous data like ocean currents, pressure gradients, and dynamic changes to water temperature that affect sound velocity profiles.

Operational Relevance

- **Increases Survivability of Nuclear and Diesel-Electric Submarines:** QAOA tackles patrol route selection as a combinatorial optimization problem, where each leg of the patrol route is a decision variable. QAOA builds a quantum circuit that encodes all feasible patrol paths. It integrates a cost function designed to minimize the cumulative detection risk, using weighted penalties for acoustically exposed zones, probable ASW intercepts, or high ambient noise areas. The algorithm outputs a probabilistically optimized patrol path that maximizes stealth while maintaining mission coverage. VQE models the undersea battle space as a complex quantum system, particularly useful when incorporating continuous variables. VQE computes the "lowest energy state" of a quantum Hamiltonian, representing the path of least probability of acoustic detection. It identifies "stealth corridors" where environmental conditions naturally mask a sub's signature. Simulations can adapt in quasi-real time to enemy ASW repositioning or environmental shifts (e.g., changing temperature profiles due to weather fronts).

Combined Operational Advantage

Feature	QAOA	VQE
Tactical Routing Optimization	✓	—
Continuous Environmental Modelling	—	✓
Response to Dynamic ASW Movements	✓	✓
Acoustic Signature Minimization	Partial	Full
Mission-Specific Constraints (e.g., Loiter Time, Launch Readiness)	✓	✓

- **Improves ISR from Undersea Assets:** QAOA tackles the problem of ISR path planning as a constrained combinatorial optimization challenge, where trade-offs must be made between sensor coverage area, time on target, acoustic and electromagnetic detectability, threat proximity, mission constraints and many other variables. A cost function is formulated where each route segment is weighted by risk (e.g., proximity to active sonar zones) and reward (e.g., ISR gain per minute). QAOA constructs a quantum circuit that explores millions of combinations of ingress/egress and loiter trajectories. It then outputs the minimum-risk, maximum-ISR path, optimized for real-time environmental and tactical conditions.

- **Reduces Predictability of Submarine Operations:** QAOA excels at solving non-deterministic routing problems where the goal is to maximize stealth and mission value while introducing controlled randomness to break adversarial prediction models. QAOA treats route generation as an optimization problem with multiple dynamic variables. Instead of producing a single optimal path, QAOA generates a high-fidelity set of optimal randomized patrol permutations, each viable under current mission constraints. Each route is scored using a cost function that blends operational effectiveness and detectability randomness a controlled departure from predictability.

  VQE enhances route unpredictability by modelling how natural environmental variability (e.g., thermoclines, seabed composition, salinity layers) affects sonar performance and detection windows in nonlinear, location-specific ways. VQE computes low-energy "eigenstates" that correspond to acoustically favourable stealth corridors. It matches submarine movement options to these transient safe zones, which fluctuate over time with environmental conditions. These corridors are rarely fixed and often inaccessible to classical prediction tools due to their complexity and volatility.

Modelling Multi-Domain Battlefields in Real Time

Modern warfare is inherently multidimensional, involving synchronized operations across land, air, maritime, space, and cyber domains. The need for real-time, adaptive modelling of these integrated battle spaces has far outpaced the capabilities of classical simulation architectures, which rely on serial or narrowly parallelized computation. These conventional systems struggle to process complex interdependencies such as how a jamming event in the cyber domain affects GPS-guided artillery in the land domain, or how a kinetic satellite strike alters air mission routing especially under time-critical constraints.

Quantum computing presents a revolutionary alternative by exploiting two quantum mechanical principles: superposition and entanglement. Superposition allows quantum bits (qubits) to exist in multiple states simultaneously. This means a quantum processor can evaluate a vast superposition of battlefield states, encompassing numerous possible configurations of force postures, weather changes, electronic interference, and adversary manoeuvres all in parallel. Entanglement enables deep correlations between qubits, which is critical for modelling cross-domain dependencies. For instance, the operational status of satellite-based ISR systems may be entangled with the effectiveness of air-ground coordination or logistics planning. This interconnectivity is embedded in the quantum model's Hamiltonian, which encodes the system's constraints, objectives, and dynamics.

Technically, quantum algorithms such as the Quantum Approximate Optimization Algorithm (QAOA) and Variational Quantum Eigen solvers (VQE) can be applied to simulate and optimize battlefield scenarios with hundreds of interacting parameters. These include asset allocation under uncertainty (e.g., how best to reassign EW aircraft after satellite communications are jammed), real-time adversarial war gaming (e.g., simulating the second- and third-order effects of enemy troop surges), mission planning in contested A2/AD environments (anti-access/area denial zones), route optimization under cyber disruption, including multi-modal supply chain recalculations in degraded networks.

Classical military simulation and planning tools frequently use the Monte Carlo method, a technique based on running large numbers of probabilistic simulations to assess risk, project outcomes, or optimize responses under uncertainty. This might take days to explore a limited set of possible futures, while a quantum computer models thousands or millions of potential futures in parallel, enabling near-real-time adjustment of plans based on unfolding battlefield data. This delay introduces command latency, a critical vulnerability in time-sensitive decision-making such as coordinated missile defence under electronic attack, dynamic ISR asset redeployment during contested air superiority, logistics rerouting in A2/AD (anti-access/area denial) zones.

Operational Relevance

- **Enhancing Joint Operations Planning:** In classical simulation frameworks, modelling joint operations is often constrained by sequential processing and limited data integration across domains. As a result, planners may miss critical interdependencies such as how the degradation of one domain (e.g., satellite assets in space) could cascade through others (e.g., comms and navigation in air and ground operations). Quantum simulations overcome these barriers by leveraging superposition and entanglement to evaluate all potential outcomes of a single event across multiple domains in parallel. For instance, the simulation of an anti-satellite (ASAT) attack can simultaneously calculate its impact on air operations (loss of GPS-based precision strike capabilities), ground logistics (navigation and supply chain rerouting), cyber security (desynchronization of encrypted networks), and naval assets (loss of satellite-linked ISR).

- **Modelling Integrated Force Effects:** Modelling how ISR feeds from UAVs interact with electronic jamming, kinetic strikes, and cyber intrusion campaigns is far beyond the capabilities of classical tools operating in isolation. Quantum-enabled simulation platforms, however, can ingest and process diverse, cross-domain data like UAV surveillance feeds, missile telemetry, and signals intelligence and simulate how each input interacts with the others under combat conditions. These models consider the cumulative (e.g., overlapping fields of fire) and emergent effects (e.g., adversary confusion due to multi-modal attacks) of joint-force actions. This helps to identify conflict amplifiers (e.g., cyber-attack triggering a kinetic response), enables de-confliction of overlapping force vectors (e.g., friendly UAV paths and artillery arcs), and offers a real-time simulation of multi-layered deterrence.

- **Preparing for Grey-Zone and Asymmetric Threats:** Grey-zone warfare where adversaries operate below the threshold of conventional conflict using non-state actors, economic pressure, disinformation, and cyber sabotage requires planning tools that are probabilistic, adaptive, and multi-layered. Quantum battlefield models can incorporate irregular warfare variables providing a realistic planning tool for hybrid and grey-zone conflicts where traditional simulation tools fall short.

  Traditional war-gaming tools struggle to capture these asymmetric and non-linear threats. In contrast, quantum simulations can model behavioural patterns of irregular forces (e.g., smuggling networks or urban insurgents), spread of disinformation via social media and its impact on local populations or troop morale, economic effects of blockade operations or infrastructure disruption. This prepares forces for hybrid scenarios involving both kinetic and information operations while assisting in shaping pre-crisis indicators of grey-zone activities.

Predicting Enemy Reactions to Doctrinal Shifts

Warfare evolves not only through new technologies, but through shifts in doctrine how those technologies are deployed, how forces are structured, and how adversaries are expected to respond. The challenge lies in anticipating these responses before they manifest on the battlefield. Classical simulations often fall short in this regard, struggling with the high-dimensional, uncertain nature of enemy behaviour across diverse scenarios. Quantum-enhanced simulations, however, enable planners to explore a wide range of adversarial reactions to doctrinal changes such as the deployment of autonomous drone swarms, space-based missile tracking, or integrated cyber-electromagnetic operations.

Quantum computers, via quantum parallelism, operate on superpositions of states allowing them to process exponentially many configurations at once. When applied to adversarial modelling, this capability allows quantum systems to simulate a large number of plausible responses of an adversary to a new weapon system (e.g., hypersonic missile deployment), multiple doctrinal adaptations (e.g., integrated cyber-kinetic responses), geopolitical consequences (e.g., alliance strengthening or economic retaliation), and cross-domain ripple effects (e.g., changes in space or cyber posture due to a naval doctrinal shift). Quantum algorithms such as the Quantum Approximate Optimization Algorithm (QAOA) or Quantum Bayesian Networks are tailored to manage NP-hard, high-dimensional strategic landscapes. These tools allow defence analysts to encode numerous interrelated factors to simulate the adaptive behaviour of adversaries not as a fixed response but as an evolving system.

Operational Relevance: Adversarial Reaction to a New Doctrine

- **Enables Red-Teaming of Emerging Capabilities:** Traditional red-teaming involves simulating an adversary's behaviour to test the resilience of new doctrines, systems, or tactics. However, this method is often limited by cognitive bias, time constraints, and the linear nature of classical modelling. In a complex, high-tempo battle space where adversaries evolve rapidly technologically and doctrinally defence planners require tools that can model not just known threats, but how an adversary might adapt to the unknown, Quantum computing enables this shift by leveraging quantum-enhanced simulations that explore a vast set of doctrinal permutations and strategic adaptations in parallel. Through algorithms such as the Quantum Approximate Optimization Algorithm (QAOA) or quantum-enhanced game theory frameworks, planners can run red-team scenarios that evolve dynamically in response to Blue Force innovations.

Quantum computers excel at combinatorial optimization solving problems with many interdependent variables and constraints. This makes them ideal for:

❑ Modelling how different adversarial doctrines might evolve in response to Blue capabilities (e.g., how the introduction of hypersonic platforms affects enemy radar postures and missile defence layering) when introducing next-gen systems such as autonomous combat vehicles or space-based ISR platforms. Quantum red-teaming can simulate how peer adversaries might counter through anti-satellite weapons, cyber sabotage, or doctrinal denial operations.

❑ Exploring new threat-concept architectures (e.g., how near-peer adversaries respond to low-earth orbit surveillance proliferation) before rolling out major doctrinal shifts (e.g., Multi-Domain Operations or Integrated Deterrence models). Military planners can run quantum simulations to identify how adversaries might attempt to exploit inter-service seams or logistical chokepoints.

❑ Simulating hybrid warfare reactions (e.g., increased use of proxies or disinformation following deployment of AI-enabled battlefield networks). Quantum-enabled red-teaming can model how adversaries react differently to unilateral vs. coalition actions, enabling better coordination and threat anticipation across allies.

• **Supports Strategic Planning against Hybrid and Non-Traditional Threats:** Hybrid and non-traditional threats ranging from cyber-sabotage and misinformation campaigns to proxy warfare and economic coercion pose a critical challenge to modern military doctrine in grey-zone conflicts where adversaries exploit ambiguity and innovation. Quantum computing introduces a paradigm shift by allowing military planners to simulate and analyse high-dimensional, probabilistic scenarios where outcomes are not singular or binary, but distributions of behaviour across political, military, informational, economic, and societal (PMESII) domains. Through quantum-enhanced game theory, Monte Carlo-like simulation at quantum scale, and quantum machine learning (QML), planners can model how hybrid actors, both state and non-state, might adapt to doctrinal shifts, and anticipate their strategic escalatory pathways.

Based on Quantum algorithms such as the Quantum Approximate Optimization Algorithm (QAOA) and quantum-enhanced Bayesian networks can model multi-agent interactions where different threat actors coordinate informally, such as insurgent networks acting under state sponsorship. Also, deception and uncertainty incorporating adversarial use of fake signals, false flags, or social engineering attacks can be detected. Using non-linear feedback loops where actions in one domain (cyber) rapidly influence others (civil unrest, diplomatic fallout) can enable right action at right time. Resilience modelling can assess the adversary's adaptive capacity in response to doctrinal or technological shifts, like the deployment of autonomous weapons or grey-zone ISR platforms.

- **Reduces Risk in Doctrinal Innovation and Procurement:** Strategic investments in doctrine and capability carry high stakes. Using Quantum Approximate Optimization Algorithms (QAOA) and quantum-enhanced strategic game theory, planners can explore a vast array of plausible enemy reactions *before* resources are committed to new doctrines or procurement pathways. Quantum simulations enable synthetic testing of new doctrines such as multi-domain swarming or AI-human teaming. By modelling how peer competitors might respond to the introduction of new platforms (e.g., space-based missile defence or rail guns), quantum systems reduce the likelihood of investing in capabilities that spark arms races or are rapidly neutralized. Quantum models will also ensure alignment between emerging doctrines and the actual utility of procured systems across war-fighting domains. For example, if a new amphibious assault concept is being developed, quantum tools can stress-test it against adversarial shore-defence innovations or maritime grey-zone tactics.

Logistics and Resource Optimization in Combat Scenarios

A robust and agile logistical backbone is the heart of every successful military campaign. Traditional logistics planning systems, typically based on classical combinatorial optimization methods like linear programming or heuristic search, struggle with the computational complexity that arises when hundreds of interdependent variables must be balanced simultaneously. These systems exhibit NP-hard scaling, where solution times increase exponentially with problem size making real-time adaptation impractical in the fog of war. Quantum Approximate Optimization Algorithm (QAOA) offers a paradigm shift to this challenge.

QAOA is designed to solve constrained combinatorial optimization problems by leveraging quantum superposition and interference to search over large solution spaces more efficiently than classical methods. Technically, QAOA encodes a cost function that reflects mission objectives (e.g., shortest safe re-supply route, optimal convoy timing, minimal exposure to hostile fire) into a quantum Hamiltonian. It then applies alternating sequences of Phase Separation Operators to mark desirable (low-cost) solutions based on real-time battlefield constraints and then mixing operators to explore new candidate solutions and avoid local minima. This quantum-classical hybrid optimization loop allows the system to converge on near-optimal decisions quickly, even when the supply chain problem involves time-sensitive variables (fuel decay, perishability), geospatial constraints (mountain passes, river crossings), security threats (IED zones, UAV exposure corridors), communication or bandwidth limitations (in degraded environments).

With QAOA running on a quantum edge processor or through secure quantum Cloud access, the system can evaluate thousands of these configurations in superposition, instantly returning the optimal load-out and route for that moment automatically adjusting as the threat picture evolves.

Advantages Over Classical Optimization

Capability	*Classical Approach*	*Quantum QAOA Approach*
Speed	Hours for high-dimensional input	Near real-time (seconds to minutes)
Scalability	Exponential slowdown with new variables	Polynomial or sub-exponential performance in certain problem classes
Adaptivity	Requires re-initialization for each change	Naturally suited for dynamic environments
Resilience in degraded networks	Needs consistent uplink and bandwidth	Can operate at edge with hybrid set-ups

Operational Relevance –

- **Improved Mobility and Sustainment in High-Tempo Combat Zones:** In fast-moving operations such as multi-pronged offensives, rapid counterattacks, or expeditionary deployments, the ability to deliver the right resources to the right units at the right time is mission-critical. Quantum systems, leveraging QAOA, encode this entire decision space into quantum bits (qubits), allowing the simulation of millions of routing, allocation, and prioritization scenarios simultaneously. Rather than trying one option at a time, the quantum processor evaluates all feasible combinations and converges on the most optimal ones under real-time battlefield constraints.

Tactical Objective	*Quantum-Enabled Outcome*
Maintain tempo in blitzkrieg-style manoeuvres	Resupply units with zero delay through dynamic corridor optimization
Minimize convoy losses in hostile terrain	Adaptive routing based on quantum evaluation of threat maps
Sustain expeditionary forces beyond traditional lines	Simulate and deploy pop-up depots and supply drones via quantum-aided prioritization
Avoid bottlenecks during joint force coordination	Model inter-service supply dependency and resolve allocation conflicts in real time

With a quantum-enabled logistics core (potentially embedded at theatre command or forward headquarters), planners can:

- ❏ Immediately simulate re-routing options when a chokepoint becomes impassable
- ❏ Adjust supply priorities (e.g., divert medical kits to a newly injured unit)
- ❏ Assess dynamic trade-offs: faster re-supply vs. exposure to ambush
- ❏ QAOA-powered simulations can provide near-instant answers to these "what-if" logistics dilemmas, allowing commanders to rehearse and pre-emptively reinforce alternative resupply modes, redundancy nodes, or forward logistics elements.

- **Simulating Supply Degradation and Adaptive Logistics:** In combat operations, the sustainability of forces depends not only on planning for ideal conditions but also on anticipating degradation; bridges may be blown off, runways cratered, or ports mined. Traditional logistics simulations work within fixed assumptions and must be manually reconfigured when battlefield conditions change. Quantum computing fundamentally alters this paradigm. With QAOA and quantum probabilistic models, planners can simulate a vast spectrum of degradation scenarios in parallel. A quantum system can model not only what happens if a key re-supply route is compromised but also the cascading second- and third-order effects across supply depots, forward operating bases, and dependent units.

Imagine an operations planner who can run thousands of "war games within the war game," showing how the loss of one bridge affects artillery units 80 km away, or how a 3-hour airlift delay can cascade into fuel shortages for a mechanized column two days later. This capability is the quantum equivalent of predictive logistics war-gaming on demand, under real-time constraints.

- ❑ Attrition Forecasting: Quantum systems can assign probabilities to infrastructure loss or convoy interdiction, allowing pre-emptive rerouting or buffer stock positioning

- ❑ Strategic Redundancy Planning: Commanders can simulate alternate logistics architectures such as mobile logistics nodes, drone supply drops, or maritime resupply relays before they are needed

- ❑ Time-Sensitive Prioritization: Algorithms dynamically re-rank supply priority as conditions evolve (e.g., pushing trauma kits ahead of vehicle spares if enemy shelling intensifies).

Terrain and Weather Simulation for Tactical Superiority

In modern warfare, environmental unpredictability can derail even the most meticulously planned operation. From sudden sandstorms grounding UAVs to rain-swollen rivers halting armour movement, terrain and weather are dynamic, uncontrollable variables that significantly influence tactical and strategic success. Traditional forecasting systems, while improved with satellite and AI data, still rely on classical computational limitations often producing delayed or inaccurate models under high-stress conditions. With quantum computing, the battlefield itself becomes legible in ways never before possible. This transforms environmental uncertainty into operational intelligence. Whether rerouting convoys around impassable terrain or forecasting radar blackouts during an electronic warfare operation, quantum terrain simulation redefines the very foundation of terrain dominance and tactical timing.

By leveraging quantum parallelism and high-dimensional modelling, quantum systems can simulate complex atmospheric systems and terrain interactions with unprecedented resolution and speed. Quantum algorithms, particularly quantum-enhanced Monte Carlo methods and hybrid machine learning models, can predict how terrain features will evolve under specific weather conditions accounting for real-time satellite inputs, topographic complexity, and fluid dynamics across large battle zones.

Terrain and weather modelling are inherently complex, nonlinear, and high-dimensional problems. Traditional models rely heavily on classical computing using numerical methods like finite element methods (FEM) or computational fluid dynamics (CFD) to simulate atmospheric dynamics and terrain interactions. However, these classical models struggle with scalability when applied to fast-changing battlefield scenarios involving multiple variables and localized microclimates. They are also limited by the curse of dimensionality, which restricts their ability to simulate multiple interdependent environmental and operational variables in real time. Quantum computing introduces exponential representational capacity through superposition and entanglement, allowing terrain and weather phenomena to be encoded in quantum states and evaluated across an enormous solution space in parallel. This allows military decision-makers to simulate dynamic environmental conditions with:

- Greater spatial and temporal resolution,
- Faster convergence times, and
- More accurate forecasts under uncertainty.

Specifically, quantum-enhanced Monte Carlo methods, quantum variational algorithms, and quantum Boltzmann machines are suited for modelling:

- Fluid dynamics of air and water (for storms, flooding, or amphibious landing support),
- Topographical changes (such as erosion, mudslide risk, or ground firmness),
- Electromagnetic propagation across terrain under variable humidity or temperature.

- **Quantum Monte Carlo for Probabilistic Terrain Forecasting:** Quantum Monte Carlo (QMC) methods outperform classical sampling by reducing the number of samples required to achieve statistical convergence. In military contexts, this means commanders can run probabilistic simulations of how terrain conditions will evolve over the next 6, 12, or 24 hours under stochastic weather inputs (e.g., rainfall variability, wind shear, snow melt). For example, modelling landslide likelihood in mountainous deployments or fording feasibility of rivers during monsoon combat operations.

- **Quantum Machine Learning for Sensor-Driven Terrain Mapping:** Quantum-enhanced neural networks, especially Quantum Convolutional Neural Networks (QCNNs) and Quantum Kernel Estimators, can rapidly fuse satellite imagery, UAV-derived LIDAR data, and real-time IoT sensor streams to generate up-to-date terrain heat maps. These maps adapt live to battlefield changes such as craters from artillery fire or structural collapse in urban warfare allowing units to reroute convoys or re-position artillery with optimal terrain advantage.

- **Hybrid Quantum-Classical Weather Simulations:** Using quantum-classical hybrid algorithms, meteorological simulations can be embedded directly into operational planning timelines. For instance, weather windows for UAV sorties or para-drop insertions can be optimized in near real-time by forecasting:

  - ❑ Wind tunnel dynamics at low altitudes,

  - ❑ Rainfall effects on electro-optical sensors,

  - ❑ Dust interference on laser range-finders in desert zones.

Simulating Swarm Behaviour and Counter-Swarm Tactics

The rise of drone swarms, autonomous, decentralized, and often AI-enabled, is re-defining battlefield dynamics. These swarms can coordinate over multiple axes of attack, adapt in-flight, and overwhelm conventional air defence systems through sheer volume and unpredictability. Simulating such complex, nonlinear group behaviour is computationally intensive due to the vast number of variables and decision pathways involved. Classical systems can only model a fraction of potential swarm formations or behaviours at once. Leveraging quantum parallelism and algorithms like the Quantum Approximate

Optimization Algorithm (QAOA) and Quantum Monte Carlo, military planners can simulate millions of inter-agent interactions simultaneously, exploring diverse attack vectors, decision rules, and countermeasures in real time. This provides unmatched capability to predict, understand, and disrupt both friendly and adversarial swarm operations.

Swarm warfare represents a disruptive leap in tactical operations, wherein large numbers of low-cost, semi-autonomous systems such as drones coordinate their actions dynamically to achieve strategic goals. These systems behave according to local rules and shared objectives, often without centralized control. This distributed autonomy and emergent behaviour present formidable challenges to both modelling and defence.

From a computational standpoint, simulating swarm dynamics requires evaluating a combinatorially explosive number of agent interactions, especially when including adversarial countermeasures and real-time environmental feedback. Classical methods, such as agent-based modelling or rule-based simulations, quickly reach scalability limits when dealing with even a few hundred drones operating in contested air space. Quantum computing offers a powerful alternative due to its capacity to handle complex combinatorial optimization and high-dimensional state spaces.

- **Quantum Parallelism for High-Fidelity Simulations:** Quantum computing breaks classical computing's sequential data processing bottleneck through a phenomenon called quantum parallelism, which enables simultaneous evaluation of exponentially many outcomes across vast combinatorial landscapes. At the heart of quantum parallelism is the principle of superposition the ability of a quantum bit (qubit) to exist in multiple states at once. When applied to simulation, this allows quantum systems to evaluate millions of possible configurations, be it drone trajectories, troop deployments, logistics paths, or electronic warfare environments in parallel, rather than one at a time. Unlike Monte Carlo simulations that require repeated sequential sampling of probability spaces, quantum simulations span entire state spaces concurrently, reducing compute time from days to minutes.

Technical Implications

Classical Simulation	Quantum Simulation (with Parallelism)
Sequentially evaluates one configuration at a time	Explores all possible configurations simultaneously using superposition
Requires high-performance computing clusters for large-scale war games	Uses fewer qubits to simulate exponentially larger state spaces
Limited by memory and CPU constraints in modelling multi-variable environments	Can encode complex interdependencies (terrain, logistics, enemy behaviour) into entangled qubit systems
Hours or days to generate a multi-path simulation output	Minutes or less with exponentially greater detail and precision

- **Quantum Machine Learning for Behaviour Prediction:** Quantum Machine Learning (QML) harnesses the principles of quantum computing to accelerate learning from high-dimensional, sparse, or noisy datasets. In swarm warfare, this translates to a game-changing ability to model, learn from, and anticipate the behaviours of large-scale, autonomous drone formations in real time. Swarm systems, whether adversarial UAVs, UGVs, or maritime drones, operate using decentralized decision-making and self-organizing behaviours, often governed by reinforcement learning or rule-based AI. QML models can identify subtle patterns, strategic shifts, or emergent behaviours by processing multiple variables (e.g., velocity, formation density, sensor emissions, and tactical responses) across massive solution spaces faster than classical ML systems. QML for behaviour prediction shifts counter-swarm warfare from reactive to proactively anticipatory. By

Technical Advantage over Classical ML

Capability	Classical ML	Quantum ML (QML)
Training on high-dimensional data	Computationally intensive	Scales efficiently using quantum feature spaces
Time to convergence on reinforcement models	Often slow, data-intensive	Exponentially faster via amplitude amplification
Simulation of probabilistic strategies	Limited by memory & compute	Handles stochastic swarm behaviours natively
Detection of non-linear correlations	Requires complex feature engineering	Naturally exploits entanglement & superposition

reducing the decision latency between swarm detection and neutralization, quantum-enhanced learning systems give military commanders a decisive edge in the age of autonomous, AI-driven adversaries.

QML offers advantages in multiple ways such as:

- ❏ **Predicting Adversary Swarm Formations & Intent:** QML models can classify swarm behaviours (e.g., encirclement, saturation attack, decoy dispersion) and predict transitions between them based on early-stage indicators like inter-drone spacing, EM signature modulation, or directionality patterns. This enables pre-emptive countermeasure deployment such as repositioning jamming arrays or deploying interceptor drones at choke points.

- ❏ **Learning-Based Counter-Swarm Optimization:** QML enhances counter-swarm AI systems that need to adapt on the fly deploying electronic warfare bursts, directional EMPs, or laser-targeting systems. It supports reinforcement learning loops where counter-tactics improve based on continuous feedback from simulated and real-world swarm encounters.

- ❏ **Enhancing Anti-Swarm War-gaming:** Quantum generative models (e.g., quantum Boltzmann machines or variational circuits) can create thousands of synthetic swarm threat scenarios, including behaviours not yet observed critical for next-gen training simulators and red-teaming tools

- ❏ **Swarm vs. Swarm Simulation:** QML enables modelling interactions between friendly and enemy swarms, including emergent collective behaviours, communication disruption, and deception tactics. This supports doctrine development for autonomous swarm engagements

- **QAOA for Defensive Strategy Optimization:** QAOA is a hybrid quantum-classical algorithm tailored for solving complex combinatorial optimization problems. In the context of counter-swarm warfare, QAOA can evaluate millions of possible defensive configurations such as placement of EW assets, interceptor launch sequences, and radar reorientations under dynamic constraints like terrain, energy consumption, line-of-sight degradation, and real-time adversarial behaviour. Swarm attacks typically present a multi-agent threat, characterized by unpredictability, decentralization, and real-time adaptation. Classical optimization methods often struggle with the time-sensitive, multi-variable nature of defending against such swarms, especially when required to balance hard constraints (e.g., available munitions) and soft goals (e.g., minimizing exposure of friendly assets). QAOA's quantum-enhanced exploration of the solution space helps generate near-optimal defensive responses far more quickly than classical solvers.

QAOA offers a tactically transformative leap in how to defend against autonomous, coordinated drone threats. Its capacity to rapidly compute optimized, scenario-specific defence strategies makes it indispensable for both kinetic response planning and electronic warfare dominance in the swarm era. By integrating QAOA into frontline C2 (Command and Control) systems, defence forces can operate with unmatched agility, leveraging quantum speed to neutralize AI-driven adversaries in real time.

QAOA Advantages in Counter-Swarm Optimization

Operational Function	*Challenge*	*QAOA-Enabled Capability*
Threat Engagement Planning	High-speed, multi-vector swarm manoeuvres	Real-time optimization of response paths
Resource-Constrained Defence	Limited missiles, bandwidth, or power in-theatre	Prioritized asset allocation and kill-chain optimization
Layered Defence Integration	Coordinating kinetic, EW, and decoy systems	Multi-domain strategy fusion
Countermeasure Sequencing	Optimal ordering of jamming, firing, re-positioning	Minimization of time-to-effect and maximization of impact
Collateral Risk Minimization	Close proximity to civilian areas or allied forces	Soft constraints embedded in cost function

Operational Relevance

❏ **Dynamic Threat Engagement Planning:** As swarm configurations evolve mid-flight (e.g., split-encirclement, feint-diversion, saturation), QAOA can re-optimize countermeasures in near real time. The system constantly evaluates multiple layers of defence (kinetic, cyber, EW) and recommends strategic resource reallocation to protect critical assets.

❏ **Prioritization Under Resource Constraints:** When munitions, power, or bandwidth are limited, QAOA can optimize which targets to intercept, where to deploy energy-intensive jammers, how to sequence responses to maximize disruption and survival. This is crucial in expeditionary or siege scenarios where logistics are degraded.

❏ **Multi-Objective Optimization:** QAOA allows the simultaneous balancing of multiple military objectives such as minimize collateral damage, maximize enemy attrition, and maintain radar signature stealth, preserve high-value assets. Such trade-offs can be encoded directly into the cost function of the QAOA algorithm

❏ **Swarm Prediction and Counter-Strategy Loop:** Integration with QML, QAOA can form a feedback loop QML predicts swarm behaviour → QAOA generates optimal counter-plan → new ISR data refines prediction → loop continues.

2.5 Quantum Algorithms in War Gaming and Strategic Forecasting

Testing the Impact of Emerging Technologies in Combat Models

The acceleration of military innovation ranging from hypersonic glide vehicles and swarming autonomous drones to satellite-disrupting jammers and post-quantum encrypted communications poses a strategic dilemma such as how to test, assess, and doctrinally integrate these technologies before they are operationally deployed. Quantum computing provides an unprecedented capability to simulate these technologies at scale, within complex, contested battle environments. Unlike traditional modelling systems, quantum simulators can evaluate thousands of interactive variables across multi-domain, multi-theatre conflict spaces, incorporating not only the raw performance metrics of emerging technologies, but their strategic impact how they shift force postures, destabilize deterrence, or change escalation ladders.

• **Quantum Simulation of High-Dimensional Conflict Environments:** Traditional military simulations are bound by classical computational limits they struggle with exponential complexity and are often confined to discrete or sequential domain modelling. Quantum computers, in contrast, exploit quantum parallelism, the ability to evaluate exponentially many configurations simultaneously using qubits and superposition. Quantum parallelism refers to a quantum computer's ability to evaluate multiple possible input states simultaneously by leveraging the superposition property of qubits. In classical systems, each possible scenario (e.g., drone formation A, hypersonic flight path B, or EW countermeasure C) must be evaluated one at a time. In quantum systems, a set of n qubits can represent 2n possible states simultaneously, which enables the system to explore an entire space of strategic or operational configurations in parallel, rather than sequentially.

This is crucial for modelling emerging technologies like hypersonic glide vehicles (HGVs) with variable trajectories, speeds, and evasion patterns, AI-enabled drone swarms with decentralized, adaptive decision nodes, satellite jamming systems that cause cascading ISR and C2 disruptions, post-quantum communications that interact differently under contested EM environments. Simulating how these variables interact when emerging technologies are introduced (e.g., what happens when you deploy a stealth drone swarm during a geomagnetic storm with EW interference?) results in a combinatorial explosion of possible scenarios. Quantum parallelism enables simultaneous evaluation of all these interdependent variables, generating an ensemble of "possible futures" in a single operation far outpacing any classical supercomputer. This results in holistic battle space modelling where emerging technologies are tested under joint-

Operational Impact

Aspect	Classical System Limitation	Quantum Parallelism Advantage
Simulation Speed	Serial processing of scenarios	All plausible configurations tested at once
Emerging Tech Interdependence	Cannot easily model cross-domain interaction	Models kinetic–cyber–space–EW integration naturally
Uncertainty & Volatility	Linear risk modelling	Probabilistic simulation of cascading effects
Scenario Breadth	Limited to narrow, pre-defined test cases	Covers millions of unexpected edge cases
Capability Integration	Difficult to test integration of multiple novel techs	Co-models impact of layered tech introductions in hybrid warfare context

force conditions involving land, air, sea, space, and cyber simultaneously.

- **Variational Quantum Algorithms for Nonlinear Behaviour Modelling:** Technologies like hypersonic or autonomous drones exhibit nonlinear performance characteristics (e.g., dynamic instability, machine learning-based behaviour). Classical simulations often rely on deterministic or discretized models. Quantum algorithms such as Variational Quantum Eigen solver (VQE) approximate energy states and behaviour outcomes of complex physical systems (e.g., fluid dynamics of hypersonic airframes or satellite thruster manoeuvres); Quantum Neural Networks (QNNs) train on sparse ISR data or synthetic threat profiles to forecast the probabilistic impact of emerging tech under uncertainty. This results in improved modelling of how HGVs or drone swarms behave under joint-force resistance, contested GPS, or counter-AI tactics.

  Essentially VQAs are hybrid quantum-classical algorithms designed to solve optimization and eigenvalue problems in high-dimensional, nonlinear environments. They combine quantum circuit (the ansatz) parameterized with tuneable gates and classical optimizer that iteratively updates the parameters to minimize a loss or cost function. These algorithms are well-suited for Noisy Intermediate-Scale Quantum (NISQ) computers and excel at finding optimal solutions in non-convex, nonlinear landscapes where traditional methods fail.

Technical Advantage of VQAs

Capability	*Military Relevance*	*VQA Contribution*
Parameter Space Exploration	Modelling how a tech (e.g., anti-GPS spoofing) performs under varying threat levels and terrain	Quantum ansatz explores a high-dimensional parameter space efficiently
Nonlinear System Identification	Understanding how multiple tech layers (EW + drone + AI comms) interact	VQAs can represent nonlinear Hamiltonians via parameterized circuits
Adaptive Training of Combat AI	Creating RL agents for testing drone swarms or autonomous UGVs	QNNs powered by VQAs adapt to nonlinear feedback environments
Stress Testing Under Hybrid Warfare Conditions	Simulating cascading failure from satellite disruption + cyber-attack	VQAs model compounding interdependencies among systems

- **Quantum Machine Learning (QML) for Strategic Impact Analysis:** Beyond performance metrics, quantum systems enable predictive modelling of second-order and third-order effects, such as how the introduction of satellite jammers might affect deterrence posture, whether adversaries will escalate in response to AI-enabled ISR deployments. Quantum Machine Learning (QML) is the fusion of classical machine learning techniques with quantum computing capabilities. It leverages quantum properties such as superposition, entanglement, and quantum interference to learn from data in high-dimensional, non-linear environments far faster and more expressively than classical AI.

  Quantum Machine Learning (QML) a fusion of classical machine learning techniques with quantum computing capabilities leverages quantum properties such as superposition, entanglement, and quantum interference to learn from data in high-dimensional, non-linear environments far faster and more expressively than classical AI. Since, emerging technologies do not operate in isolation their effects cascade across force posture, doctrinal shifts, deterrence equilibrium, coalition behaviour and more. Classical AI struggles with these long-range, entangled causalities. QML offers the ability to learn and forecast these cascading effects using historical, simulated, and real-time combat data from multiple domains.

  QML models use quantum-encoded features representing technological attributes (range, speed, autonomy, and stealth), adversary posture (reaction patterns, alliance movements), environmental conditions (domain constraints, terrain, and EW saturation) and historical behavioural patterns. These are processed

QML Pipeline for Strategic Impact Testing

Stage	*Quantum Role*
Data Encoding	Quantum feature maps encode military data into Hilbert space
Training	Quantum neural nets learn from synthetic or real combat simulations
Inference	QML predicts outcomes like adversary repositioning or coalition fragmentation
Interpretation	Outputs are mapped to military decision metrics risk indices, escalation alerts

by Variational Quantum Circuits, Quantum Neural Networks (QNNs), or Quantum Boltzmann Machines, which discover hidden correlations, forecast strategic shifts, recommend adaptive decisions.

- **Quantum Monte Carlo Simulations for Escalation Scenario Testing:** At the core of Quantum Monte Carlo (QMC) is the representation of a complex strategic environment using quantum state functions. Each input variable such as troop presence, weapons deployment, political pressure, alliance posture, and economic sanctions is mapped to a quantum bit (qubit) or a configuration of qubits. For example, Qubit group A encodes military readiness across four regions, Qubit group B encodes adversary political tolerance to provocation, Qubit group C encodes ISR coverage density in a theatre. These quantum states interact with each other under defined probabilistic rules, mimicking the interdependency of real-world military and political systems. In classical Monte Carlo, systems are evolved over many randomly sampled paths to explore possible outcomes. In QMC, this is enhanced by quantum superposition enabling simultaneous evaluation of multiple paths. Quantum paths represent different "future" possible escalatory responses from adversaries, shifts in political tone, or changes in military posture. These are evolved using quantum probability amplitudes, capturing the interference patterns of multiple decision chains and something classical systems cannot model.

Escalation is often non-linear and recursive; one actor's response triggers a counter, which affects a third party, and so on. QMC then assigns probability weights to each escalatory branch allowing interactions between events (e.g., a cyber-attack influences naval movements) thereby evolving the system until it reaches a conflict terminal state, de-escalation, or stalemate, for example [Maritime Deployment] → [Satellite Interference] → [Alliance Activation] → [Cyber Retaliation] → [Limited Strike]. Each transition has associated probabilities dynamically computed by the QMC engine.

After thousands (or millions) of quantum iterations, QMC uses importance sampling and wave function collapse to extract likelihood of each escalation path while averaging time-to-escalation from the first move across critical nodes (decisions or postures most likely to trigger conflict).

QMC Component	Role in Escalation Scenario Modelling
Quantum State Encoding	Models' variables like troop readiness, diplomatic pressure, tech deployments
Quantum Path Integrals	Simulates simultaneous possible futures across interacting systems
Stochastic Transition Matrix	Governs how events probabilistically lead to escalation or de-escalation
Quantum Interference Modelling	Captures non-linear feedback between decisions across domains
Collapse & Sampling	Extracts most probable paths and risk-weighted outcomes

Operational Relevance and Strategic Utility

- **Assessing ROI on Emerging Technologies:** Quantum-enabled simulations allow decision-makers to test not just whether a new system works, but whether it produces strategic overmatch, deterrence leverage, or mission assurance under real-world constraints. By running probabilistic combat outcomes with and without the technology, quantum test beds inform resource allocation, R&D prioritization, and early fielding strategies.

- **Supporting Capability Development and Procurement:** By integrating evolving threat models (e.g., peer use of quantum radar or AI jamming) into scenario-based evaluations, quantum simulation supports capability requirement generation and procurement validation. This reduces the risk of investing in technologies that may become obsolete or strategically irrelevant due to adversary countermeasures or domain convergence.

- **Validating Tech Integration into Combat Doctrines:** Quantum models can simulate how new technologies impact command-and-control structures, force dispersion, deterrence thresholds, and joint fire coordination. For example, introducing autonomous ISR drones in a contested A2/AD environment may change how strike missions are tasked or how SIGINT is distributed across echelons. Quantum systems allow doctrinal innovation to precede hardware deployment, ensuring seamless integration and minimal friction in the field.

Key Applications of Quantum Combat Test Beds

Emerging Tech	*Quantum Simulation Advantage*
Hypersonic Missiles	Simulate interception probability, time-to-target dynamics, and deterrence thresholds.
Autonomous Drone Swarms	Model swarm behaviour vs. anti-swarm defences; assess latency and bandwidth requirements.
Satellite Jamming Systems	Evaluate cascading impact on GPS, ISR, and SATCOM-dependent systems in joint operations.
Post-Quantum Comms	Test resilience against quantum-enabled SIGINT and denial-of-service attacks.
Directed-Energy Weapons	Assess line-of-sight dependency, energy availability, and risk of collateral electromagnetic effects.

Quantum-Powered Threat Assessment and Early Warning Systems

In the contested maritime domain, anticipating threats before they materialize is crucial for maintaining naval superiority and deterrence. From stealth submarine incursions and spoofed commercial shipping patterns to cyber-injected AIS data and electromagnetic deception, modern naval threats are often characterized by high uncertainty, weak signals, and complex interdependencies. Classical AI systems struggle in environments where data is incomplete, ambiguous or non-linear which is typical in naval ISR. This is where Quantum Machine Learning (QML) and quantum-enhanced pattern recognition offer a transformative advantage. Quantum-powered early warning tools transform strategic ambiguity into operational clarity. In an era of grey-zone tactics, cyber-submarines, and electromagnetic deception, militaries that deploy quantum-enhanced threat assessment will not just react faster; they will shape the battle space in advance.

Fleet-Level Decision Support

Decision Scenario	*QML/Quantum Pattern Benefit*
Submarine hunt in cluttered littoral waters	Detects subtle hydrodynamic anomalies, even in sonar-dead zones
Evaluating threat from convoy with mixed cargo	Analyses container manifests, maritime trajectories, and frequency shifts
Pre-emptive manoeuvring in response to grey-zone ops	Flags coordinated cyber, EW, and fleet re-positioning patterns
Crisis escalation detection (pre-war)	Integrates satellite heat signatures, comms changes, and ship alerts

QML-powered early warning systems act as decision augmentation tools in the Combat Information Centre or regional command HQs. They deliver high-confidence threat flags based on pre-detection modelling before any kinetic signature or explicit alert is registered.

By fusing QML into Command and Control (C2) and Maritime Domain Awareness (MDA) architectures, naval forces can shift from:

- Reactive → Predictive

- Sensor-based detection → Pre-sensor strategic inference

- Sequential ISR workflows → Quantum-enhanced, concurrent threat modelling.

This evolution reduces decision latency, enhances force survivability, and increases strategic surprise mitigation, making quantum-powered early warning a doctrinal imperative in next-generation maritime warfare.

- **Quantum Machine Learning (QML) for Strategic Threat Inference:** QML applies the principles of quantum computing to enhance machine learning models. Unlike classical ML systems, which are limited by high computational cost and feature dimensionality, QML can process vast, multi-domain datasets with non-linear correlations and complex interdependencies exactly the kind that typify adversarial military behaviour in today's hybrid battle space.

  Quantum Feature Encoding helps QML maps input data (such as ELINT patterns, troop movement logs, or geospatial indicators) into quantum Hilbert spaces, where relationships between features become more distinguishable due to the exponentially large representation capacity of quantum states, for example encoding environmental and sensor data into quantum circuits to detect submarine noise anomalies camouflaged under sea turbulence. Variational Quantum Classifiers (VQCs) which are essentially quantum-classical hybrid models learn to classify threat scenarios such as distinguishing military logistics build-ups from civilian movements. VQCs optimize quantum circuits to minimize classification errors across high-dimensional inputs (e.g., radar + HUMINT + AIS spoof data). Unlike classical SVMs or neural networks, VQCs can handle non-linearly separable and entangled features, which are common in deception-heavy environments.

Quantum Support Vector Machines (QSVMs) use quantum kernels to detect subtle deviations in known operational patterns helping identify emergent behaviours like coordinated grey-zone activities or strategic feints. QSVM can help in distinguishing routine maritime logistics from pre-positioning manoeuvres in a contested strait using vessel route changes, comms silence, and EM spectrum shifts. Quantum Generative Adversarial Networks (QGANs) simulate adversary behavioural profiles by learning from historical conflict data and generating future variants. These can be used to hypothesize unknown tactics or spoof detection strategies.

Function	QML Contribution
ISR fusion and anomaly detection	Learns subtle threat indicators hidden in noisy sensor and satellite data
Doctrine shift anticipation	Predicts changes in enemy operational patterns using behaviourally enriched training data
Early warning and escalation modelling	Correlates cyber intrusions, diplomatic tone shifts, and unit redeployments
Disinformation and info-war threat detection	Identifies synthetic influence operations or botnet-based narrative shaping

Operational Military Applications

- ❑ Detect early indicators of undersea incursions by analysing patterns of sonar echoes, marine traffic changes, and surface wave anomalies.

- ❑ Identify spoofed commercial vessels or "ghost ships" by training on subtle changes in AIS broadcasts or voyage behaviour using quantum-enhanced anomaly detection.

- ❑ Forecast cyber-kinetic escalations by correlating signals from cyber intrusions with changes in regional naval deployments.

In future war-fighting scenarios, speed of inference equals speed of dominance. QML reduces "detection-to-decision" lag from hours to seconds and enables the pre-emptive neutralization of threats that would otherwise escalate undetected. This shifts strategic posture from:

"Wait for the threat to materialize"

to

"Detect the threat signature before it becomes a headline."

- **Quantum-Enhanced Pattern Recognition for Strategic Threat Inference:** Quantum-enhanced pattern recognition uses quantum computational power particularly its superposition, entanglement, and interference capabilities to identify subtle, low-probability but-high-impact patterns across complex datasets. These may include sensor feeds, cyber logs, diplomatic communications, satellite imagery, and battlefield telemetry, all processed in parallel and correlated with high fidelity.

Quantum systems based on Quantum Feature Space Mapping project input data into Hilbert spaces of exponentially high dimensionality. This transformation makes complex or hidden relationships in the data more linearly separable and computationally tractable, even for seemingly unrelated variables, correlating maritime vessel loitering near dual-use infrastructure with simultaneous cyber reconnaissance on port security systems. Quantum Interference-Based Similarity Detection enables high-precision comparison of data patterns. When multiple data features are encoded into quantum states, the resulting interference patterns highlight areas of constructive similarity and destructive noise ideal for identifying incipient threat signatures. This results in isolating pre-conflict signal patterns in ISR data from multiple failed state actors to predict the next potential flashpoint.

Inspired by Grover's Algorithm, quantum-enhanced search can amplify the likelihood of rare but critical matches ideal for rare-event detection like WMD precursor shipment routing, unauthorized satellite uplinks, or cyber kill chain activities. Quantum systems can then analyse

Use Case	Quantum Pattern Recognition Benefit
WMD precursor detection	Detect unusual shipping manifest combinations from multiple supplier chains
Cyber-physical warfare correlation	Correlate malware telemetry with physical command post positioning
Force mobilization detection	Identify logistic staging patterns that mimic civilian supply chains
Influence operations by early warning	Detect coordinated inauthentic behaviour across foreign-language digital platforms
Grey-zone manoeuvre detection	Uncover slow, distributed, pattern-of-life shifts (e.g., unmanned maritime swarms or militia arming)

entangled variables features that are classically independent but quantum-correlated such as energy grid anomalies and digital influence campaigns, satellite drift and cyber targeting behaviour, financial outflows and strategic military movements. This allows commanders to infer threat posture shifts based on hidden linkages that classical AI may overlook.

Quantum-enhanced pattern recognition systems are most effective when deployed as decision support modules in joint intelligence operations centres (JIOCs), strategic early warning cells, cyber fusion centres, naval and air domain awareness hubs. These systems continually learn from adversary activity and improve over time, transforming pattern recognition into adaptive situational awareness that can alert before kinetic onset.

With adversaries increasingly masking aggression behind legal ambiguity, commercial infrastructure, or information fog, quantum-enhanced pattern recognition provides the technological edge to:

❏ Pre-emptively detect adversarial build-up or intent

❏ Shorten sensor-to-shooter timelines

❏ Reduce false positives in early warning systems

❏ Enhance credibility and readiness in deterrence postures.

Decision Support for Joint Maritime Operations with Allied Forces

In modern maritime warfare, victory increasingly depends on the ability of multinational forces to operate as a cohesive, synchronized coalition. Joint maritime operations, especially in high-tempo conflict zones like the Indo-Pacific or North Atlantic, involve intricate coordination of multi-national fleets, air support, logistics chains, and intelligence networks. Quantum computing introduces a transformational capability by enabling real-time, high-dimensional optimization of coalition force deployments, communication synchronization, and operational tasking.

Quantum algorithms can process complex interdependencies among allied forces such as fleet positioning, intelligence-sharing protocols, logistics compatibility, and overlapping rules of engagement far more efficiently than traditional systems. This will empower commanders with decision support systems that propose optimal collaborative actions, factoring in each nation's operational constraints, capabilities, and geopolitical sensitivities. Consider a multilingual AI staff officer embedded in a joint task force able to instantly translate not just language, but also operational doctrines, fleet readiness statuses, and national command caveats providing commanders with cohesive, executable battle plans that align all allied assets without operational friction. Quantum-enabled systems compress weeks of multinational planning into actionable options in minutes.

- **Improved Interoperability across Navies:** Modern coalition maritime operations face acute interoperability challenges. Differences in tactical doctrines, C4ISR architectures, encryption standards, and logistics chains introduce delays in threat detection, task assignment, and weapon engagement coordination. These delays are often measured in critical minutes or seconds, which can decide the outcome of anti-submarine warfare (ASW) hunts, missile defence intercepts, or contested littoral operations.

  A Quantum Command Decision Suite (QCDS)-based decision-making suite will use quantum-accelerated joint operational modelling to run real-time "what-if" coalition combat simulations while factoring in the following complexities that classical HPC struggles to handle at operational tempo –

  - ❑ **National Doctrine Constraints.** QML models will encode decision-making rules for each navy's engagement criteria, ROEs, and escalation policies

  - ❑ **Sensor Integration Variances.** Quantum-enhanced pattern recognition will be able to harmonize disparate sonar, radar, EO/IR, and satellite feeds across platforms of different nations, overcoming classification mismatches

  - ❑ **Encrypted Data Sharing.** Quantum-secure channels (QKD) could allow near-instantaneous transfer of operationally critical sensor data without risking exposure to adversary cyber-espionage

  - ❑ **Logistical Dependencies.** Quantum Monte Carlo–based logistics models dynamically re-route supply chains (e.g., fuel, ammunition, maintenance) to prevent operational pauses during high-tempo operations.

Quantum Command Decision Suite – Interoperability Data Flow for Improved Interoperability across Navies

Stage	Function	Quantum Capability Used	Operational Example	Strategic Benefit
1. Multi-Navy Data Ingestion	Collect raw sensor, comms, and intel data from allied navies (radars, sonars, EW, satellite feeds)	Quantum Secure Communication (QKD) for tamper-proof, real-time data exchange	NATO SNMG pulls ASW sonar tracks from the USA, U.K., and German ships in a task group	Eliminates intel latency and prevents data spoofing
2. Doctrine Harmonization Layer	Map differences in national doctrines, ROE (Rules of Engagement), and sensor classification thresholds	Quantum Knowledge Graph + Hybrid Classical-Quantum Reasoning	Harmonizes Japanese P-1 ASW patrol doctrine with U.S. P-8A operating patterns	Reduces misinterpretation of threat severity and rules
3. Quantum Data Fusion	Merge multi-sensor, multi-format inputs into unified threat picture	Quantum Bayesian Inference for probabilistic correlation	Fuses French sonar contacts with Indian SIGINT intercept to confirm submarine identity	Improves accuracy of threat detection
4. Tasking Simulation & De-confliction	Run large-scale joint mission simulations accounting for doctrine, asset positioning, logistics	Quantum Monte Carlo & Quantum Optimization	QUAD carrier groups de-conflict air-sea strike windows to avoid fratricide	Maximizes strike efficiency and coordination speed
5. Quantum-Enhanced Pattern Recognition	Detect hidden adversary tactics from cross-navy datasets	QML-trained models on historical combat+ live data	Identifies probable wolf-pack submarine formation in Pacific	Early detection of coordinated enemy manoeuvres
6. Adaptive Command Recommendations	Deliver optimized, real-time joint orders to all allied units	Quantum Reinforcement Learning	Suggests re-tasking Australian frigate to plug a sonar coverage gap mid-operation	Dynamic, data-driven course correction
7. Continuous Secure Feedback Loop	Push updates from frontline assets back to the joint command	QKD + Quantum Blockchain for immutable updates	AUKUS anti-submarine screen continuously updates positional data	Maintains situational awareness without compromise

The QCDS will be able to run a multi-nation, quantum-optimized patrol assignment in seconds, integrating live sonar contacts, predicted submarine paths from QML models, and fuel state data from each platform. The output: non-overlapping patrol sectors with optimally positioned relay nodes for persistent sonar coverage and real-time re-tasking if an adversary submarine changes course. This quantum-enabled interoperability will allow coalition navies to function as if they were one integrated fleet, without politically sensitive data overexposure or doctrinal compromise, preserving both national sovereignty and operational synergy.

Prioritizing Threat Responses Based on Strategic Objectives

In a modern battle space, the number of simultaneous threats can exceed human decision bandwidth and classical computational triage capabilities. Leveraging the immense parallel processing power of quantum computing, such systems can simultaneously model dozens or hundreds of complexes, fast-changing threat scenarios. They do not simply flag what is visible; they rank threats based on projected impact, timing of arrival, and alignment with overarching strategic objectives. They weigh each threat's impact potential (e.g., lethality, political consequences), timing (how soon it will materialize), and strategic value (how it aligns with or undermines mission objectives). Unlike classical systems that sequentially score threats, a quantum system can model and rank them in parallel across a massive possibility space, even under uncertainty and incomplete information. Considering multiple factors such as command radar screen during a joint force operation where dozens of returns appear at once some are high-altitude drones conducting reconnaissance, others are hypersonic glide vehicles closing in on critical infrastructure, and some are electronic warfare decoys designed to confuse sensors a conventional radar operator sees them all as blips of similar urgency. But a quantum-enabled *"tactical radar brain"* instantly identifies which blips are decoys, which are lethal, and which are part of a larger diversionary tactic, then prioritizes interception sequencing accordingly. It is the difference between *"reacting to every noise in the dark"* and *"striking precisely where the real danger lies."*

- **Optimized Resource Allocation**: In a multi-domain operation spanning air, land, sea, cyber, and space with finite resources response time is critical. Rule-based prioritization or classical optimization software will be too slow or too simplistic with multiple threats emerging simultaneously. Quantum computing fundamentally changes this calculus by applying specialized algorithms that can evaluate millions of deployment permutations in parallel, delivering an optimized allocation plan in seconds.

One such approach is the Quantum Approximate Optimization Algorithm (QAOA), which models the allocation problem as a complex combinatorial optimization challenge. In operational terms, QAOA acts like a "digital war room," rapidly testing and ranking deployment scenarios for limited interceptors against multiple incoming missile salvos, factoring in probabilities of interception, potential collateral damage, and strategic consequence. For resource balancing across simultaneous fronts, Quantum Annealing as implemented in systems like D-Wave, offers a near-instantaneous "energy landscape search" to identify the global optimum. This can be likened to a naval fleet commander dynamically deciding which ships should reinforce a contested sea lane versus escorting a high-value logistics convoy, balancing immediate threats with long-term operational sustainment.

Finally, for scenarios where uncertainty dominates such as cyber defence against a coordinated multi-vector intrusion, Variational Quantum Eigen solver (VQE)-based methods can model risk-adjusted payoff landscapes. This enables a command centre to "pre-clear" the most strategically advantageous deployments by simulating evolving threat states and resource readiness, similar to pre-positioning rapid reaction units in areas where insurgent activity is *most likely* to have a decisive effect.

By integrating QAOA, Quantum Annealing, and VQE into operational decision loops, military planners can move beyond reactive asset allocation into predictive, impact-driven resource deployment, ensuring that finite resources are always committed to the threats that matter most to mission success, not merely the ones that appear first or loudest.

How QML Enhances Optimized Resource Allocation

Component	Description	Example in Military Context	Relevant QML Algorithms
Multi-Domain State Encoding	Encode the operational state (threat type, location, timing, intent, and capability) from all domains into a quantum state representation.	Encode incoming UAV swarm, naval strike group movement, and cyber intrusion status simultaneously.	Quantum Feature Encoding (Amplitude Encoding, Angle Encoding)
Strategic Objective Weighting	Assign dynamic weights to threats based on mission impact, not just severity. QML models learn these weightings from prior mission data and simulations.	If cyber-attack on satellite uplink will disable precision targeting, it's weighted higher than a low-risk UAV swarm.	Quantum Neural Networks (QNNs) trained on simulated combat outcomes
Real-Time Resource-Threat Matching	Use quantum optimization to assign limited interceptors, cyber teams, or naval batteries to threats based on weighted priorities.	Matching three available cyber response teams to seven ongoing intrusion attempts for maximum mission impact reduction.	Quantum Approximate Optimization Algorithm (QAOA)
Adaptive Reallocation Under Change	QML continuously recalculates threat-resource pairings as conditions evolve, considering both confirmed and probabilistic threat data.	If one naval missile battery is destroyed, resources are instantly reallocated to cover exposed sectors.	Quantum Reinforcement Learning (QRL)
Uncertainty Management	Quantum superposition allows simultaneous evaluation of multiple "what-if" scenarios without full enumeration.	Evaluate outcomes if a satellite is jammed *and* a cyber intrusion occurs at the same time.	Variational Quantum Classifiers (VQC) for probabilistic classification of threat impact scenarios

- **Decision Acceleration in Complex Battle Spaces:** In high-intensity, multi-front engagements, the greatest threat to mission success is often not enemy firepower, but decision paralysis. Quantum computing offers a decisive advantage by running multiple operational outcome simulations in parallel, drastically compressing the time needed to identify the most viable course of action. The Quantum Amplitude Estimation (QAE) algorithm can evaluate the likelihood of success for competing strategies, whether it's committing an armoured reserve to a flanking manoeuvre or ordering a high-risk cyber counterstrike by rapidly calculating success probabilities based on known and projected battlefield variables.

For engagements where the battle space itself is shifting, such as a carrier strike group under simultaneous missile, drone, and submarine attack, the Variational Quantum Eigen solver (VQE) can model evolving threat states in real time, effectively "forecasting" how the tactical picture will change depending on the commander's next move. This enables pre-emptive actions instead of reactive scrambling.

Quantum Algorithms for Battlefield Decision-Making Contexts

Operational Challenge	*Quantum Algorithm*	*Military Parallel*	*Operational Effect*
Optimized Resource Allocation – Assigning scarce assets (interceptors, cyber teams, naval missile batteries) to the most decisive threats	Quantum Approximate Optimization Algorithm (QAOA)	War-gaming to determine the optimal deployment plan across multiple fronts	Maximizes mission impact by allocating limited resources to threats with the highest outcome influence
	Grover's Search Algorithm	Rapid target prioritization from large ISR datasets	Speeds up finding high-value targets in vast intelligence feeds
	Quantum Annealing	Selecting best strike packages from multiple available units	Identifies resource combinations that yield the highest operational success rate
Decision Acceleration in Complex Battle Spaces – Avoiding decision paralysis in multi-front engagements	Quantum Amplitude Estimation (QAE)	Estimating probability of success for each course of action under uncertainty	Provides commanders with rapid, probability-ranked action choices
	Variational Quantum Eigen solver (VQE)	Predicting evolving threat landscapes (e.g., missile swarm + submarine + cyber intrusion)	Anticipates how battlefield dynamics will shift based on each potential decision
	Quantum Walk Algorithms	Recon patrol selecting fastest and safest infiltration route	Finds optimal decision pathways even in highly complex, branching scenarios

When the complexity is so great that the number of possible actions explodes beyond human evaluation joint special operations, unmanned swarm management, and space asset defence occurring simultaneously the Quantum Walk algorithms are the best. By mapping decision pathways like reconnaissance patrol routes through a complex urban environment, they identify the shortest, safest, and most strategically effective action sequence, even when traditional AI or human planners would stall. The result is decision superiority, not merely reacting faster, but choosing better under pressure. By embedding QAE, VQE, and Quantum Walk methods into the operational decision loop, commanders can turn a chaotic, fog-of-war scenario into a quantum-informed, high-clarity battle space, where each choice is both time-efficient and strategically optimal.

Chapter Three

Quantum Communications and Operational Security

"Quantum Communications and Operational Security," provides a clear operational brief on how quantum technologies will safeguard joint-force communications, protect sensitive intelligence, and strengthen information warfare posture from seabed to space. It explains, in practical terms, how quantum effects make interception detectable and keep classified traffic confidential even against adversaries equipped with advanced computing. The chapter ties these capabilities to real missions—subsurface patrols, carrier strike operations, distributed air missions, and land-force maneuvers—where resilient, low probability of intercept links and tamper evident keying directly affect tempo, survivability, and command continuity.

The chapter's core technical section introduces Quantum Key Distribution (QKD) and shows how widely taught protocols such as BB84 generate encryption keys while flagging any eavesdropping at the physical layer. It maps QKD to concrete military topologies—fiber on fixed bases and headquarters, free space links for ship to ship and air to air, and satellite relays for beyond line of sight command. Officers are given practical integration guidance: how to layer QKD over current IPsec/MACsec networks, exchange keys through standard key management interfaces, and adopt hybrid designs that pair QKD with post quantum cryptography to overcome range, weather, and deployment constraints.

"Quantum Cryptography and National Security" then sets the policy and doctrine frame for protecting classified data and C4ISR networks in a world of "harvest now, decrypt later" adversaries. Readers are taken through the systems engineering elements required for quantum secure networks—optical paths, trusted nodes, emerging quantum repeaters, key distribution services, and rigorous red team validation under electronic warfare conditions. A focused segment on quantum secured Command and Control (C2) shows how QKD enabled command paths improve compromise visibility, preserve continuity of operations, and harden against jamming and deception, with specific naval patterns for ship to ship, ship to shore, and satellite links under electronic attack.

The chapter closes with a concise survey of global quantum communication programs and their military implications, covering major initiatives in the United States, China, and Russia across national backbones, space based QKD, and metropolitan pilot networks. It translates these developments into risk and readiness terms: accelerating timelines, rising potential for technological surprise, and the need for immediate action on doctrine, interoperability, and coalition planning. Finally, it highlights Indian demonstrations in fiber and free space QKD as a credible path to indigenous capability and maritime security, underscoring the imperative for joint training, acquisition, and operational test campaigns to field quantum secure communications at scale.

3.1 Quantum Key Distribution (QKD) for Secure Communications

- **How QKD Ensures Secure Communications in Military Operations**

Quantum Key Distribution: Securing Signals at the Speed of Light – Because the cost of interception is mission failure

Understanding the Physics behind QKD: Why Eavesdropping Fails

Heisenberg as a Sentinel: When Physics Guards the Line – You can't listen in without leaving fingerprints

The reliability and integrity of communication links can determine the success or failure of military operations campaign. Quantum Key Distribution (QKD) provides a fundamentally new way to secure those channels by embedding the laws of physics directly into the encryption process. QKD is different from secure radios or encrypted channels. Instead of relying only on mathematics, it relies on the laws of physics to protect communications. This makes it a game-changer for modern battle networks.

At the heart of QKD are two principles of quantum physics:

- **Signals can't be measured without leaving a mark.** The tiny light particles (photons) carry the key change the moment someone tries to observe them. It's like breaking the wax seal on a confidential envelope; you can't do it without being noticed.

- **Signals can't be copied.** Unlike regular data, which can be duplicated, a quantum signal cannot be cloned. It's similar to a one-of-a-kind physical key: if the enemy tries to make a duplicate, the key breaks.

Imagine a secure courier delivering sealed, tamper-proof orders to forward units. If the seal is broken even slightly, the receiving commander instantly knows the integrity of the message has been compromised no matter how skilfully the intruder had tried to hide it. QKD works in the same way. The "seal" is the fragile quantum state of each photon, and breaking it is impossible without leaving an unmistakable signature. Because of these properties, any interception attempt instantly disturbs the signal. For commanders, this means you don't just find out after your network is compromised; you get an immediate alert the moment the line is tapped.

Quantum Key Distribution		
Concept	**Analogy**	**Operational Benefit**
Quantum key distribution – Uses quantum mechanics to distribute encryption keys	Sending a locked box with a key inside	Allows secure communication over insecure channels
Qubit – Quantum bit that can exist as 0,1 both simultaneously	Spinning coin that's both heads and tails until observed	Enables secure key exchange using the uncertainty principle
Quantum state measurement – Eves dropping detection	If the seal of the box is broken, you know that someone has tried to open it	Prevents undetected interception of the encryption key

Fig. 3.1: QKD

Operational Relevance:

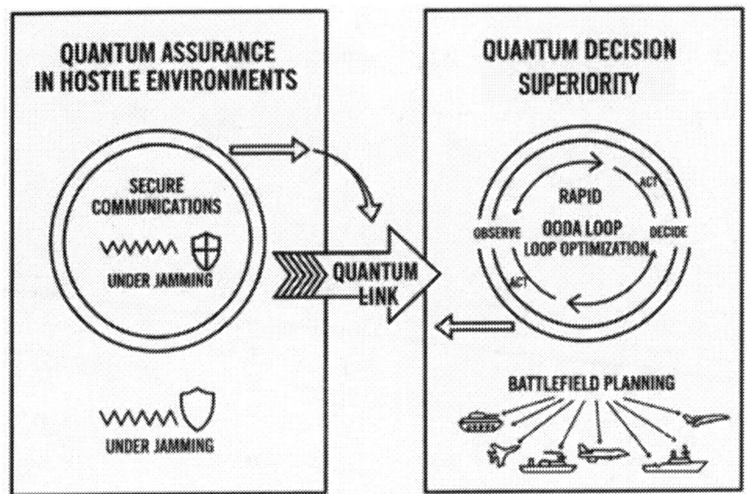

Fig. 3.2: Quantum Advantage: Assured Operations and Faster Decisions

- **Assurance in Hostile Environments:** In a cyber-contested battle space, classical systems face the constant risk of network infiltration, malware implants, and stolen encryption keys. By contrast, Quantum Key Distribution (QKD) ensures that even if adversaries manage to penetrate the network, they cannot decrypt intercepted data without triggering immediate intrusion alarms. Under electronic warfare (EW) conditions, where jamming, spoofing, and signal interception undermine trust in communications, QKD offers resilience by distributing keys via photons making them immune to classical interception methods. Spoofing becomes impossible as adversaries cannot forge a valid quantum key. In denied communications zones, such as behind enemy lines or in anti-access/area denial environments, classical relays and satellites may be compromised or destroyed. QKD mitigates this risk through pre-established networks using satellite constellations or fibre backbones, enabling commanders to deploy with a cache of pre-secured keys that preserve coordination even when conventional communication channels are degraded. At the command level, QKD delivers four critical assurances in terms of guaranteed authenticity, where orders can be executed without doubt of compromise; resilience, through fallback to alternative secure nodes if one communication path is denied; interoperability, by creating a single quantum-secure backbone for joint and multi-domain operations; and trust in degraded environments, ensuring commanders retain secure coordination even when adversaries dominate the electromagnetic spectrum.

Environment	Threats to Classical Systems	QKD Assurance Advantage	Tactical Payoff
Cyber-Contested Battle space	Key theft, malware implants, network infiltration	Intrusion-free key generation; no stored keys to steal	Secures C2 even if enemy owns parts of the network
Electronic Warfare Zone	Jamming, spoofing, interception of signals	Quantum keys immune to spoofing; tamper instantly flagged	Preserves trust in targeting, navigation, and coordination orders
Denied/Disrupted Zones	Loss of satellites, fibre cuts, A2/AD interdictions	Pre-distributed quantum keys and space-based QKD	Ensures continuity of secure comms under enemy disruption
Joint Force Operations	Multiple service networks with varied resilience	QKD as a unified secure backbone	Seamless secure interoperability between Army, Navy, Air Force, Cyber, and Space forces

- **Strategic Decision Superiority:** With QKD-secured links, commanders maintain absolute confidence in the integrity of their most sensitive communications from nuclear command-and-control to special operations tasking even in the face of quantum-enabled adversaries. Unlike classical encryption, which risks future decryption or spoofing, QKD guarantees that any interception attempt is immediately exposed, preserving the authenticity of orders, the confidentiality of covert operations, and the stability of crisis signalling. This ensures leadership can act decisively, sustain deterrence credibility, and project resilience across multi-domain operations without fear of compromise.

Dimension	Classical Risk	Quantum-Assured Advantage	Operational Implication
Nuclear Command & Control (NC2)	Future quantum computers may retroactively decrypt stored traffic, exposing nuclear directives or authentication codes.	QKD keys are information-theoretically secure, immune to brute force even by quantum adversaries.	Maintains credibility of nuclear deterrence and ensures adversary cannot inject false launch/recall orders.
Special Operations Tasking	Intercepted tasking orders could reveal infiltration routes, extraction timings, or covert alliances.	Any interception attempt introduces measurable disturbances, alerting commanders in real time.	Guarantees clandestine ops remain uncompromised even when staged deep in hostile territory.

Dimension	Classical Risk	Quantum-Assured Advantage	Operational Implication
Theatre-Level Command (Multi-Domain Ops)	Adversaries may spoof or replay high-level directives, creating confusion across land, air, sea, space, and cyber domains.	Quantum-secure channels authenticate each message as originating from legitimate command authority.	Prevents adversary-induced friction, enabling seamless synchronization of joint operations.
Strategic Diplomacy & Crisis Negotiation	Diplomatic cables or back-channel crisis comms risk exposure to interception or leaks.	QKD provides tamper-proof encryption, ensuring confidentiality of crisis management dialogues.	Preserves escalation control by protecting sensitive signalling between capitals.
Doctrine & Deterrence Signalling	If adversaries suspect command networks can be penetrated, they may miscalculate and escalate.	Assurance of unbreakable C2 integrity projects resilience and reduces incentive for pre-emptive strikes.	Reinforces deterrence by denial and ensures adversaries respect red lines.

QKD vs. Classical Key Exchange: Eliminating the Human Weak Link

Keys Born in Quantum Silence: No Intercept, No Decrypt – Human operators are no longer soft targets

In the chaos of modern battlefields, communication superiority often determines whether forces seize the initiative or cede it. Traditional encryption methods, while robust today, remain vulnerable to both electronic warfare interference and the looming threat of quantum-enabled adversaries who could decrypt intercepted keys. QKD replaces this fragile model with an architecture where keys are generated using quantum particles and validated through the laws of physics themselves. This creates channels that are not just encrypted, but intrinsically tamper-evident, removing the possibility of undetected interception.

For tactical edge units, whether special forces teams, forward observers, or unmanned systems, relaying ISR feeds the value is clear. Secure QKD links with headquarters ensure that real-time battlefield instructions cannot be spoofed or corrupted. If an adversary attempts to intrude, the disturbance is immediately detected, allowing commanders to terminate the channel before false or compromised data enters the operational cycle. The analogy is akin to an encrypted radio set that automatically alerts the operator if someone tries to eavesdrop on the frequency, enabling secure continuity of command even under fire. From an electronic warfare perspective, QKD also provides resilience. Traditional signals can be jammed or masked, forcing reliance on alternative channels or couriers, which introduce delays and risks. QKD-backed systems are less susceptible to such manipulation because the very act of interference corrupts the key exchange, immediately flagging the intrusion. In joint operations where multinational forces, cyber units, and kinetic elements must synchronize, this guarantees that encrypted data streams, from drone reconnaissance to artillery fire missions, remain both authentic and uncompromised in real time.

Operational Relevance

QKD in battlefield communications closes one of the most critical vulnerabilities of classical systems: the possibility of hidden compromise. It ensures that combatants can operate with clarity (knowing messages are genuine), confidentiality (adversaries cannot read them), and continuity (communications persist even under hostile EW pressure). In the words of a NATO commander, the advantage is clear: *"With QKD, we don't just transmit orders; we transmit certainty."*

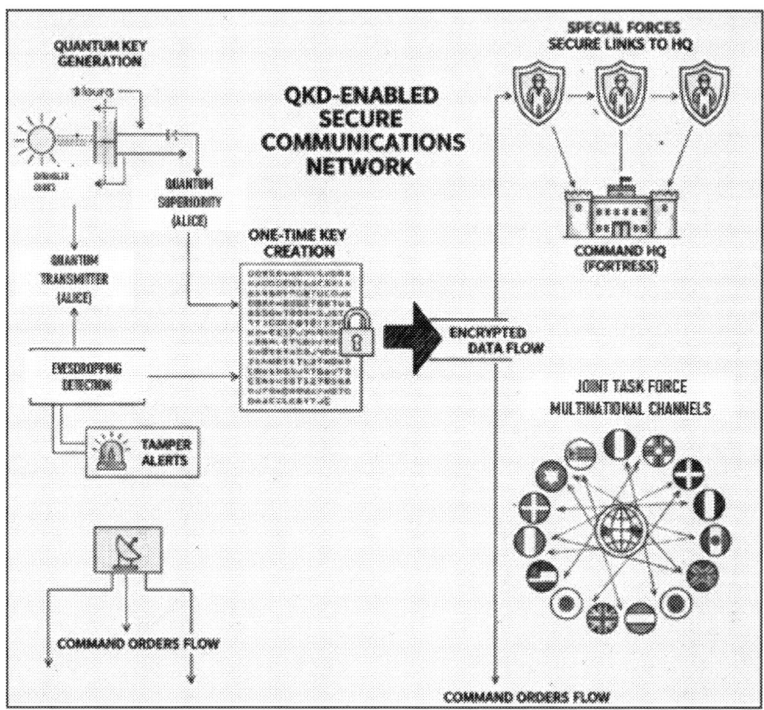

Fig. 3.3: QKD Communications: Physics-Secured Command Assurance

QKD vs. Classical Battlefield Communications

Dimension	Classical Comms	QKD-Enabled Comms
Core Principle	Relies on mathematically complex encryption (RSA, ECC) vulnerable to quantum decryption in future.	Keys generated using quantum particles; secured by physics, not maths.
Key Vulnerability	Keys can be intercepted, stored, and decrypted later ("harvest now, decrypt later").	Any eavesdropping disturbs the quantum state, instantly exposing intrusion.
Dependence on Infrastructure	Requires pre-shared keys, third-party certs, and secure courier's human weak links.	Self-generated one-time keys; removes dependence on external trust chains.
Electronic Warfare (EW) Resilience	Signals can be jammed, spoofed, or modified without detection until too late.	Attempts to jam or tamper corrupt the key exchange and trigger automatic alerts.
Tactical Edge Use	Forward units.risk compromised comms in denied areas; delayed re-supply or fire support.	Special forces and ISR units maintain tamper-proof links with HQ, ensuring real-time integrity.
Operational/ HQ Use	Strategic HQ relies on layered comms with latent risk of key compromise.	Joint task forces and multinational HQs assured of uncompromised channels, enabling synchronized operations.
Command Confidence	Orders could be intercepted, altered, or replayed without detection.	Orders delivered with clarity, confidentiality, and continuity removing doubt.

Satellite-Based QKD: Beyond the Line of Sight

Quantum from Orbit: Securing the Sky Link – When the theatre of war spans continents, space becomes the conduit

In the 21st-century battle space, secure communications cannot remain confined to terrestrial limits. Fibre-based QKD offers unmatched security on the ground, but its reach is constrained by distance and terrain. Satellite-based QKD extends this advantage into the strategic domain, allowing secure distribution of quantum keys across continents and oceans. By harnessing space assets, militaries can establish tamper-proof communication links well beyond the line-of-sight, ensuring that the chain of command remains unbroken even in globally dispersed operations.

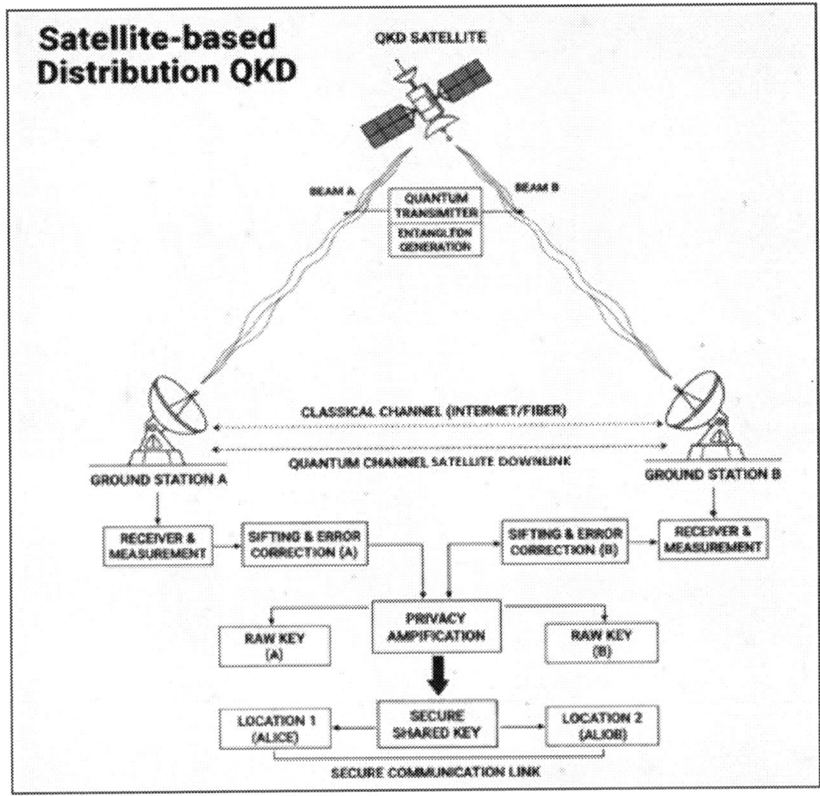

Fig. 3.4: Satellite QKD: Beaming Secure Keys From Space

Consider the difference between a supply convoy moving along contested roads versus an air bridge that bypasses enemy choke points entirely. Terrestrial QKD over fibre is akin to the convoy, reliable but geographically limited and vulnerable to disruption. Satellite-based QKD is the air bridge immune to ground interdictions, capable of projecting secure keys across thousands of kilometres, and ensuring continuity of supply regardless of the terrain below. Similarly, space-based QKD sidesteps the vulnerabilities of physical fibre routes, enabling strategic reach without exposure to local interdiction or sabotage. Satellite-based QKD provides militaries with secure long-range communications vital for joint and coalition operations, where commands are spread across theatres and continents. It creates resilient pathways in denied or remote environments where terrestrial infrastructure is either compromised or absent, such as during high-altitude operations in Ladakh, naval deployments in the Indian Ocean, or forward presence in expeditionary campaigns. Even in scenarios where adversaries have degraded ground communications through cyber, kinetic, or EW attacks, space-based QKD ensures continuity of command and control. By freeing secure communications from the limitations of geography, satellite QKD becomes a force multiplier, safeguarding both deterrence and war-fighting readiness.

Operational Relevance

In a crisis scenario, an Indian naval carrier strike group operates deep in the Indian Ocean, tasked with deterring hostile submarine activity near vital sea lanes. Midway through deployment, adversaries sever undersea communication cables and launch cyber-attacks on regional ISPs, attempting to isolate the task force from mainland command. In a classical set-up, the fleet would risk operating "blind" or fall back on vulnerable, easily intercepted fallback comms.

Classical SATCOM vs. Space-QKD

Dimension	*Classical SATCOM*	*Space-QKD (Quantum Key Distribution via Satellites)*	*Operational Relevance*
Intercept Ability	Encrypted keys exchanged using classical mathematics (RSA, ECC) vulnerable to interception, brute force, or future quantum decryption.	In keys secured by quantum mechanics, any interception attempt collapses the quantum state and is immediately detectable.	Prevents adversary SIGINT units from silently harvesting keys and decrypting orders at leisure.
Resilience in Contested Environments	Susceptible to cyber intrusions, jamming, spoofing, and compromised ground stations.	Quantum-secured links remain inviolable; adversary interference is evident, not covert.	Preserves command-and-control integrity in high-intensity EW or cyber warfare.
Infrastructure Dependence	Relies on terrestrial ground stations, undersea cables, and civilian/dual-use satellite networks, all potential single points of failure.	Independent of terrestrial fibre; direct key delivery from satellite to military terminals.	Ensures secure connectivity for forces in remote or denied areas without relying on vulnerable civilian infrastructure.
Key Lifespan & Security Horizon	Current keys (RSA/ECC) risk obsolescence under future quantum computing attacks, potentially compromising archived traffic.	QKD keys are unconditionally secure (information-theoretic); cannot be retroactively broken.	Protects classified operational traffic for decades, safeguarding historical mission data.
Strategic Relevance	Provides robust but ultimately *conditional* security, dependent on adversary's computing limits.	Delivers *absolute* assurance of key secrecy under the laws of physics, not mathematics.	Elevates deterrence credibility: adversaries cannot plan on decrypting command traffic, even with quantum advantage.

Instead, the task force maintains seamless connectivity with the Strategic Forces Command through satellite-based QKD channels. Secure keys are continuously refreshed via quantum signals transmitted from an overhead satellite, immune to interception. Commanders exchange operational orders, targeting updates, and sensor data with complete assurance of confidentiality and integrity. Despite the adversary's best attempts to degrade communications, C2 continuity is preserved, allowing the strike group to coordinate ASW (anti-submarine warfare) operations and maintain deterrence. The lesson for commanders: Space-QKD ensures that even when adversaries sever, jam, or compromise terrestrial networks, strategic communication links remain untouchable, guaranteeing command without compromise.

Detecting and Responding to Communication Breaches in Real Time

Invisibility Has a Tell: Instant Breach Detection with Quantum — Interception is no longer silent

One of the most critical vulnerabilities in classical key exchange is that adversaries can often intercept or copy cryptographic keys without immediate detection. This is akin to an enemy reconnaissance patrol slipping behind friendly lines, mapping positions unnoticed and only later calling in artillery fire. By the time the compromise is realized, sensitive information may already have been exploited, operational deception achieved, and battlefield momentum lost.

- **QKD systems detect eavesdropping attempts instantly:** QKD however, is built on the laws of quantum mechanics, which guarantee that any intrusion leaves behind a trace. The core principle is that quantum states cannot be measured without disturbance. When an adversary attempts to observe or copy the photons transmitting the key, their measurement inevitably alters the photon's polarization or phase. This disruption manifests as an increase in the Quantum Bit Error Rate (QBER), a measurable indicator of eavesdropping. In protocols like BB84, legitimate users (traditionally called Alice and Bob) exchange photons polarized in random bases. After transmission, they publicly compare a subset of their measurements. If the channel is uncompromised, the QBER remains within a very low threshold (typically <1%). If an eavesdropper ("Eve") attempts interception, the no-cloning theorem ensures she cannot copy the photon without introducing detectable errors: the QBER rises sharply (e.g., >10%), signalling a breach. Military operators monitoring the QKD system can thus detect an attack within seconds to minutes of occurrence.

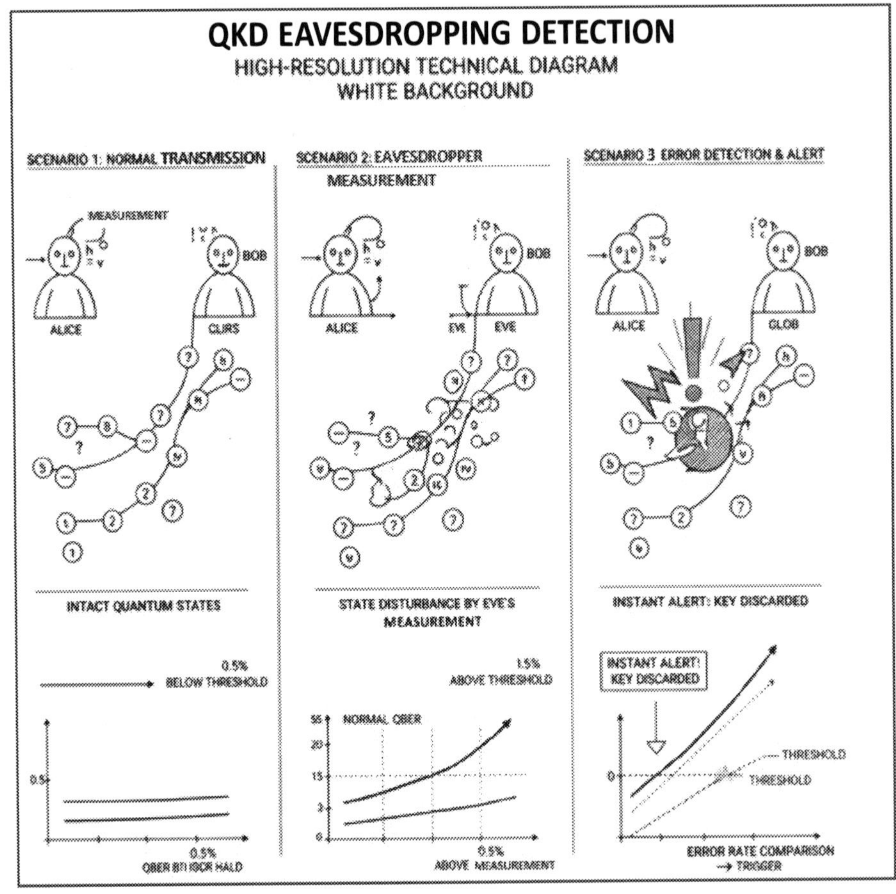

Fig. 3.5: *Instant Intrusion Detection: How QKD Exposes Eavesdroppers*

For military communications, this translates into real-time cyber situational awareness. As soon as the QBER exceeds safe thresholds, operators can terminate the compromised link to prevent leakage, activate pre-designated contingency networks (fibre, satellite, or mobile QKD relays), trigger deception counter-measures, ensuring adversaries cannot inject false traffic or corrupt fire-control orders. In effect, QKD integrates detection and response at the speed of physics. It doesn't rely on human operators to notice anomalies or forensic teams to analyse after-action reports; it provides an immediate breach alert. For commanders, this is the communications equivalent of radar detecting incoming missiles: you may not stop the probe, but you know it is happening, and you can react before it causes operational damage.

Breach Handling – Classical vs. Quantum Key Distribution

Dimension	*Classical Key Exchange (RSA, Diffie–Hellman, etc.)*	*Quantum Key Distribution (QKD)*
Detection Time	Weeks to months (often post-facto, via forensic analysis or adversary activity noticed later).	Seconds to minutes (via real-time rise in Quantum Bit Error Rate).
Attack Stealth	Adversary can passively copy traffic without alerting operators.	Any interception disturbs quantum states; cannot be hidden.
Operator Workload	High – requires constant monitoring, intrusion detection systems, and manual forensic investigation.	Low – detection is built into the physics of the channel; alerts are automated.
Operational Consequences	Keys may already be compromised when breach is discovered; adversary can inject false orders or decrypt previous traffic.	Breach attempts are detected immediately; operators can cut link or switch to contingencies before critical data leaks.
Strategic Risk	Adversary gains intelligence undetected; could compromise campaign plans.	Adversary presence is revealed instantly; quantum link acts as both secure channel and early warning sensor.

- **QKD Minimizes Data Loss and Prevents Misinformation Injection:** Since in QKD, any attempt at eavesdropping is instantly reflected in the QBER once QBER crosses a pre-defined military threshold, the key exchange session is aborted before keys are finalized. This ensures that no compromised keys are ever used to encrypt traffic, meaning the adversary gains zero usable intelligence. However, in classical systems, adversaries may siphon off partial keys undetected, enabling brute-force or replay attacks later while in QKD, partial key interception is impossible without detection, because quantum states collapse under observation. This "all-or-nothing" property means the adversary either gets detected or gets nothing there is no middle ground.

Military QKD systems can be configured with rapid re-keying protocols. When a breach is flagged, new quantum keys are generated on alternate channels or via redundant satellites. This prevents even a short disruption from turning into a critical intelligence leak. Operationally, it's equivalent to an armoured brigade switching to a pre-laid alternate command net the moment its primary frequency is jammed. In classical systems, once an adversary compromises encryption keys, they can not only read but also inject traffic, e.g., sending false target coordinates or cancellation orders with QKD. Since keys are never compromised in the first place, adversaries cannot gain authentication rights to craft false messages. This preserves command authenticity, which is critical in high-speed, high-stakes operations such as missile defence or multi-domain fire coordination.

Classical Key Exchange vs. QKD – Handling Data Loss & Misinformation

Dimension	Classical Key Exchange (e.g., RSA, Diffie-Hellman)	Quantum Key Distribution (QKD)
Detection of Key Breach	Breaches often remain undetected until after exploitation (days to months).	Instant detection via rise in Quantum Bit Error Rate (QBER).
Partial Key Leakage	Possible adversary may capture segments of the key and use brute force or side-channel exploits later.	Impossible; any interception collapses the quantum state, yielding no usable key material.
Use of Compromised Keys	Keys may continue to be used without operator knowledge, exposing ongoing comms.	Compromised session is aborted immediately, ensuring no bad key is ever used.
Adversary Ability to Inject False Traffic	Once keys are compromised, adversary may spoof commands, send false target data, or cancel orders.	Impossible adversary never obtains usable keys, so they cannot craft authentic traffic.
Data Loss from Breach	High adversary may exfiltrate large volumes before discovery.	Minimal to zero session stops at first sign of tampering.
Operator Workload	High requires incident response, log reviews, forensic checks, and possibly re-keying under duress.	Low automated abort and re-keying, with clear alerts for command decision.
Operational Consequences	Risk of misfire, fratricide, or command paralysis if false orders are injected.	Preserves command integrity and tempo, ensuring only valid, authenticated orders flow.

QKD is typically paired with one-time pad encryption or quantum-hardened symmetric ciphers. Any tampering or misinformation attempt results in authentication failures, instantly alerting operators. This gives not only confidentiality but also integrity assurance, a guarantee that the order received is exactly the order sent.

Integration with Existing Defence Networks: From Fibre to Field

Quantum-Enabled Networks: Seamless Security in Every Byte—Upgrading without uprooting your infrastructure

One of the strategic advantages of QKD is its ability to integrate into existing defence communication infrastructures without requiring a wholesale replacement of systems. Modern militaries already rely on robust fibre-optic backbones connecting command centres, forward operating bases, naval ports, and intelligence hubs. QKD can be layered directly onto these optical networks, enabling secure key exchange between trusted nodes with minimal disruption to current workflows. This makes QKD a force multiplier rather than a disruptive overhaul. A joint task force can establish secure base-to-base links over terrestrial fibre, extend QKD-based protection to ship-to-shore communications where naval task groups interface with port commands, and even push drone-to-operator control links through hybrid quantum-classical channels. Importantly, QKD supports hybrid deployment, where quantum-derived keys are combined with classical encryption, allowing gradual adoption while ensuring interoperability with existing command, control, communications, computers, intelligence, surveillance, and reconnaissance (C4ISR) systems. For commanders, this approach reduces both technological risk and operational friction: units can field-test QKD in controlled corridors while retaining classical fallback, ensuring mission continuity. Over time, as quantum-secure infrastructure scales, forces can phase out reliance on vulnerable classical key exchanges, achieving seamless security without operational disruption.

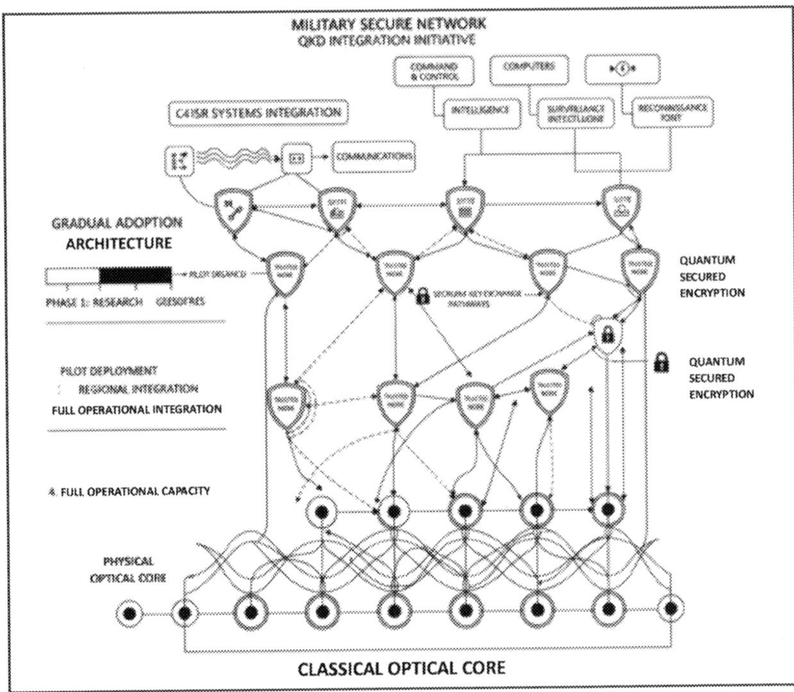

Fig. 3.6: Hybrid QKD Integration: Securing C4ISR Networks Seamlessly

Classical Defence Networks vs. Quantum-Integrated Defence Networks

Dimension	Classical Defence Networks	Quantum-Integrated Defence Networks (QKD-Enabled)
Adaptability	Dependent on fixed encryption standards; slow to adapt against emerging threats.	Layered onto existing fibre/optical links; adaptable to both legacy and new systems with minimal disruption.
Interoperability	Requires uniform key protocols; integration across services and allies often cumbersome.	Supports hybrid quantum-classical encryption; interoperable across joint and coalition forces during phased adoption.
Operational Risk	Vulnerable to undetected interception and decryption (present or future quantum attacks).	Eavesdropping instantly detectable; commanders can switch to contingency comms or halt transmission, reducing exposure.
Future Readiness	Relies on cryptographic algorithms that may be obsolete against quantum adversaries.	Future-proof against quantum-enabled decryption; scalable from base-to-base to ship-to-shore and drone-to-operator comms.

- **Enabling Secure Base-to-Base, Ship-to-Shore, and Drone-to-Operator Communications:** One of the most compelling strengths of QKD is its ability to extend secure communication across the diverse operational theatres of land, sea, and air. Unlike a classical key exchange, which relies on mathematical hardness assumptions, QKD derives its security directly from the laws of quantum mechanics. This distinction matters operationally because it allows commanders to secure communications end-to-end, whether transmitting over fixed optical links between bases, through maritime SATCOM channels from ships to coastal HQs, or via airborne relays to unmanned systems.

  QKD can be layered over terrestrial fibre-optic infrastructure, enabling ultra-secure links between command centres, airbases, and forward operating bases. Using polarization or phase encoding, single-photon transmissions provide encryption keys that are unbreakable and continuously refreshed. This ensures that strategic command directives and real-time ISR feeds cannot be intercepted or decrypted, even by adversaries with future quantum computers. QKD can also be integrated with free-space optical links and SATCOM relays, providing secure communication channels between ships at sea and coastal command facilities. Unlike traditional SATCOM, where signals can be intercepted or jammed, QKD links introduce quantum keys that collapse if tampered with. Naval task groups gain jam-resistant and interception-proof communications, crucial during carrier strike operations, amphibious landings, or contested sea lanes.

  QKD-enabled free-space optics allows secure downlinks from drones, UAVs, and satellites directly to operators. Unlike classical links, where keys may be re-used or compromised, QKD refreshes keys at quantum rates, making data exfiltration impossible without detection. This will prevent adversaries from hijacking drones, injecting false sensor data, or spoofing command signals preserving ISR integrity and ensuring drones remain extensions of the commander's will, not the enemy's tool.

QKD across Operational Domains

Domain	Typical Threats to Communications	Operational Risks (Classical Exchange)	QKD Protections	Operational Advantage
Base-to-Base (HQ ↔ FOB, Command Centres)	• Fibre-optic tapping • Insider interception • Future quantum decryption of stored traffic	• Undetected key compromise • Orders could be altered or replayed • Long-term confidentiality risk	• Intrusion attempts instantly detected (quantum no-cloning principle) • Keys refreshed continuously in real-time • Resistant to "store now, decrypt later"	• HQ directives and ISR feeds remain uncompromised • Long-term secrecy assured even against adversary quantum computers
Naval (Ship ↔ Shore, Ship ↔ Ship)	• SATCOM interception jamming & spoofing signal relay compromise	• Adversary could decode fleet movements • False signals may be injected • Task force coordination at risk	• QKD over free-space optical and satellite links Keys collapse if tampered with (eavesdropper visible) • Integration with hybrid SATCOM-quantum systems	• Jam/intercept-proof fleet communications Secure strike group coordination • Protects blue-water ops in contested sea lanes
Air/Drone (Drone ↔ Operator, UAV ↔ Base)	• Command link hijacking • Data exfiltration from UAV sensors GPS/spoofing injection attacks	• Enemy could seize drone control • ISR feeds manipulated or falsified • Loss of mission-critical assets	• QKD-enabled free-space optics • Keys refreshed per transmission Sensor-to-operator encryption immune to interception	• Drones remain loyal to operators, not adversaries • ISR integrity preserved • Prevents misinformation injection during air ops

- **Hybrid Quantum-Classical Encryption: A Bridge to Quantum-Secure Forces:** QKD is not designed to immediately replace every classical encryption system in defence networks. Instead, it is most effective when deployed in a hybrid mode, where QKD-derived keys are layered on top of or combined with existing classical encryption algorithms such as AES-256 or elliptic-curve cryptography. In such architectures, QKD continuously generates fresh, provably secure symmetric keys, which then feed into conventional cryptographic systems. This ensures that even if adversaries develop quantum computers capable of breaking RSA or ECC, the underlying keying material remains quantum-secure. The hybrid approach allows militaries to upgrade incrementally, reducing the operational shock of large-scale infrastructure changes.

For the armed forces, hybrid quantum-classical encryption is vital because it allows:

❏ **Gradual Rollout Across Theatres:** Bases, fleets, and air wings can adopt QKD progressively, beginning with strategic command-and-control nodes before extending to tactical assets like drones and mobile command posts.

❏ **Interoperability with Allies:** Coalition operations often require compatibility with partners who may not yet be quantum-enabled. Hybrid systems ensure encrypted interoperability without exposing the entire network to legacy vulnerabilities.

Adoption Matrix: Classical vs. Hybrid vs. Pure Quantum

Dimension	*Pure Classical Encryption*	*Hybrid Quantum-Classical Encryption*	*Pure Quantum (QKD-Only)*
Readiness	Fully mature, universally deployed in defence networks; but increasingly vulnerable to quantum threats.	Deployable today on existing fibre/optical channels; works alongside current systems; operationally feasible in near-term deployments.	Experimental at scale; requires dedicated quantum infrastructure and global QKD satellites; not fully combat-ready.
Operational Risk	Highly vulnerable to "harvest now, decrypt later" and future quantum decryption; long-term secrecy compromised.	Low QKD secures keying material while classical ensures continuity; layered defence reduces catastrophic failures.	Medium unbreakable in theory, but susceptible to technical fragility (line-of-sight limits, environmental disruptions) and lacks mature failover.
Transition Speed	No transition required; status quo. But standing still increases vulnerability over time.	Moderate to fast incremental rollout across bases, fleets, and drone C2; minimal disruption to operators.	Slow requires wholesale replacement of current crypto and comms infrastructure; high training overheads.
Cost	Lowest sunk cost in existing infrastructure. But hidden long-term costs of insecurity are enormous.	Medium leverages existing infrastructure, adding QKD modules; spread cost across gradual adoption cycles.	Highest full infrastructure rebuilds, new satellites, hardened optical links, and specialized terminals.
Future Readiness	Obsolete against quantum adversaries; systems compromised once adversary reaches scale.	Future-proofed protects against quantum threats while keeping current systems alive; acts as bridge to full quantum.	Ultimate goal maximal long-term security; strategic dominance in a quantum battle space.

❑ **Reduced Training and Logistical Load:** Operators continue to work with familiar encryption systems, while QKD silently strengthens key distribution in the background. This reduces the immediate need for large-scale re-training or reconfiguration of battlefield equipment.

❑ **Assured Continuity of Secure Communications:** In case QKD links are temporarily degraded (e.g., due to weather impacts on free-space optical channels), communications automatically fall back on classical encryption without exposing the network to catastrophic compromise.

- ## STRATEGIC APPLICATIONS OF QKD IN ENCRYPTION-RESISTANT MILITARY NETWORKS

Quantum Fortresses: Building Military Networks That Don't Crack Securing the command chain in a post-decryption era

Quantum Hardened Command-and-Control (C2) Networks

Unbreakable Chains of Command – When every order is encrypted at the speed of light

The backbone of any military operation is the Command-and-Control (C2) network, the structure through which intent, orders, and situational updates flow. Its integrity determines whether commanders can synchronize combat power across dispersed formations. By integrating QKD into C2 pipelines, commanders gain the ability to secure orders and situational reports with keys that adversaries cannot copy or predict. In effect, the "chain of command" becomes quantum-hardened every link, from strategic headquarters to tactical units, is underpinned by continuously refreshed, quantum-secure keys. For the operator in the field, the system appears seamless; messages are encrypted and decrypted with familiar devices. Any attempt at interception collapses the quantum state, immediately alerting the system to intrusion.

Just as a column of armoured vehicles maintains integrity by using secure recognition signals that cannot be forged, a QKD-enabled C2 network ensures that only authentic, uncompromised orders propagate across the force. This prevents adversaries from injecting false commands or delaying genuine ones an especially critical safeguard during high-tempo operations in contested electromagnetic environments.

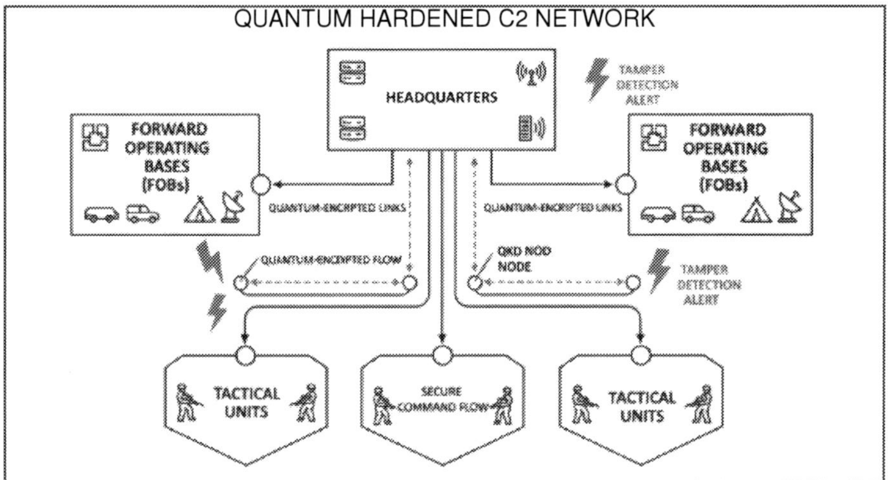

Fig. 3.7: Quantum-Hardened C2: Unbreakable Command Networks

The operational relevance of QKD-hardened C2 is most evident in multi-domain warfare. Coordinating an air strike package, naval task force manoeuvre, and ground offensive requires synchronized timing and unquestioned trust in the authenticity of orders. In hostile environments where adversaries employ electronic warfare, cyber infiltration, and deception campaigns, QKD ensures that commanders retain decision superiority. Orders remain encrypted "at the speed of light," immune to both current and future decryption threats, preserving unity of effort under the most demanding combat conditions.

- **Hardening Command-and-Control (C2) Networks:** Quantum keys generated via fibre or free-space optical channels are ingested into the Key Management System, which replaces or supplements existing keying material (e.g., EKMS, NATO KMI).These keys are securely loaded into link encryptors (e.g., TACLANE, KIV-series, or their Indian equivalents), ensuring all IP-based and voice traffic across the C2 backbone is wrapped in quantum-secured encryption sessions. Unlike classical key distribution, which may re-key daily or weekly, QKD can supply high-entropy, continuous key streams, allowing near-real-time re-keying. This drastically reduces adversary windows for brute-force or traffic analysis attacks. Even if parts of the network are compromised, quantum keys are distributed with end-to-end secrecy, preventing lateral movement by adversaries.

- **Protecting Strategic Assets and Platforms:** Quantum keys are distributed from secure ground nodes to strategic platforms, then used to refresh one-time-pad (OTP) or AES session keys in crypto modules for highly sensitive command messages. Because QKD detects eavesdropping, adversaries cannot spoof or silently tamper with launch orders. A detected disturbance in the quantum channel would invalidate the key, forcing an immediate fall-over to contingency keying material. GEO and LEO satellites equipped with QKD payloads can distribute symmetric keys directly to command bunkers, missile silos, and mobile platforms extending quantum-assured reach into strategic operating theatres. Keys generated via QKD ensure that both authentication tokens (verifying commander's identity) and encryption for the order message remain uncompromised, even by quantum-enabled adversaries.

- **Securing Multi-Domain Operations (MDO):** QKD nodes can be configured to provide compartmentalized key sets, allowing selective sharing of cryptographic material among coalition partners while preserving national control of sensitive links. Quantum keys feed into domain-specific KMS pipelines. e.g., Air Operations Centre (AOC), Maritime Operations Centre (MOC), Cyber Ops Command each receiving domain-isolated keys to reduce spill over risk. QKD enhances the security of cross-domain solutions (CDS) that bridge classification levels (e.g., SECRET to TOP SECRET networks), ensuring that even if data must traverse multi-level gateways, the encryption keys remain quantum-resilient. Tactical UAVs, satellites, and maritime patrol aircraft can receive refreshed quantum keys to securely uplink ISR feeds into joint command centres, denying adversaries the chance to intercept imagery or sensor data.

- **Survivability in Denied and Contested Environments:** Ship-to-ship, aircraft-to-aircraft, or ship-to-aircraft links can be established using optical terminals capable of quantum photon transmission. These work independently of GPS or SATCOM. Vehicle-mounted or UAV-mounted QKD terminals can establish line-of-sight secure key exchanges in expeditionary scenarios (e.g., forward-deployed brigades in EW-contested zones). In case fibre-optic undersea cables are tapped or satellites are denied, commanders can rely on direct optical QKD links as emergency re-key channels for tactical crypto devices. When jamming increases error rates, the system adapts by shortening session lifetimes and pulling additional keys from QKD buffers to sustain secrecy even under heavy interference.

Strategic Applications of QKD in Encryption-Resistant Military Networks

Application	Concept	Technical Detail	Operational Relevance
Hardening Command-and-Control (C2) Networks	QKD provides tamper-proof keys for encrypting command channels.	Continuous key refresh at quantum-secured rates; keys injected directly into C2 encryption devices.	Preserves chain-of-command integrity; prevents spoofed or intercepted orders under cyber/EW attack.
Protecting Strategic Assets and Platforms	Secures nuclear, naval, and air defence data links.	QKD keys secure fire-control, early warning, and deterrence systems against quantum decryption.	Ensures credibility of deterrence and second-strike forces; prevents crypto-targeted paralysis of assets.
Securing Multi-Domain Operations (MDO)	Enables secure, synchronized cross-domain communications.	Keys distributed via fibre, FSO, or satellites into Link-16/22, SATCOM, and IP-based systems.	Guarantees mission synchronization across air, land, sea, cyber, and space in contested EM environments.
Survivability in Denied & Contested Environments	Maintains comms when satellites/jamming/cyber are compromised.	Free-space optical QKD secures line-of-sight comms for mobile HQs, carriers, and forward bases.	Provides resilient comms during satellite denial, EW saturation, or cyber intrusion.
Future-Proofing Strategic Infrastructure	Anticipates quantum threats to classical cryptography.	QKD delivers information-theoretically secure keys immune to future quantum computers.	Extends lifecycle security of platforms (to 2050+), avoiding costly retrofits and crypto obsolescence.

Protecting Satellite and UAV Communications

Shielding Eyes in the Sky – When quantum keys protect our most critical aerial assets

With modern military operations increasingly depending on satellites and unmanned aerial vehicles (UAVs) for command, surveillance, targeting, and battle damage assessment, these assets extend the commander's reach deep into contested regions and provide real-time intelligence feeds. In an era where adversaries are actively developing quantum-enabled decryption capabilities, traditional cryptographic measures may not guarantee the integrity of satellite-ground or UAV-command channels.

QKD for Protecting Satellite & UAV Communications

Objective	Threat	How QKD Mitigates	Operational Relevance
Secures satellite–ground and UAV–command links from quantum-enabled adversaries	Quantum-capable adversaries intercept control/telemetry channels; decryption of classical keys; command insertion.	QKD generates one-time, physics-secured keys that cannot be cloned, ensuring encrypted + authenticated command links.	Maintains exclusive command authority over UAVs and satellites; denies adversary any chance of decryption.
Prevents hijacking, spoofing, or jamming of unmanned systems	Hijacking (false commands), spoofing (fake ground station signals), jamming (control link denial).	Hijacking: Commands rejected without quantum-authenticated key. Spoofing: False signals fail key validation. Jamming: Even fallback channels protected with quantum-secured frequency hopping.	Protects unmanned ISR, strike, and EW platforms from takeover; adversary limited to disruption only, never control.
Essential for ISR dominance	ISR satellites and UAVs are prime targets for data theft, signal corruption, or disruption.	QKD ensures confidentiality + integrity of ISR data streams; prevents adversary manipulation of reconnaissance feeds.	Guarantees trustworthy situational awareness, enabling secure targeting, surveillance, and operational decision-making.

- **QKD in Satellite–UAV Links:** QKD directly addresses this vulnerability by creating tamper-proof encryption keys shared between satellites, ground stations, and UAV controllers. QKD uses single photons transmitted between a satellite (or UAV) and a ground station to establish encryption keys. Each photon is encoded with a quantum state (polarization, phase, or time-bin), forming a one-time random sequence. By the laws of quantum mechanics, any interception attempt by an adversary disturbs the photon's state, producing an observable error rate (Quantum Bit Error Rate – QBER). This property ensures that commanders know immediately if a link has been compromised. Unlike classical cryptography (which can be retroactively decrypted if the cipher text is stored), QKD ensures forward secrecy. Keys generated today cannot be reconstructed tomorrow, even by a future quantum computer. Adversaries attempting man-in-the-middle attacks on satellite uplinks or UAV command channels would introduce measurable anomalies in QBER, alerting operators in real time. Thus, interception attempts are detectable during the mission, not after compromise. Once generated, quantum keys are fed into standard encryption algorithms (e.g., AES or OTP – One-Time Pad). The command uplink (orders to satellite/UAV) and telemetry downlink (status, ISR feeds) are both encrypted with these fresh, physics-secured keys. Even if adversaries capture the RF signal, the cipher text is information, theoretically secure it contains no exploitable structure for decryption. QKD has already been demonstrated in orbit (China's *Micius* satellite, 2017) for long-distance secure key distribution. Military application extends this principle to geostationary and LEO satellites supporting C2, ISR, and PNT (Position, Navigation, Timing). With QKD, adversaries cannot hijack navigation signals or spoof ISR downlinks without being detected instantly. UAVs are highly vulnerable to link spoofing or hijacking (e.g., adversaries sending false commands). By using QKD-secured uplinks, only the authenticated ground station can issue valid encrypted commands. Any injected or replayed signal without the correct quantum-derived key is discarded as invalid. Similarly, ISR video/data downlinks are encrypted with quantum keys, preventing adversaries from intercepting or altering live intelligence feeds.

- **QKD Protects Against Hijacking, Spoofing, and Jamming:** QKD can overcome unauthorized command insertion to prevent hijacking of UAVs when an adversary intercepts the UAV control link and injects malicious commands to redirect a drone and crash land in hostile territory. UAV and control station share fresh, one-time quantum keys that are unforgeable. Commands are encrypted and authenticated using Message Authentication Codes (MACs) derived from these quantum keys. If an adversary tries to insert false commands, the UAV immediately rejects them because they lack the valid quantum-authenticated signature. Spoofing prevention by injecting false signals when an adversary impersonates the ground station, tricking the UAV into thinking hostile commands are authentic such as GPS spoofing, telemetry spoofing, etc., can be prevented by QKD. The UAV continuously validates incoming commands with the quantum-secured authentication key. Spoofed signals, even if identical in waveform and modulation, fail cryptographic validation because the adversary cannot reproduce the correct quantum key. Attempts at replay attacks (re-sending an old valid packet) also fail because QKD ensures forward secrecy with time-synchronized key updates; old keys are immediately discarded. UAV will only accept commands encrypted/authenticated with the live quantum key. Everything else is automatically discarded. When denial of service attempts are made by the adversary by flooding the control frequency band with noise, preventing UAV from receiving commands or sending telemetry, QKD does not stop jamming (a physical-layer denial); it ensures any fallback control channel is encrypted and authenticated with quantum keys. QKD does not stop jamming (a physical-layer denial); it ensures any fallback control channel is encrypted and authenticated with quantum keys. This raises the cost of jamming from brute-force noise flooding to sophisticated EW campaigns with limited effectiveness.

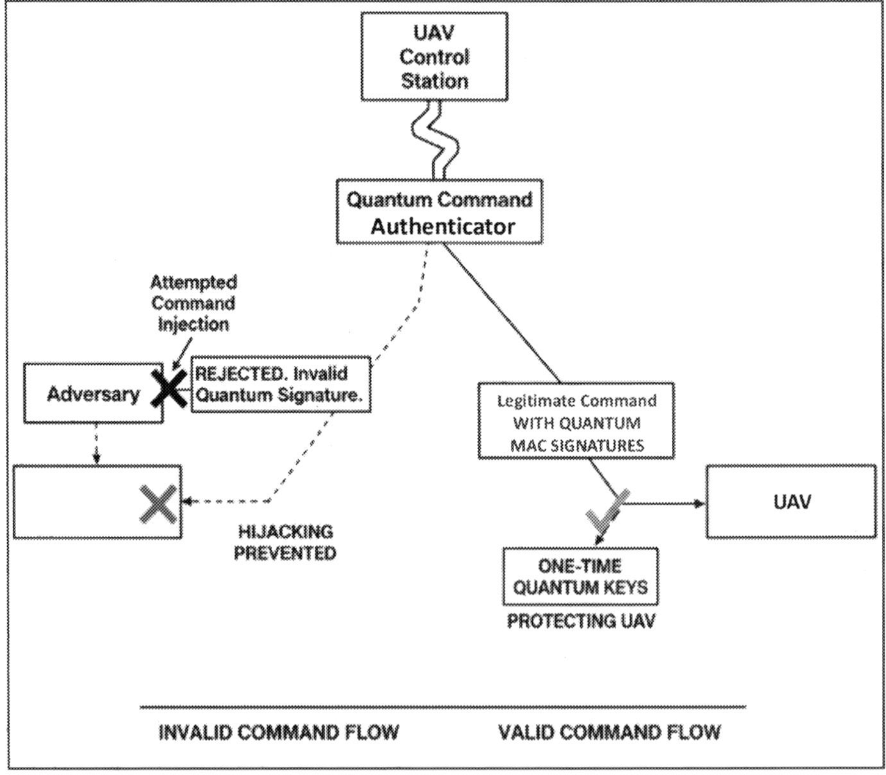

Fig. 3.8: Quantum-Secured UAVs: Preventing Command Hijacking

QKD for UAV & Satellite Link Protection

Threat Vector	Description (Military Context)	QKD Mitigation Mechanism	Operational Effect
Hijacking Prevention (Unauthorized Command Insertion)	Adversary attempts to inject malicious commands into UAV or satellite control channel to redirect, crash, or weaponize the platform.	QKD delivers unique one-time quantum keys for command authentication. Only commands signed with valid QKD-derived keys are accepted; all others are rejected.	Ensures exclusive control authority; prevents loss of mission-critical ISR/strike assets to hostile takeover.
Spoofing Prevention (False Signal Injection)	Enemy mimics ground station or uplink signals to mislead UAVs/satellites into accepting fake instructions or corrupted telemetry.	QKD-enforced mutual authentication between platform and ground station; spoofed signals fail key validation and are automatically discarded.	Maintains signal authenticity, ensuring operators trust only genuine command & telemetry channels.
Jamming Mitigation (Denial of Service Attempts)	Adversary floods communication spectrum to disrupt UAV or satellite control links and ISR data flows.	While QKD cannot stop RF energy denial, it secures fallback channels via QKD-derived encrypted frequency-hopping patterns and supports rapid re-keying after jamming attempts.	Preserves mission continuity; adversary limited to temporary disruption without data compromise or hostile redirection.

Tactical Field Networks with Quantum Secure Backbones

Quantum at the Tactical Edge – When forward bases need fortress-grade encryption

Operational Need	QKD Function	Military Analogy	Operational Relevance
Forward-deployed secure command posts	Deployable QKD nodes integrated into mobile HQs and field command shelters.	Mobile radar units extending coverage into forward areas.	Ensures forward commanders have encryption-resistant comms for orders, targeting, and logistics.
Resilience in contested or degraded environments	Continuous refresh of cryptographic keys at the quantum level.	A constant re-supply convoy that never runs dry.	Protects tactical networks against adversary quantum decryption, jamming, or intrusion.
Protection against cyber exploitation and geo-location	Prevents adversary interception, exfiltration, or location tracking.	Camouflage and deception at the quantum level.	Denies enemy cyber units the ability to geo-locate or compromise field HQs, ensuring survivability.

- **Deployable QKD nodes can be integrated into mobile command posts or field HQs:** QKD nodes can be made compact and ruggedized for field deployment, enabling secure key exchange not only in fixed installations but also in mobile command vehicles, forward operating bases (FOBs), or temporary field headquarters. These nodes can connect via satellite uplinks, fibre spools, or free-space optical channels to higher headquarters, ensuring that even at the tactical edge, commanders operate within a quantum-secured communications framework.

Units manoeuvring through contested zones now deploy quantum-secured communication nodes housed within ruggedized, hardened shelters or purpose-built armoured vehicles. These mobile quantum installations maintain their protective encryption capabilities even when units are rapidly displaced or redeployed across dynamic operational areas, ensuring continuous secure communications regardless of tactical positioning.

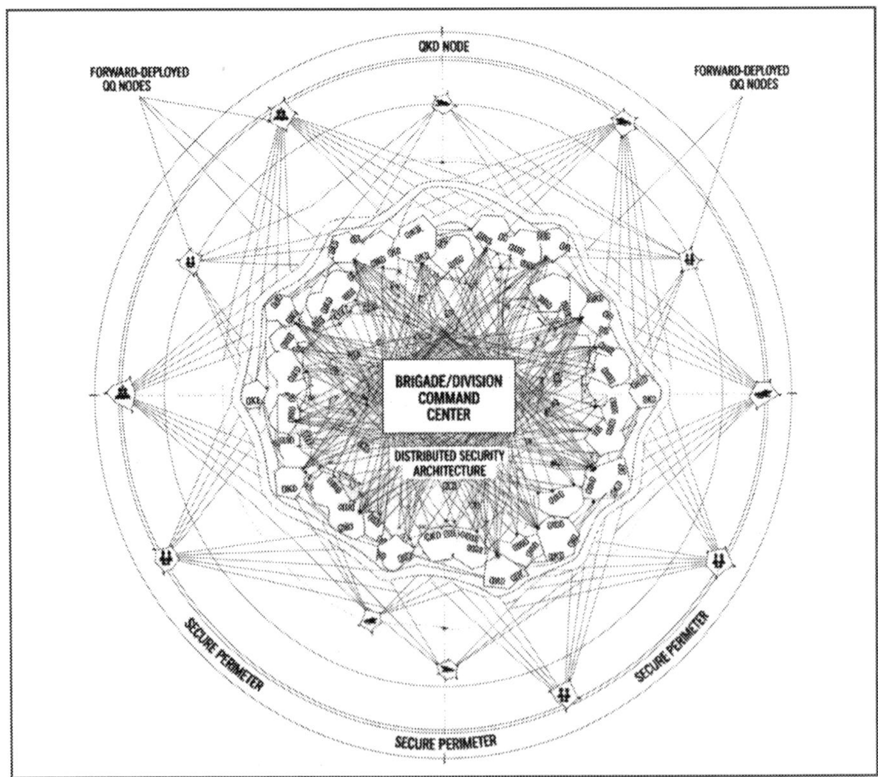

Fig. 3.9: Tactical QKD Mesh: The Encryption-Fortified Bubble

The integration of mobile headquarters with QKD networks creates unprecedented operational security for command-and-control functions. These quantum-protected command posts can issue tactical orders, receive critical intelligence, surveillance, and reconnaissance (ISR) feeds, and coordinate complex fire support missions with complete confidence that adversary forces cannot intercept or decrypt their communications. This security remains intact even when operating under intense near-peer electronic attack scenarios, where conventional encrypted communications would be vulnerable to sophisticated jamming or cryptanalytic efforts. The quantum encryption provides an unbreachable communication layer that maintains operational effectiveness despite advanced enemy electronic warfare capabilities.

Forward-deployed QKD nodes can be interconnected to form comprehensive secure tactical communication networks, creating what military strategists describe as an encryption-fortified operational "bubble" surrounding brigade or division-level command elements. This mesh network architecture ensures redundant communication pathways and distributed security, preventing single points of failure that could compromise entire operational communications. The quantum key distribution system's fundamental security principle that quantum keys cannot be stolen, copied, or intercepted in transit without immediate detection provides an additional layer of operational security. Even if adversary cyber warfare units successfully overrun or compromise individual communication nodes, they cannot exfiltrate quantum encryption keys or re-use them for subsequent operations, maintaining the integrity of the broader secure communication network and protecting classified operational information from enemy exploitation.

- **Secures battlefield communications even in contested or degraded environments:** QKD provides continuous generation of one-time-use encryption keys that are physically tamper-evident. Unlike conventional Public Key Infrastructure (PKI) systems that can be undermined by brute-force cryptanalysis, compromised key repositories, or delayed resupply of key material, QKD creates encryption keys on demand, refreshed in near-real time. This ensures resilient and uncompromisable encryption even when adversaries employ cyber, electronic warfare (EW), or kinetic means to degrade battlefield communication networks.

Where traditional systems suffer under conditions of jamming, spoofing, or cyber infiltration, QKD-equipped tactical nodes can detect hostile interception attempts through elevated Quantum Bit Error Rates (QBER). This provides built-in intrusion; detection operators know immediately if their channel is under attack. Even in contested spectrum environments or when physical infrastructure is damaged, secure communication remains intact as long as a minimal optical or satellite relay is operational.

QKD for Battlefield Communications in Contested Environments

Threat Condition	*Conventional Vulnerability*	*QKD Countermeasure*	*Commander's Advantage*
Electronic Jamming (EW saturation of RF spectrum)	Radios/networks experience denial of service; fallback channels may be insecure.	QKD enables secure optical or satellite key distribution; encryption remains uncompromised despite RF degradation.	Maintains secure C2 continuity even under EW attack.
Spoofing & False Signal Injection	Adversary can insert falsified orders, redirect units, or feed misinformation into tactical nets.	Quantum Bit Error Rate (QBER) anomaly alerts instantly flag interception/ spoof attempts.	Commanders gain intrusion awareness and deny enemy disinformation ops.
Cyber Intrusion/ Key Repository Compromise	Centralized key stores can be hacked or exfiltrated, exposing encryption.	QKD generates on-demand one-time keys via physics-secured channels; no stored keys to steal.	Ensures zero-key compromise, even when rear echelons are penetrated.
Infrastructure Degradation (fibre cuts, relay loss)	Communications collapse until alternate networks are restored; risk of unsecured fallback.	Deployable QKD nodes with free-space optics/ satellite links re-establish secure comms rapidly.	Provides resilient fallback pathways for forward-deployed forces.

- **Prevents data exfiltration or location tracking by enemy cyber units:** In classical networks, cyber adversaries can intercept packets, extract metadata, and analyse traffic flows to infer content, source, or geo-location of tactical nodes. Even if payloads are encrypted, traffic analysis and key compromise can expose sensitive operational data (e.g., unit positions, command patterns). QKD eliminates this risk by using quantum-generated one-time keys that cannot be copied or stored. Any attempt to eavesdrop on the quantum channel changes the quantum states, producing detectable anomalies (Quantum Bit Error Rate spikes).

With QKD, adversaries cannot "harvest-now, decrypt-later" since keys are non-reusable and quantum-protected. Even if they collect cipher text today, it will remain mathematically and physically unreadable in the future. Adversary cyber units often track key exchanges, handshake signals, and routing metadata to geo-locate HQs, mobile command posts, or UAV control stations. QKD reduces this surface since the quantum channel does not broadcast metadata in usable form; attempts to probe it reveal themselves instantly. Commanders can maintain stealth of force disposition, protect the integrity of operational plans, and deny adversaries the ability to map C2 hierarchies or movement corridors through cyber reconnaissance.

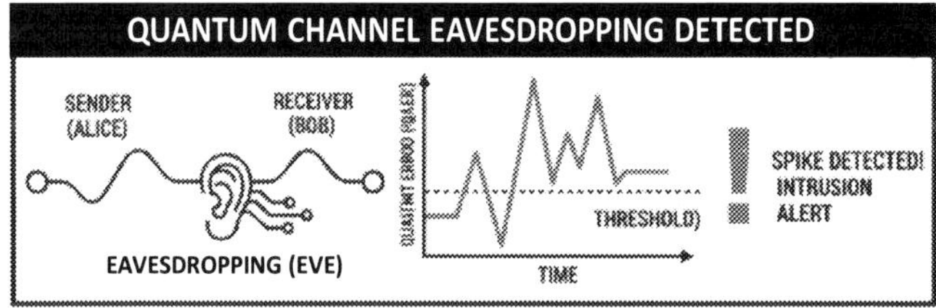

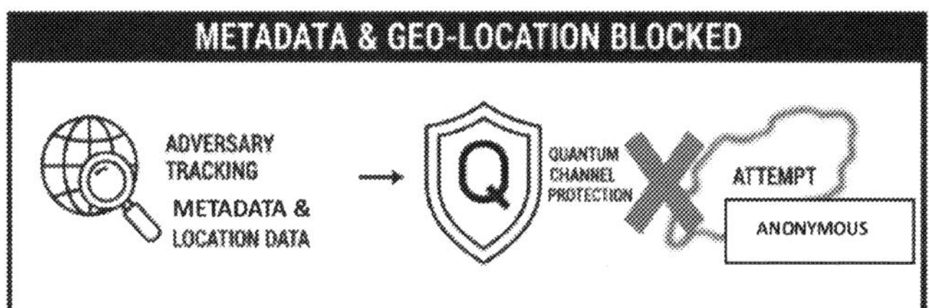

Fig. 3.10: Future-Proof Security: QKD Against Tomorrow's Threats

QKD for Cyber Exfiltration & Tracking Denial

Threat	*Vulnerability in Conventional Comms*	*QKD Countermeasure*	*Commander's Edge*
Data Exfiltration by Cyber Units	Encrypted traffic can be harvested, stored, and decrypted later with quantum computers ("harvest-now, decrypt-later" threat).	Quantum keys are one-time, generated on demand; eavesdropping causes detectable quantum errors (QBER). No usable key material is ever exposed.	Prevents loss of battle plans, orders, or sensor intelligence. Assures commanders that sensitive C2 remains unreadable for adversaries now and in the future.
Location Tracking via Traffic Analysis	Adversaries exploit hand-shake signals, routing metadata, and comms frequency patterns to geo-locate command posts, UAV stations, or forward HQs.	QKD channels do not leak exploitable metadata. Any attempt to probe causes immediate anomaly detection.	Protects force disposition secrecy. Denies adversaries the ability to map HQ positions or infer movement corridors.
C2 Hierarchy Mapping	Adversaries infer command structure by analysing traffic volumes, timing, and key exchange patterns.	QKD enforces symmetric, ephemeral keying with no re-usable patterns, obscuring link analysis attempts.	Preserves operational deception and prevents the enemy from targeting command nodes.

Quantum Resilience in Cyber-Contested Environments

Holding the Line in the Quantum Fog – Where cyber-attacks end at the quantum threshold

In modern battlefields, communications are under constant siege from cyber intrusion, electronic warfare, and adversarial attempts at decryption. Traditional cryptographic systems, while robust today, remain vulnerable to quantum-enabled adversaries who may harvest and later decrypt sensitive transmissions. QKD offers a fundamentally different form of resilience, shifting the contest from software-based intrusion to the immutable laws of physics. This resilience is decisive in cyber-contested environments where near-peer rivals and asymmetric actors alike attempt to deny commanders secure information dominance.

- **QKD systems provide early-warning of tampering or cyber infiltration:** One of the most decisive advantages of QKD in the military domain is its ability to serve not only as a secure encryption tool but also as an intrusion detection system at the physical layer of

Doctrinal Element	Explanation	Military Analogy	Operational Relevance
Early Warning of Cyber Infiltration	QKD detects hostile interception by registering abnormal quantum error rates (QBER).	Tripwires and motion sensors around a forward operating base; enemy probing creates instant alarms.	Provides commanders with immediate indication of hostile cyber activity; prevents silent compromise of mission-critical comms.
Immunity to Man-in-the-Middle & Brute Force	Keys are generated from unclonable quantum states, preventing impersonation or brute-force decryption.	Constantly shifting, unforgeable radio call signs; only two authentic stations can produce them.	Ensures orders, ISR feeds, and drone command links remain authentic and resistant to forgery or hijacking.
Strengthened Posture vs. Near-Peer & Asymmetric Threats	QKD creates physics-enforced resilience, unbreakable by classical or quantum computing.	Reinforced concrete fortress vs. sandbags impervious to sustained or precision attacks.	Preserves secure C2 (command & control) under sustained cyber/EW pressure, maintaining operational continuity in contested theatres.

communication. Unlike classical cyber security measures, which often detect compromises only after data has been exfiltrated or systems corrupted, QKD exposes hostile tampering attempts in real-time. This is because quantum states used for key exchange cannot be measured, copied, or altered without introducing a detectable disturbance. Any adversary's attempt to eavesdrop on or intercept the quantum channel manifests as a rise in the Quantum Bit Error Rate (QBER), providing an immediate warning signal to operators.

This capability transforms the security equation. In contested or denied environments, where cyber infiltration may be continuous and stealthy, QKD gives an unambiguous early-warning mechanism. This ensures that compromised channels can be shut down or re-routed before sensitive information is exposed and decision cycles are preserved. As commanders can trust the authenticity of the communication channels, enemy attempts to remain covert in cyberspace are defeated, forcing them into detectable behaviour. In essence, QKD does not just encrypt the line; it ensures the line itself becomes a sensor against enemy intrusion. This dual role secure transmission and proactive detection makes QKD uniquely suited for future battlefields where cyber infiltration will be as constant as artillery fire.

QKD as Early-Warning Sensor

Aspect	Details
Doctrinal Function	QKD acts as a digital trip wire detecting intrusion attempts at the quantum layer before message compromise occurs.
Mechanism	Any adversary interception of quantum states causes a rise in Quantum Bit Error Rate (QBER), triggering an automatic tamper alert.
Military Analogy	Comparable to trip wires and perimeter sensors around a forward operating base: the enemy cannot probe without exposing themselves.
Operational Advantage	Provides real-time intrusion detection; commanders know instantly if a link is under attack.
Commander's Use Case	Enables immediate channel shutdown or re-routing to maintain secure C2. Protects decision superiority by ensuring message authenticity.
Threat Context	Counters stealthy cyber infiltrations from near-peer adversaries and irregular actors alike. Prevents covert intelligence-gathering.
Force Preservation Impact	Protects battlefield decision cycles from disruption, ensuring orders are not corrupted or delayed by hidden cyber-attacks.

- **Immunity to man-in-the-middle and brute-force decryption:** QKD provides inherent immunity against two of the most dangerous attack vectors in military communications: *man-in-the-middle (MITM) interception* and *brute-force cryptographic decryption*. Unlike classical encryption systems where an adversary might impersonate trusted nodes or deploy supercomputing brute-force attacks, QKD leverages the laws of quantum mechanics. Any attempt to impersonate a legitimate endpoint in a QKD link requires the adversary to reproduce or measure quantum states. This is physically impossible without introducing measurable errors, thus exposing the deception instantly. Likewise, brute-force decryption, no matter the computational resources, cannot succeed, as QKD relies on *information-theoretic security*, not mathematical hardness.

In a cyber-contested battle space, near-peer adversaries such as China or Russia are developing advanced AI-driven intrusion capabilities and quantum computing systems capable of breaking RSA/ECC-based military encryption. If orders to air defence units or nuclear command nodes are susceptible to such breaches, the entire deterrence

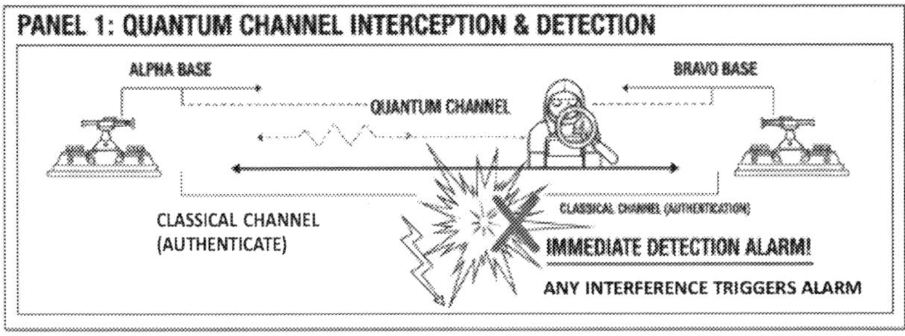

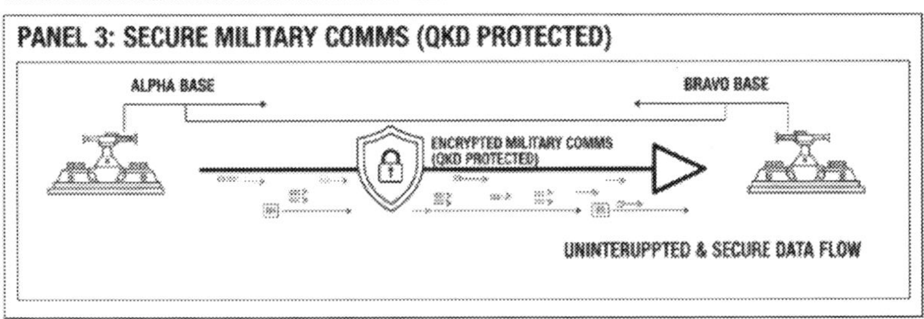

Fig. 3.11: Unbreakable by Design: QKD's Dual-Layer Defense

posture collapses. By employing QKD, armed forces ensure that even a quantum-capable adversary cannot decrypt command traffic or inject false orders. This immunity preserves decision dominance, preventing hostile forces from corrupting C2 integrity or causing "friendly fire by deception."

QKD transitions the military from a model of *defence-in-depth* (where compromise is possible but delayed) to one of *absolute denial* of adversary access. No adversary, regardless of computing power, can sit invisibly inside the channel or peel away encryption by brute force. For commanders, this translates into battlefield confidence, the assurance that what is sent is what is received, unaltered and unseen.

QKD – Immunity to MITM & Brute-Force

Doctrinal Aspect	Officer-Focused Guidance
Threat Addressed	Man-in-the-middle interception, adversary spoofing, and brute-force decryption by classical/quantum computers.
Concept	QKD leverages quantum states that cannot be copied or measured without disturbance, ensuring that any adversary interference exposes itself through error rates. Brute-force is irrelevant; security is physical, not mathematical.
Military Analogy	Like a courier whose message self-destructs if intercepted and simultaneously signals HQ of tampering, enemy cannot impersonate the courier or break the lock.
Operational Relevance	Secures C2 traffic in cyber-contested battle space, denies adversaries the ability to inject false orders or decrypt sensitive operational directives. Ensures decision dominance under quantum-capable threat conditions.
Doctrinal Impact	Shifts from defence-in-depth (delayed compromise) to absolute denial of access. Reinforces trust in secure orders during high-tempo or nuclear-related operations.
Field Employment	Deploy QKD-secured links between command posts, missile batteries, naval task forces, and nuclear command nodes. Monitor error rates as indicators of attempted interception.
Officer Takeaway	With QKD, the enemy cannot sit inside your network unseen or break in with brute force. Orders remain authentic, unaltered, and unencryptable assured security by design.

- ## INTEGRATION OF QKD WITH EXISTING MILITARY COMMUNICATION INFRASTRUCTURE

Quantum Backbone: Integrating Next-Gen Security into Today's War-fighting Networks – Upgrading without disrupting, securing without compromising

Retrofit of Legacy Systems with Quantum Interfaces

Old Iron, New Shields – Breathing quantum life into battle-tested hardware

One of the most pressing challenges for the communication infrastructure whether radios, ship-borne communication suites, or field routers is that they have been designed to operate over decades-long lifecycles. These legacy systems were not built with quantum-era threats in mind, yet they remain mission-critical and cannot be discarded or fully replaced without significant disruption and cost. QKD offers enabling retrofit solutions; modular QKD interfaces can be layered over existing encrypted systems, transforming them into quantum-resilient assets without altering their operational workflows. QKD modules function as an add-on security layer. Instead of replacing classical encryption hardware, QKD uses tamper-proof encryption keys to legacy devices, which then continue to use their existing encryption protocols for data transmission. In this manner, QKD strengthens rather than replaces the current system, ensuring that even older communication networks benefit from future-proof security against quantum adversaries. QKD modules shield legacy radios, routers, or ship systems from interception, ensuring that these platforms continue to serve effectively in the quantum threat environment.

- **Layering QKD Over Existing Encrypted Communication Systems:** Traditional military communication systems rely on classical key exchange protocols such as RSA or Diffie-Hellman to generate and distribute encryption keys making them vulnerable to future quantum attacks. QKD modules do not replace the encryption algorithms or hardware already in use; rather, they supply the encryption systems with quantum-generated keys. Here is how layering works in practice:

  - ❑ **Quantum Layer:** The QKD module generates keys through quantum channels (fibre-optic or satellite-based free-space links). Any eavesdropping attempt disturbs the quantum states and is immediately detected.

  - ❑ **Interface Layer:** The keys are then securely passed from the QKD module into the legacy encryption system through a standardized interface (often PCIe cards, hardware security modules, or external key injectors).

  - ❑ **Classical Layer:** The legacy system (radio, router, satellite terminal) continues to encrypt and transmit data using AES or other military-grade encryption but with fresh, quantum-secure keys.

- **Phased Retrofit of Quantum Key Distribution into Legacy Military Communication Networks**

  Army Signal Corps – Tactical and Operational Networks

  Phase 1: Pilot Integration at Brigade/Division HQs

  - ❑ Deploy QKD modules as overlays on strategic backbone links between corps HQs and theatre commands

- Integrate with army radio networks (HF/VHF/UHF) by connecting QKD-generated keys into secure field routers and encryption boxes

Phase 2: Extension to Tactical Edge

- Enable mobile signal units to field QKD-enabled key injectors that refresh tactical radios in real time
- Provide tamper-proof command-to-unit communications during high-intensity operations

Doctrinal Alignment: Fits within the Army's doctrine of "Assured Tactical Communications", emphasizing survivability of command and control (C2) under electronic warfare pressure.

Naval Communication Commands – Ship-to-Shore and Fleet Operations

Phase 1: Shore-Based Gateways

- Retrofit naval shore stations with QKD modules for secure key distribution to deployed fleets over satellite uplinks
- Establish QKD-secured operational orders between Fleet HQ and task force flagships

Phase 2: Fleet-Wide Rollout

- Integrate QKD interfaces into shipboard communication suites (satcom, HF/VLF links) without requiring re-design of the combat management system
- Use inter-ship free-space optical QKD to secure task force comms, ensuring resilient blue-water operations

Doctrinal Alignment: Supports "Maritime Domain Awareness and Secure Fleet C2" doctrine, ensuring operational orders remain uncompromised even in heavily contested environments.

Air Force C2 Networks – Airborne Platforms and Ground Stations

Phase 1: Ground-to-Air Command Links

- Introduce QKD retrofits in air operations centres (AOCs), securing uplinks to AWACS, UAV control stations, and strategic bomber commands
- Layer quantum keys over existing SATCOM and line-of-sight data links

Phase 2: Airborne Network Extension

- ❏ Enable airborne relay aircraft (AWACS, tankers) to carry QKD modules to distribute quantum-secure keys in flight
- ❏ Future-proof drone swarms and manned-unmanned teaming (MUM-T) by ensuring operator-drone channels are tamper-proof

Doctrinal Alignment: Reinforces the Air Force emphasis on "Command of the Air through Resilient Networks", where survivability of the C2 chain directly impacts mission assurance.

Deployment via Fibre and Free-Space Optical Channels

Through Wire and Air – Where photons carry the code of command

At fixed installations such as command headquarters, naval bases, and air operations centres, QKD can be deployed over existing fibre-optic backbone networks. This allows quantum keys to be continuously distributed between command nodes, ensuring the integrity of secure C2 channels. Much like laying hardened field telephone cables in previous wars ensured secure wired communication between bunkers, fibre-based QKD leverages the infrastructure already in place, but with the added guarantee that any interception attempt is instantly detectable. This is especially critical in hardened, high-value facilities where continuity of secure command is non-negotiable.

- **Free-Space Optical QKD for Mobile and Satellite-Denied Operations:** Fibre connectivity cannot extend to every battle space. Modern operations often demand secure communication where physical infrastructure is absent, degraded, or under threat. In such environments, whether manoeuvring armoured brigades, naval task forces at sea, or expeditionary units operating in austere terrain Free-Space Optical (FSO) QKD offers a resilient solution. FSO QKD uses narrow laser beams to transmit quantum states (photons) through the open atmosphere between two line-of-sight terminals. Each beam carries the fragile quantum bits that form unbreakable encryption keys. Unlike radio-frequency or satellite channels, which radiate broadly and can be jammed or intercepted, FSO links are inherently directional and low probability of intercept/detection (LPI/LPD) significantly reducing adversary exploitation opportunities.

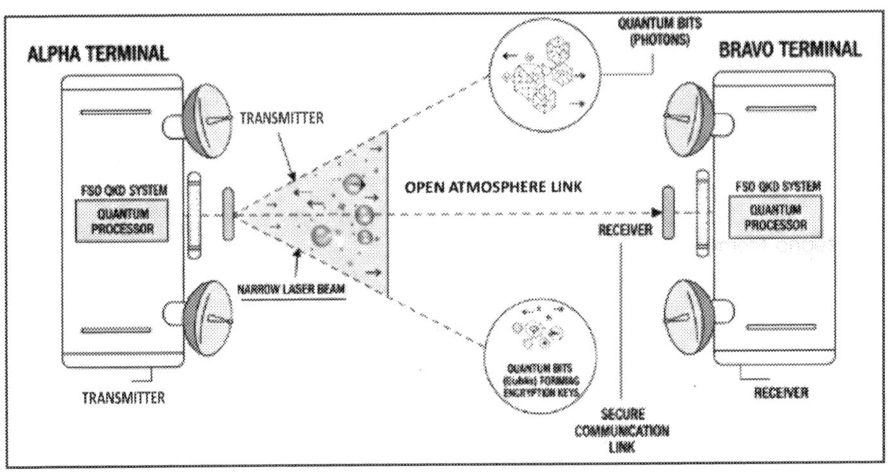

Fig. 3.12: FSO QKD: Laser Beams Carrying Unbreakable Keys

FSO QKD systems can be mounted on command vehicles, mobile shelters, shipboard masts, or UAV platforms, enabling secure key exchange while on the move. Emerging stabilization technologies allow systems to maintain optical alignment even with platform vibration or sea states. Since the photons carry only the encryption key and not the operational data stream, FSO QKD can be overlaid on existing tactical communication networks. The secured keys can then drive classical encryption across radios, SATCOM, or line-of-sight microwave systems already in use. Should space assets be jammed, blinded, or destroyed, FSO QKD provides a ground- or ship-based alternative for maintaining secure links, ensuring C2 continuity without reliance on space-based infrastructure. Unlike RF-based systems, optical beams do not contribute to the electromagnetic order of battle, reducing the chance of electronic warfare targeting or SIGINT detection.

FSO QKD can be compared to the naval signal lamps once used for fleet manoeuvring; a narrow, focused beam visible only to the intended recipient. Where Morse code once carried manoeuvre instructions across a silent sea, today's FSO QKD beams silently distribute quantum keys that can secure the entire fleet" digital communications. Just as a ship outside the beam's line of sight could never intercept the signal lamp, an adversary outside the narrow quantum beam path cannot obtain the keys without detection.

Compatibility with Encryption Key Management Systems (KMS)

Quantum Keys, Command Certainty – Plug-and-play protection in the crypto core

One of the biggest hurdles in adopting new cryptographic technologies across armed forces is not the science, but the practicality of integration. Military organizations already rely on sophisticated Key Management Systems (KMS) to distribute, store, and rotate cryptographic keys for radios, satellite terminals, and secure data links. From a doctrinal perspective, KMS is the "supply chain of trust", a logistics system not for ammunition or rations, but for cryptographic lifeblood.

QKD strengthens this chain by feeding physics-guaranteed keys directly into existing KMS frameworks. Rather than requiring new workflows or entirely novel cryptographic devices, QKD modules can be configured to supply keys into the same secure loaders, fill devices, or crypto-management suites already familiar to communications detachments. This means that operators do not have to relearn encryption handling procedures, the same keying equipment, the same fill routines; the same accountability measures remain intact. QKD simply enhances the trustworthiness of the keys themselves, making them immune to interception, duplication, or brute-force decryption.

Operationally, this integration is vital. It means that a brigade signal company can replace its key material under combat conditions without worrying about adversary compromise. A naval task force can refresh encryption across an entire battle group without physically transferring keys by courier. An air operations centre can rapidly re-key its tactical data links if compromise is suspected, with assurance that the replacement keys have not been intercepted. The reduction in training overhead and the preservation of familiar workflows ensures adoption can be rapid and seamless, while the high-assurance replacement of keys strengthens confidence in mission-critical systems, even under cyber or electronic warfare pressure. In short, QKD does not discard the military's existing crypto-infrastructure; it plugs into it like a new power source, one rooted not in man-made complexity, but in the fundamental laws of quantum mechanics. This ensures that the chain of command remains encrypted, authenticated, and uncompromised, even against the most advanced adversaries.

QKD links are to be employed in two complementary modes: fibre-based connections for securing fixed command installations (HQ, command centres, naval bases) and FSO links for protecting deployed or mobile nodes such as carrier strike groups, expeditionary headquarters, and airborne command posts. Once established, QKD modules are to generate quantum-assured keys and inject them directly into the KMS, appearing to operators as a standard key feed without altering existing workflows. Keys are then to be distributed through established channels, secure loaders, over-the-air rekeying, or wired fills into tactical radios, Link-16/22 terminals, SATCOM systems, and IP-based encryptors. Field units will continue to operate familiar crypto-enabled equipment, with the assurance that compromised keys can be replaced immediately through QKD refresh, denying adversaries any window of exploitation. In the event that classical cryptographic algorithms are degraded or broken, commanders are to rely on the QKD-fed KMS pipeline to ensure continued confidentiality of operational orders and tactical communications under conditions of cyber or electronic attack.

3.2   Quantum Cryptography and National Security

• **Quantum cryptography's role in safeguarding classified data and securing defence communications**

Photon-Proof Security: Defending Secrets in the Age of Quantum Intrusion – Where encryption meets the edge of physics

Unbreakable Encryption Using Quantum Key Distribution (QKD)

Beyond Codebooks – Securing messages with the laws of nature, not man-made algorithms

In modern warfare, secure communication is not just an enabler; it is the very spine of command and control (C2). Classical encryption systems, no matter how advanced, ultimately rely on human-devised algorithms and computational limits. With the advent of quantum computing, those limits are destined to collapse. QKD represents a doctrinal leap. It secures messages not by human ingenuity, but by the immutable laws of physics. QKD relies on photons to transmit encryption keys. According to quantum mechanics, any attempt to measure a photon's quantum state alters it irreversibly. This creates an automatic alarm system: if an adversary tries to intercept, anomalies (bit errors, increased noise) are instantly visible to both sender and receiver. This provides a built-in *early-warning system* against espionage and cyber infiltration. Unlike traditional systems where adversaries can eavesdrop silently, here the "trip wire" is woven into the medium itself. Commanders know when to terminate or re-route communications to protect mission integrity.

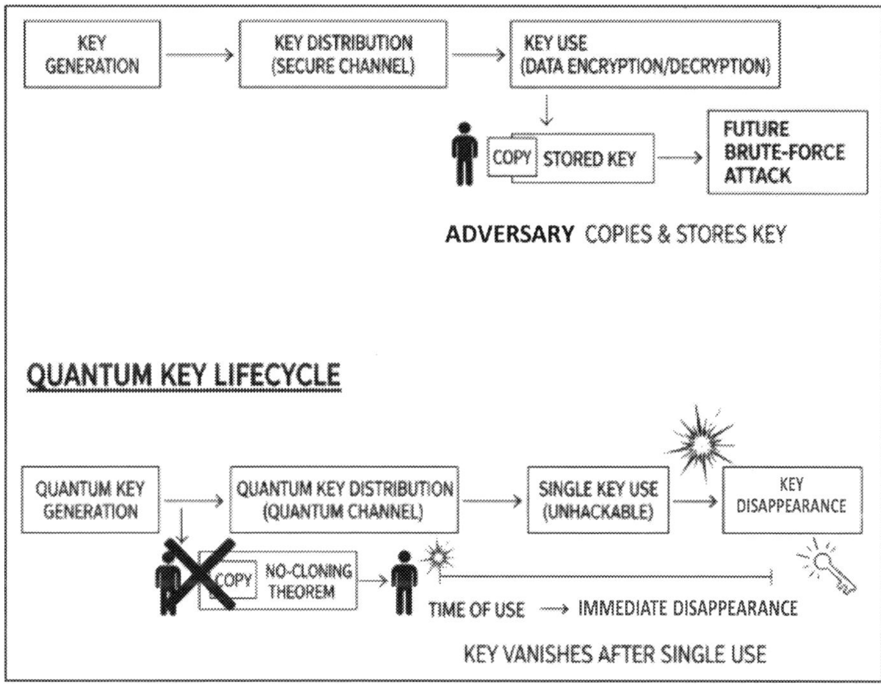

Fig. 3.13: The No-Cloning Advantage: Keys That Vanish After Use

QKD keys are governed by the no-cloning theorem, which states quantum information cannot be perfectly copied. Unlike digital encryption keys that can be stolen and stored for later brute-force or quantum-powered cracking ("harvest now, decrypt later"), QKD keys vanish once used and cannot be replayed. No-cloning theorem protects against both present and future threats. Even adversaries with massive quantum computing power cannot retroactively decrypt QKD-secured traffic. For military archives, battle orders, intelligence logs, and nuclear readiness directives this ensures they remain classified for ever, not just for 10–20 years. The ultimate application of QKD lies in securing strategic communications pathways: between political leadership, service headquarters, and strategic/nuclear assets. In such cases, compromise is not an intelligence loss. It risks unauthorized launches, disrupted deterrence, or breakdown of national C2.

Doctrinal Block	Concept	Military Analogy	Operational Relevance
Interception Detection	Any eavesdropping attempt disturbs quantum states, instantly exposing adversary presence.	Like a secure field line rigged with trip wires, any tampering causes an alarm.	Provides a built-in early-warning system against cyber and SIGINT infiltration; allows commanders to abort or re-route compromised channels.
Uncopyable Keys	Keys follow the no-cloning theorem; they cannot be copied, stored, or cracked.	Comparable to one-time cipher pads written in self-erasing ink; once read, they disappear for ever.	Neutralizes "harvest now, decrypt later" threats; archives, orders, and readiness directives remain classified permanently.
Strategic Command Assurance	Ensures secure backbone for HQ-to-nuclear/strategic asset communications.	Like PALs (Permissive Action Links) preventing unauthorized weapon use.	Prevents adversary command injection; safeguards deterrence credibility and continuity of command under cyber or EW attack.

- **Instant Interception Detection:** QKD exploits the principle that quantum states collapse when observed. When an adversary tries to intercept or copy these quantum signals, the very act of observation disturbs the quantum state. This disturbance is instantly noticeable at the receiving end, alerting defenders that the channel is compromised. Commanders gain a real-time intrusion alarm for critical communication links. Instead of discovering a breach months later through forensic analysis, they know instantly if adversaries have tried to tamper with quantum-secured channels. This allows HQs to rapidly switch to backup channels, re-key systems, or re-route operational data before any compromise occurs. Knowing that detection is guaranteed discourages adversaries from even attempting man-in-the-middle operations against QKD-secured lines, shifting the cyber battlefield calculus. In nuclear C2 (Command and Control) or strategic asset coordination, this early-warning function ensures *command integrity*, preventing false orders or infiltration at the most critical level.

Aspect	Explanation	Military Analogy	Operational Relevance
Concept	Quantum states (photons) carrying keys cannot be observed without disturbance; any interception instantly alters the signal, revealing adversary presence.	Trip wire perimeter defence – the enemy cannot cross without triggering an alarm.	Provides real-time breach detection; commanders know immediately if adversaries attempt to access communications.
Doctrinal Utility	QKD transforms every enemy interception into a detectable event; secrecy and integrity are enforced by the laws of physics, not algorithmic strength.	Like mined choke points, any intrusion exposes the attacker.	Enables rapid shift to alternate channels, immediate re-keying, or re-routing orders before compromise spreads.
Strategic Relevance	Eliminates silent eavesdropping; adversaries cannot build intelligence undetected. Ensures command integrity high-value communications (HQ ↔ nuclear assets).	Equivalent to watchtowers lighting up a battlefield when an enemy probe is detected.	Strengthens deterrence: adversaries know that QKD-secured lines cannot be probed without exposure, altering the cyber contest balance.

- **Uncopyable and Uncrackable Keys:** Unlike digital keys, which can be copied undetected or eventually broken through brute-force or quantum computing power, quantum keys are unclonable by nature. The "no-cloning theorem" ensures that a key distributed via quantum channels cannot be duplicated or reverse-engineered. Unlike classical cryptographic keys, which can be copied and attacked with brute-force or quantum algorithms (e.g., Shor's Algorithm), QKD keys exist only as quantum states. Once measured or disturbed, they collapse irreversibly, making replication or theft impossible. Keys transmitted via QKD can never be compromised, even by adversaries with supercomputers or future quantum computers. This ensures communications remain secure beyond 2050 threat horizons. For nuclear command-and-control, ballistic missile submarines, or space-based assets, QKD guarantees that encryption keys protecting launch orders or satellite tasking cannot be intercepted or brute-forced. Long-term deterrence prevents adversaries from "store now, decrypt later" strategies, where sensitive military communications are harvested today for decryption in the quantum era. QKD ensures they remain unreadable forever. With guaranteed un-breakability, commanders can transmit targeting data, ISR feeds, and joint orders across services without fear of eventual compromise.

Protection of Strategic Command and Control (C2) Systems

Command Without Compromise – Safeguarding the chain of command from quantum disruption

Quantum cryptography transforms C2 from a vulnerable communication link into an incorruptible command lifeline ensuring orders are never intercepted, altered, or doubted even under conditions of maximum cyber-electronic warfare. It preserves leadership continuity, reinforces unity of command, and guarantees that military operations remain coordinated and effective under the harshest wartime pressures.

Command and Control (C2) networks forming the nervous system of military operations: Any compromise will cripple decision-making, sow confusion, and even paralyze response in critical moments. QKD ensures that C2 messages are shielded by the laws of quantum physics rather than man-made algorithms. Attempts to intercept or manipulate keys are instantly detected, rendering classical espionage or cyber infiltration futile. This elevates C2 systems from being merely "encrypted" to being tamper-proof at the most fundamental level of communication.

Doctrinal Aspect	Military Analogy	Operational Relevance
Shielding C2 from Espionage, Jamming, or Spoofing	Like invisible sentries guarding every runner carrying orders, the moment an enemy touches the message, it vanishes and alarms are raised.	Ensures only authentic, uncompromised orders flow through the chain of command, denying adversaries the ability to mislead or confuse forces.
Preventing Digital "Decapitation Strikes"	Comparable to a fortress whose gates cannot be picked or forced; no matter how much force is applied, leadership access points remain secure.	Protects critical message integrity, preserving leadership continuity even under massive cyber or electronic attack.
Reinforcing Wartime Communication Hierarchies	Like a battlefield trumpet whose sound can never be mimicked by the enemy, troops always know which orders come from their true commander.	Guarantees unbroken trust in orders, preventing hesitation, fragmentation, or paralysis in high-intensity and nuclear scenarios.

- **Shielding C2 Messages from Espionage, Jamming, or Spoofing: QKD provides a new defensive shield by making espionage functionally impossible.** Since QKD relies on the exchange of quantum states (such as photons), any attempt by an adversary to observe or intercept the key immediately alters those quantum states, triggering an alert. Unlike traditional code-breaking, where an enemy might stealthily monitor encrypted flows for months, QKD ensures that surveillance attempts are both futile and instantly detectable. Against jamming, QKD-enhanced communications can be embedded within resilient optical channels and hardened satellite links. Even if an adversary floods the spectrum with noise, they cannot replicate or spoof the quantum signals themselves. This denies the enemy the ability to block, overwrite, or confuse the flow of C2 orders.

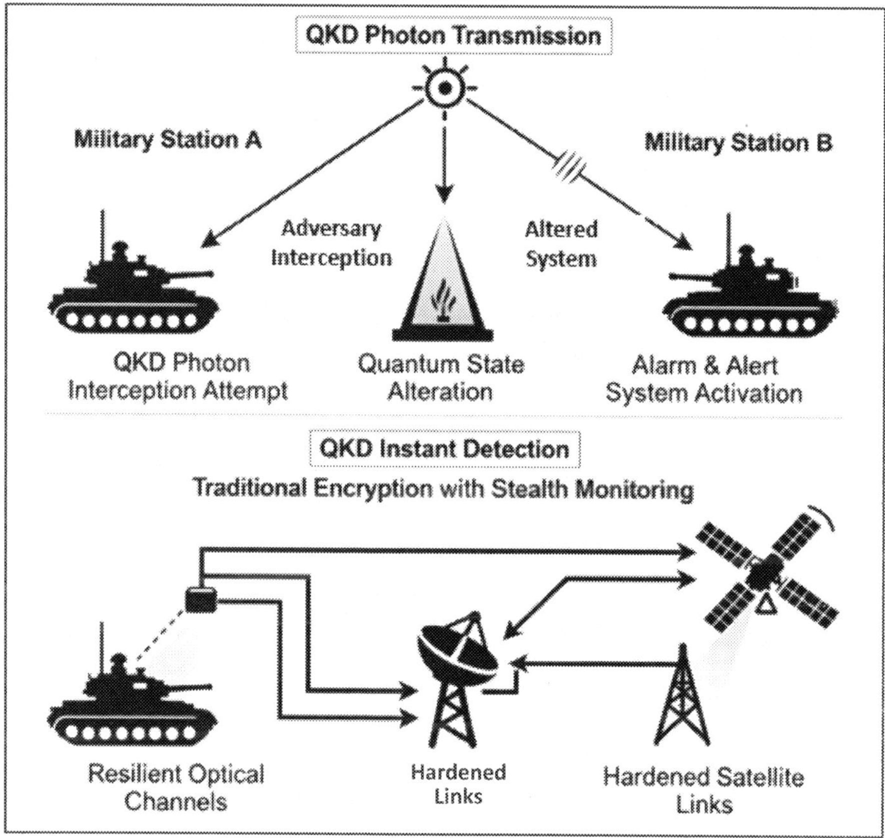

Fig. 3.14: Instant Threat Detection: QKD's Real-Time Alert System

Spoofing, the digital equivalent of impersonating a commander's voice, is also neutralized. Because QKD-derived keys are one-time-use and physically unclonable, adversaries cannot generate counterfeit encryption keys to masquerade as legitimate command authorities. Just as no enemy can mimic a commander's personal seal without detection, no hostile actor can forge QKD-secured authentication. This ensures that in the fog of war, whether in high-intensity conventional battles, cyber-electronic contested zones, or nuclear command readiness scenarios, C2 messages remain immune to espionage, resistant to jamming, and un-spoofable by adversaries. Forces can therefore act on orders with absolute confidence, knowing they originate from legitimate command authorities and remain intact despite adversarial interference.

Threat Dimension	Quantum Shielding Mechanism	Operational Relevance for C2
Espionage (Interception of C2 traffic)	QKD keys exist only as fragile quantum states; any interception attempt alters the state, instantly revealing eavesdropping.	Eliminates silent surveillance of command channels ensures adversaries cannot monitor or extract keys without immediate detection.
Jamming (Disruption of communication links)	Quantum-secured signals ride on optical/ satellite channels that cannot be replicated; even spectrum flooding cannot mimic quantum states.	Protects continuity of command orders under electronic warfare; forces remain linked to HQ despite enemy jamming attempts.
Spoofing (Forging C2 orders/ false commands)	QKD keys are one-time-use, uncopyable, and physically unclonable, preventing adversaries from fabricating authentic command signatures.	Reinforces chain-of-command legitimacy in wartime; troops can trust every order is genuine, denying adversaries the ability to sow confusion.

Defence Against Quantum-Era Espionage and Signal Interception

Silent Signals, Safe Secrets – Defeating tomorrow's spies with today's quantum shield

Quantum cryptography is not merely a defensive measure for current communications; it is a time-proof shield that protects the military decision-making chain, both in the heat of battle and across the generational arc of national security.

The emergence of quantum computing poses a direct threat to classical encryption systems. Adversaries who intercept today's encrypted military communications may remain unable to read them immediately, but once quantum capabilities mature, they could retroactively decrypt these messages, exposing sensitive operational details, intelligence reports, and strategic directives. QKD prevents this vulnerability by ensuring that encryption keys themselves cannot be copied or reconstructed even by an adversary armed with a large-scale quantum computer.

Just as the Enigma machine was rendered obsolete by Allied cryptanalysis in World War II, classical encryption risks obsolescence in the face of quantum attacks. QKD is the equivalent of deploying a cipher that even a future Ultra or quantum "code-breaking division" cannot break. With QKD, commanders can transmit orders across theatres of war confident that even if intercepted, adversaries gain nothing. This neutralizes the quantum espionage threat before it materializes, ensuring that battlefield communications remain ahead of the adversary's technological curve.

Aspect	Concept	Analogy	Operational Relevance
Neutralizes the threat from quantum-enabled adversaries cracking intercepted data	Quantum computers could eventually break today's encryption. QKD ensures keys are immune from interception or reconstruction, even by quantum attack.	Like WWII's Enigma machine being broken by Allied cryptanalysis experts, QKD is a cipher that even a future "Ultra" or quantum code-breaking unit cannot crack.	Commanders can issue orders confident that, even if intercepted, adversaries gain nothing. Eliminates the risk of quantum espionage undermining operations.
Preserves confidentiality even if data is stored and analysed decades later	Adversaries may record encrypted comms now and wait for future decryption ("store now, decrypt later"). QKD blocks this tactic permanently.	Like an enemy submarine recording every sonar ping hoping to reconstruct patrol routes later, in quantum-secured comms, the archive remains meaningless static.	Protects operational records, nuclear command traffic, and intelligence reports long-term. Ensures enduring confidentiality of mission-critical communications.
Future-proofs sensitive archives against delayed decryption attacks	Sensitive archives, war plans, nuclear codes, intelligence remain valuable for decades. Quantum-secured archives remain sealed against all future cryptanalytic advances.	Like a hardened underground bunker designed to withstand both todays and tomorrow's weapons, QKD is the digital bunker for archives.	Denies adversaries the advantage of time. Strategic doctrines, leadership directives, and classified history remain secure across generations.

- **Preserves confidentiality even if data is stored and analysed decades later:** Modern adversaries routinely capture encrypted communications on high-capacity storage systems. Even if they cannot decrypt now, they bank the traffic, anticipating that future advances (quantum computers, AI-accelerated cryptanalysis) will eventually break the ciphers. QKD relies on quantum states of photons. Any attempt to intercept or copy photons introduces observable disturbances (Heisenberg's uncertainty principle and no-cloning theorem). Keys generated through QKD are therefore secure since they cannot be copied during transmission and cannot be reconstructed later from stored traffic. Even if an adversary records the cipher text today, without the QKD-generated one-time keys, the traffic remains mathematically unbreakable for ever.

Once transmitted under QKD, a command or archive remains undecipherable decades later, even if future enemies achieve quantum supremacy. Strategic deterrence relies not only on current capability but also on the secrecy of legacy nuclear and war-fighting directives. QKD ensures these remain impenetrable, denying adversaries historical insight into command doctrine. Military archives (strategic communications logs, war diaries, satellite intelligence) remain shielded against future digital archaeology by hostile powers.

Classical Encryption (RSA, ECC, AES-256)	*Quantum-Secured Encryption (QKD + OTP)*
Security is computational: breaking it requires impractical computing power today.	Security is information-theoretic: guaranteed by quantum physics, not by assumptions about computing limits.
Vulnerable to "harvest now, decrypt later."	Immune: recorded cipher text cannot be retroactively decrypted.
Strength decreases over time as hardware/algorithms improve.	Strength remains constant, regardless of future quantum breakthroughs.

- **The strategic advantage of quantum cryptographic systems in defence contexts**

Quantum Fortresses: Securing the Edge in Digital Battlefields – Dominating the invisible frontlines of tomorrow's wars

Uninterceptable Key Exchange with Quantum Key Distribution (QKD)

No Keys Left Behind – Weaponizing quantum laws to block enemy eavesdropping

At the heart of secure communications lies the exchange of cryptographic keys. These are codes that unlock encrypted messages. Traditional key exchange methods, no matter how advanced, ultimately rely on mathematical difficulty, which can be overcome with time, computing power, or insider compromise. QKD employs the fundamental laws of quantum physics to guarantee that keys cannot be copied, cloned, or intercepted without detection. In QKD, any hostile attempt to "tap the line" alters the quantum state of the photons carrying the key, immediately revealing the intrusion. This transforms key exchange from a vulnerable handshake into an intrusion-proof security mechanism, backed not by man-made algorithms but by the unyielding physics of the quantum world.

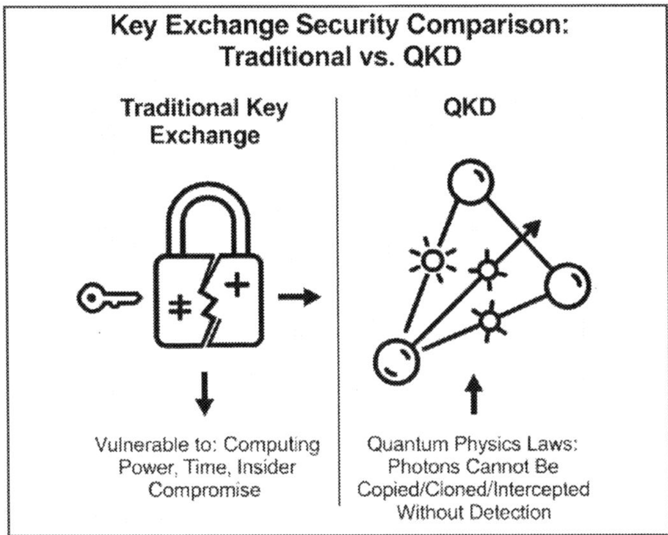

Fig. 3.15: From Math to Physics: QKD's Unbreakable Guarantee

QKD-based key exchange is not merely a technical upgrade; it is a strategic advantage. In the context of nuclear command-and-control, where absolute assurance is required that launch or stand-down orders cannot be forged, intercepted, or delayed, QKD provides a safeguard that even quantum-enabled adversaries cannot bypass. Similarly, in special operations and black ops communications, where stealth and deniability are paramount, QKD ensures that any attempt at interception is both impossible to conceal and operationally useless to the adversary. By making interception detectable and key compromise impossible, QKD eliminates the risk that adversaries will exploit stolen keys to read, replay, or fabricate commands.

Concept	Analogy	Operational Relevance
QKD ensures cryptographic keys cannot be cloned, copied, or intercepted without detection, as any eavesdropping attempt disturbs the quantum state of photons.	Like a sealed diplomatic pouch with a dye-pack: any tampering irreversibly marks the contents and alerts sender and receiver.	Guarantees uncompromised command chains, especially for nuclear control orders and special ops communications. Prevents adversaries from silently stealing or fabricating keys, closing one of the oldest gaps in military cryptography.

- **Uses quantum principles to ensure keys can't be cloned or secretly intercepted:** A no-cloning theorem forms the core principle of QKD's security: This states that an unknown quantum state (e.g., a photon encoding part of a key) cannot be copied perfectly. For an adversary attempting to intercept and duplicate photons in transit, quantum mechanics prohibits making an identical copy without disturbing the original and therefore any interception attempt alters the signal and reveals the intruder. In QKD schemes like BB84, photons are encoded in different polarization bases (rectilinear vs. diagonal). Measuring a photon in the wrong basis changes its state irreversibly. Therefore, if an eavesdropper tries to measure in transit, the measurement error propagates into the key exchange, and Alice and Bob (sender and receiver) detect higher-than-expected error rates (Quantum Bit Error Rate – QBER).

  During key reconciliation, publicly compare subsets of their measurement bases (not the actual key values): If the error rate exceeds a threshold (e.g., >11% in BB84), it indicates tampering/eavesdropping. The compromised key is discarded, and a new round is initiated. Unlike classical communication where packets can be

Concept	Analogy	Operational Relevance
No-Cloning Theorem: Quantum states (photons carrying key bits) cannot be copied perfectly, making duplication of encryption keys physically impossible.	Like trying to copy a sealed handwritten order that vanishes if duplicated, the act itself destroys the original.	Prevents adversaries from making silent copies of keys; guarantees that even if comms are monitored, encryption integrity is preserved.
Heisenberg's Uncertainty Principle: Measuring photons in the wrong basis irreversibly alters their state, introducing detectable errors (QBER).	Similar to opening a coded safe with the wrong combination; it scrambles the lock, and the tampering is immediately visible.	Enemy attempts to "listen in" create measurable errors; defenders are alerted to compromise attempts instantly.
Detection of Eavesdropping: Any interception attempt raises error thresholds; keys are discarded and re-issued if tampering is suspected.	Like a tripwire in a secure bunker, any disturbance sets off an alarm, forcing a reset of defences.	Ensures nuclear command, black ops, and strategic comms cannot be silently compromised; adversaries cannot "store now, decrypt later."

copied without trace, quantum signals cannot be tapped without disturbing them. This makes "store now, decrypt later" impossible: adversaries cannot even capture the raw key material to use against future quantum computers.

Zero-Day Resistance Against Cryptographic Exploits

Built from Physics, Not Patches – No backdoors, no exploits, just laws of nature

Conventional cryptographic systems rely on mathematical hardness assumptions. The belief that certain problems (like factoring large primes or solving discrete logarithms) are computationally infeasible. This makes them inherently vulnerable to *zero-day exploits,* hidden flaws in algorithms, unpatched software vulnerabilities, or undiscovered mathematical shortcuts that adversaries can weaponize without warning. Quantum cryptography, however, shifts the foundation of security away from mathematics and code to the immutable laws of physics. QKD is not secured by computational difficulty but by the inviolate principle that a quantum state cannot be copied or measured without altering it. This renders adversarial zero-day attacks whether algorithmic breakthroughs, insider-planted backdoors, or stealthy exploitation of software flaws irrelevant.

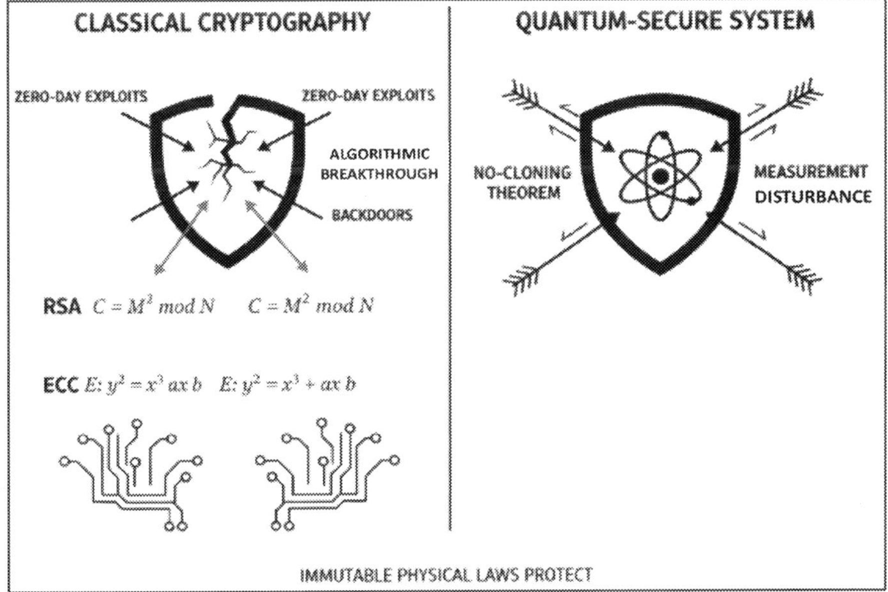

Fig. 3.16: Physical Laws vs Mathematical Vulnerabilities: The Quantum Advantage

Traditional cryptography is like a fortress built with walls of stone: strong today, but potentially undermined if an enemy discovers an unguarded tunnel or invents a new siege engine. Zero-day exploits are those hidden tunnels unknown even to the defenders until it is too late. Quantum cryptography, by contrast, is like fortifying the fortress with a moat of fire that physics itself sustains. No hidden tunnel exists, because the moat cannot be crossed without being seen and extinguished in the attempt. The defence is no longer based on human design or code audits but on physical laws that cannot be rewritten by adversary ingenuity.

For long-duration intelligence missions such as signals interception platforms, space-based surveillance assets, and embedded HUMINT networks the costliest risk is not today's compromise but tomorrow's retrospective decryption of years of sensitive data. Quantum-secured systems provide immunity against such retroactive exposure, guaranteeing that today's secrets remain sealed even if adversaries gain unprecedented computational capabilities in the future. For strategic deterrence forces, where credibility depends on absolute assurance of secure command and control, immunity from zero-day cryptographic exploits removes an adversary's ability to silently undermine launch authority or disrupt nuclear signalling channels. In coalition operations, where interoperability creates complex security dependencies, quantum cryptography ensures trust by eliminating reliance on potentially compromised third-party software updates or algorithms.

Concept	Analogy	Operational Relevance
Classical cryptography depends on mathematical hardness (RSA, ECC), leaving it vulnerable to zero-day exploits, algorithmic breakthroughs, or backdoors. Quantum systems rely on immutable physical laws (no-cloning, measurement disturbance), eliminating this attack surface.	Classical systems are a fortress of stone: strong but secretly undermined if an enemy finds a hidden tunnel (zero-day exploit). Quantum cryptography is a moat of fire maintained by physics; no secret tunnel exists, and intrusion is immediately visible.	Protects long-duration, high-value missions (satellites, SIGINT, HUMINT networks) from future decryption. Guarantees secure C2 for deterrence forces by removing exploit-based compromise risks. In coalition ops, removes dependency on potentially compromised third-party algorithms or updates. Provides future-proof security edge.

Immunity Against Future Quantum Computer Attacks

Concept	Analogy	Operational Relevance
Classical encryption (RSA, ECC) will fail once large quantum computers emerge. QKD derives keys from quantum states, which cannot be retroactively broken. Secures secrets decades into the future.	Like sealing war plans inside a vault that self-destructs if tampered with, even if the enemy builds new tools tomorrow, yesterday's data remains safe.	Protects strategic deterrence communications, nuclear command links, and intelligence archives from being decrypted decades later. Ensures continuity of deterrence credibility.

Tactical Superiority in Highly Contested Electronic Warfare Zones

Concept	Analogy	Operational Relevance
Classical comms fail in jammed, GPS-denied, or spoofed zones. Quantum channels encode information in photons, unaffected by EM noise.	Like a submarine using sonar pings inaudible to the enemy, communication continues while the adversary hears only silence.	Enables SOF units, naval task groups, drone swarms to maintain secure C2 under heavy jamming or deception environments. Extends operational reach in denied battle space.

Zero-Day Resistance against Cryptographic Exploits

Concept	Analogy	Operational Relevance
Classical crypto depends on algorithms and updates, vulnerable to zero-day exploits and backdoors. Quantum cryptography depends on immutable physics, immune to such flaws.	A classical fortress may hide secret tunnels unknown to defenders (zero-day exploits). Quantum is a moat of fire guarded by natural law; no tunnels exist, and entry is always detected.	Secures long-duration, high-value missions (satellite relays, HUMINT networks, SIGINT). Eliminates dependency on compromised vendors/updates. Provides reliability in coalition operations where software trust is uncertain.

Unclonable Keys and Interception Resistance

Concept	Analogy	Operational Relevance
Quantum states obey the no-cloning theorem: keys cannot be copied or intercepted without detection. Any eavesdropping attempt disturbs the system and is revealed.	Like special ink that changes colour if touched, an adversary cannot steal or duplicate without leaving visible traces.	Guarantees C2 integrity for expeditionary forces. Ensures adversaries cannot harvest keys silently for future use. Builds trust in joint operations by proving comms are uncompromised.

- **Quantum systems are not reliant on mathematical complexity or software updates:** Traditional cryptography (RSA, ECC, AES, etc.) secures communications by relying on the mathematical hardness of specific problems: integer factorization, elliptic curve discrete logarithm, symmetric key brute force limits which takes too long for current computers to solve from the security perspective. If adversaries discover a faster algorithm (e.g., Shor's Algorithm on a quantum computer) or exploit a weakness in implementation, the whole system collapses. A continuous cycle of software updates, patching, and algorithm replacement is required to stay ahead of attackers.

QKD does not rely on a mathematical puzzle at all. Security is rooted in principles of physics – the Heisenberg Uncertainty Principle and quantum no-cloning theorem. Any interception attempt disturbs the quantum state and reveals eavesdropping instantly. There is no mathematical problem to "solve" or algorithm to "crack"; it's law-of-nature security, not complexity-based security. Therefore, there is no cycle of "patching, updating, or migrating algorithms." Once deployed, the physics remains constant, making the system resistant to both current and future computational advances. This removes the burden of constant crypto-agility cycles (new algorithms every

Theme	Concept	Analogy	Operational Relevance
1. Immunity Against Future Quantum Computer Attacks	QKD security is rooted in physics (uncertainty principle, no-cloning), not mathematical hardness. Prevents adversaries with quantum computers from decrypting sensitive data.	Classical codes = safes eventually cracked with stronger tools. Quantum = safe that self-destructs if tampered with.	Protects long-term intelligence, nuclear C2, and strategic secrets from "harvest now, decrypt later" threats. Ensures deterrence stability.
2. Tactical Superiority in Highly Contested EW zones	Quantum links resist EW interference, spoofing, and jamming; remain secure and tamper-evident even in denied environments.	Classical radios in EW= shouting in a storm. Quantum = laser beam between scouts, invisible and tamper-proof.	Enables SOF comms, naval task groups, drone swarms to operate securely in EM-contested battle spaces. Assures connectivity when classical systems fail.
3. Zero-Day Resistance Against Cryptographic Exploits	Immune to zero-days, algorithmic breakthroughs, or backdoors. Security based on immutable physics, not patches.	Classical = armoured vehicle needing upgrades against new threats. Quantum = force field impervious to any innovation.	Removes risk of sudden crypto-collapse. Critical for long-duration missions, intelligence archives, and autonomous systems where updates are impossible.
4. Physics Over Mathematics – Independence from Updates	Quantum cryptography does not rely on complexity assumptions or periodic software updates. Security is permanent, enforced by physical laws.	Classical = locks replaced as lock-picking evolves. Quantum = door that destroys tools if tampered with, needing no upgrades.	Provides permanent, non-upgradable security for nuclear C2, forward-deployed forces, and intelligence stores where patching is infeasible.

10–15 years), ensures long-duration missions and intelligence archives (satellite imagery, HUMINT reports, nuclear command links) remain secure for decades without fear of a future algorithmic breakthrough. It also provides strategic zero-day resistance; there are no undiscovered mathematical weaknesses or hidden backdoors to exploit, security is anchored in immutable physics.

3.3 QUANTUM-SECURE COMMUNICATIONS NETWORKS

• **Design and Deployment of Quantum-Secure Military Networks**

Quantum Shield Networks – Fortifying the digital battlefield from base to battle space

Architecture of a Quantum-Secure Military Network

Fortress by Design – Embedding quantum strength into the very fabric of communication

The architecture of a quantum-secure military network is not an add-on to existing systems. It is a purpose-built defensive framework that embeds security into the very structure of communication. Unlike classical networks, where protection is often layered on as patches, a quantum-secure architecture integrates QKD at every node-to-node connection, ensuring that from the very first exchange of information, adversaries face an impenetrable barrier grounded in the laws of physics.

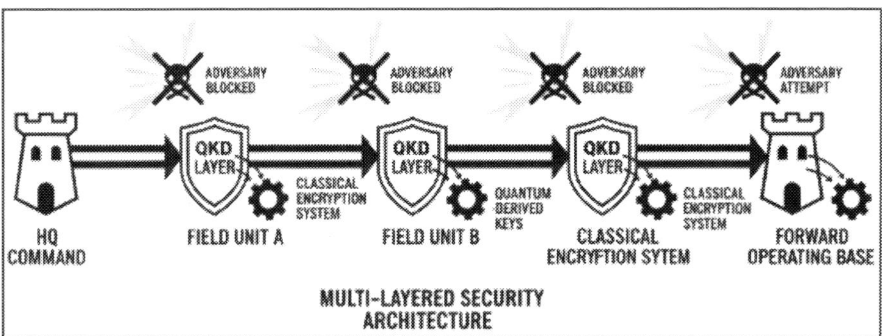

Fig. 3.17: Fortified at Every Node: QKD's Multi-Layered Shield

At its core, the architecture establishes a multi-layered shield. Each hop between network nodes is fortified by QKD, preventing undetected interception of keys. On top of this, quantum-derived keys feed into classical encryption systems, ensuring backward compatibility while eliminating the vulnerabilities of purely algorithmic security. The design further segregates sensitive pathways, dedicating isolated, high-assurance channels for nuclear command-and-control, satellite uplinks, and strategic intelligence transfers. This network is best understood as a fortified military base with concentric defences. The QKD layer acts as the outer wall with sentries who detect any breach attempt instantly. Behind it, existing classical encryption is the inner barracks and reinforced bunkers, providing additional fallback layers. Finally, the segregation of sensitive traffic is akin to restricted-access command bunkers within the base isolated, shielded, and accessible only to the highest echelons. Even if an adversary penetrates the outer perimeter, the most critical command functions remain sealed off and untouchable.

For modern militaries operating in multi-domain battlefields, such architecture ensures enduring communication superiority. Adversaries equipped with quantum computers or advanced cyber capabilities cannot silently infiltrate or decrypt traffic. Strategic command remains uncompromised during prolonged operations, even in cyber-contested or space-denied environments. The architecture's segregation principle ensures that compromise in tactical layers never escalates to strategic collapse, preserving nuclear deterrence integrity, safeguarding coalition interoperability, and maintaining uninterrupted secure C2 (command and control) across land, sea, air, cyber, and space domains.

Concept	Analogy	Operational Relevance
Integrates Quantum Key Distribution (QKD) at each node-to-node hop, ensuring undetectable interception is impossible. Combines quantum-derived keys with existing encryption to build a multi-layered shield. Segregates critical pathways for command-and-control and nuclear systems.	A fortified military base with concentric defences: QKD as the outer wall with sentries that detect breaches instantly; classical encryption as inner reinforced bunkers; segregated pathways as restricted-access command bunkers deep inside.	Ensures enduring communication superiority across contested domains. Prevents silent infiltration even by quantum-capable adversaries. Guarantees nuclear command integrity, coalition interoperability, and uninterrupted secure C2 across land, sea, air, cyber, and space.

- **Incorporates quantum key distribution (QKD) at each node-to-node hop:** At every communication hop (satellite → ground station, ship → fleet HQ, forward post → brigade HQ), a dedicated QKD channel is established. Keys are exchanged using photons whose quantum states collapse upon interception, instantly exposing eavesdropping attempts. This design ensures that even if a single link is compromised, adjacent links remain independently secure, preventing cascading breach across the network. Functions as a layered, mesh-based architecture, with QKD regenerating new keys at each relay rather than relying on end-to-end single key chains.

In a distributed theatre (carrier groups, airbases, forward operating units), local QKD relays prevent single points of failure. Even if adversaries saturate one comms node, adjacent QKD links remain uncompromised, enabling rapid re-routing and maintaining the chain of command. Allied forces can insert QKD-enabled gateways at coalition nodes, ensuring joint force secure comms without sharing master keys. Node-to-node QKD allows secure key refresh at hardened silos, submarines, or airborne command posts, ensuring Nuclear Command & Control survivability under contested conditions.

- **Multi-Layered Security Design:** Quantum-secure networks do not *replace* existing encryption, they *fortify it.* QKD produces unbreakable one-time keys, but those keys can be used to feed into classical ciphers like AES or OTP (one-time pad). This means even if one layer is probed, the second layer remains intact. The design follows a "defence-in-depth" principle. The quantum protocols secure the key distribution, while classical algorithms encrypt the actual message traffic. Together, they create redundancy adversaries must break both physics and mathematics simultaneously, an infeasible task.

  Like a fortified base with layered defences, outer minefields, inner fences, watchtowers, and a central bunker, no single breach collapses the whole. If the enemy somehow neutralizes the outer wire (classical encryption), they still face a hardened inner wall (QKD-secured keys). Each barrier buys time and preserves integrity.

- **Segregated Mission-Critical Pathways:** Segregation of mission-critical pathways ensures that communications carrying the highest national security consequences such as Nuclear Command, Control, and Communications (NC3), Special Operations Forces (SOF) tasking, strategic missile warning, and satellite downlinks. These are physically and logically separated from routine operational traffic. These channels are provisioned with dedicated QKD-secured circuits or quantum-

Concept	Analogy	Operational Relevance
Node-to-Node Quantum Key Distribution (QKD): Each hop in the network generates and exchanges keys using entangled photons or single-photon transmissions. Any interception attempt collapses the quantum state, instantly signalling compromise. Keys are refreshed continuously, making interception impossible without detection.	**Convoy re-supply under armed escort:** At every checkpoint along a supply route, fresh sealed crates of ammunition are handed over under guard. If a crate is tampered with, the seal breaks immediately and the breach is known, ensuring that only uncontaminated supplies reach the front.	**Tamper-Proof Resilience:** Guarantees that every link in the chain of command remains uncompromised. Adversaries cannot siphon keys undetected. Even in contested environments, each node remains cryptographically fresh, denying opponents a foothold to exploit.
Multi-Layered Security Design: QKD is not a replacement but a force multiplier. Quantum keys feed into classical ciphers (AES, OTP), combining physics-based unbreakability with mathematical encryption. This defence-in-depth model ensures redundancy: adversaries must break both layers simultaneously an infeasible task.	**Fortified base defences:** An outer perimeter of fences and minefields (classical crypto) protects the base, while an inner concrete bunker (QKD) holds mission-critical assets. Even if the perimeter is breached, the core remains unassailable.	**Continuity & Failsafe:** Hybrid layering ensures secure comms persist even if quantum channels are temporarily degraded. Coalition forces can interoperate by layering national ciphers on top of shared quantum keys. Strategic NC3 traffic can employ QKD-generated OTP for maximum protection, while general ops remain shielded by AES+QKD hybrid models.
Segregated Mission-Critical Pathways: Sensitive channels (NC3, SOF, nuclear command) are assigned dedicated QKD-secured circuits, separate from general operational traffic. This prioritization ensures the highest-grade quantum protection is reserved for the most consequential communications.	**Hardened command bunker vs. field HQ:** Just as generals shelter in a hardened facility during strategic operations while tactical units use mobile HQs, critical communications are routed through the most protected pathways.	**Tiered Protection:** Prevents dilution of quantum-secure bandwidth. Ensures that even in peak load or degraded network conditions, command-critical circuits retain uncompromised protection.

dedicated fibres/satellites, guaranteeing that their security posture is never diluted by the traffic or vulnerabilities of general-purpose networks. By isolating these flows, quantum bandwidth and cryptographic strength are preserved for those transmissions whose compromise would have catastrophic strategic consequences.

Even under conditions of electronic warfare, cyber-attack, or partial network degradation, leadership retains an uncompromised, hardened communication channel. This ensures adversaries cannot conduct "bandwidth starvation" or "traffic flooding" attacks to degrade critical communications by overwhelming general-purpose links. The tiered defence posture is maintained, mirroring military principles of force protection where elite guard units shield nuclear assets while general infantry secures the perimeter. Similarly, sensitive comms enjoys the most secure quantum protections, while tactical comms retains strong but shared defences.

Establishing Quantum Key Distribution (QKD) Nodes

Securing the Spine – Building a keychain of trust across operational theatres

Establishing QKD nodes involves embedding quantum-enabled key distribution infrastructure directly into the backbone of military communication networks. These nodes act as trusted gateways at critical locations such as military bases, command posts, naval fleets, and airborne command relays, ensuring secure quantum key exchanges between operational centres. QKD nodes function much like fortified supply depots strung across an extended battlefield. Just as logistics depots guarantee uninterrupted re-supply lines for fuel and ammunition, these nodes guarantee a continuous, tamper-proof supply of encryption keys to the operational force. An adversary

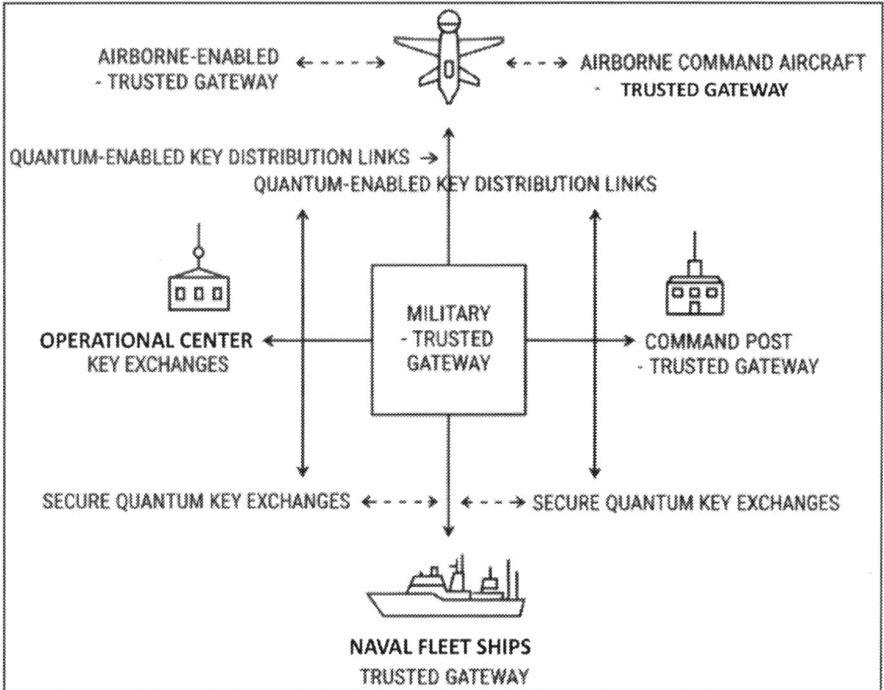

Fig. 3.18: QKD Nodes: Trusted Gateways Across All Domains

attempting to infiltrate or corrupt one of these depots would be instantly detected, similar to setting off perimeter trip wires. In this way, QKD nodes form the "supply chain of trust," extending secure communications across multiple domains land, sea, air, space, and cyber.

The deployment of QKD nodes ensures resilient and uninterrupted encryption refresh cycles even under conditions of heavy network strain, electronic warfare, or cyber-attack. For joint and combined operations, such nodes enable secure multi-theatre coordination without fear of delayed or compromised key distribution. Naval groups can manoeuvre with confidence, knowing their command links are quantum-secured across oceanic distances. Forward-deployed bases remain connected through a cryptographic lifeline immune to zero-day exploits or classical decryption. In prolonged campaigns, where command networks must endure sustained stress, QKD nodes guarantee the operational continuity of secure orders, situational awareness, and intelligence flow.

Doctrinal Element	Description
Concept	QKD nodes are deployed at critical military installations, bases, command centres, naval assets, and airborne relays to provide continuous, tamper-evident key exchange. Unlike classical cryptography, QKD derives its strength from quantum mechanics, instantly exposing any interception attempt by adversaries. These nodes form the secure backbone of the military network, ensuring reliable, intrusion-resistant encryption refresh cycles.
Analogy	QKD nodes function like fortified supply depots along an extended battlefield logistics chain. Just as depots guarantee steady re-supply of ammunition and fuel, QKD nodes guarantee a steady flow of cryptographic keys. Any tampering attempt sets off alarms, much like triggering trip wires or breaching defensive perimeters. This creates a "supply chain of trust" spanning all operational domains—land, sea, air, space, and cyber.
Operational Relevance	Ensures mission-ready encryption refresh cycles even under conditions of heavy network strain or adversarial interference. Provides uninterrupted secure command and control links for joint and coalition operations. Naval fleets, forward bases, and airborne assets maintain trusted communication lifelines immune to zero-day exploits or quantum-enabled cyber-attacks. QKD nodes thus guarantee the cryptographic resilience of operations across distributed theatres.

- **Deploying QKD Nodes at Military Bases, Command Centres, and Naval Assets:** The photon source emits single photons or entangled photon pairs as information carriers. Quantum channel interface is typically fibre-optic links for terrestrial bases, free-space or satellite-optical terminals for naval/airborne platforms while Key Management Processor *(KMP)* integrates raw photon measurements into shared symmetric keys using quantum post-processing. Tamper detection module monitors signal statistics (e.g., Quantum Bit Error Rate, QBER). Any anomaly suggests eavesdropping or jamming. QKD backbone nodes are installed in hardened bunkers or comms shelters, linked via buried fibre-optic cables with quantum repeaters (future-ready). Base nodes serve as regional key distribution centres, refreshing keys for forward-deployed outposts. Strategic nodes integrate with existing Command, Control, Communications, Computers, Intelligence, Surveillance, and Reconnaissance (C4ISR) infrastructure. Keys are distributed to subordinate units, ensuring hierarchical but intrusion-proof dissemination. Redundant quantum channels (fibre + satellite) provide dual-path resilience.

Ship-borne QKD terminals use stabilized optical telescopes to maintain free-space quantum links with satellites or coastal QKD stations. Submarines rely on buoyed relay antennas or surfaced optical uplinks for scheduled QKD refresh during surfacing windows. Naval strike groups can operate with a "floating QKD spine," where the carrier acts as the primary node and distributes refreshed keys to escorts. For continuous intrusion-resistance, each photon transmission collapses if intercepted, producing detectable QBER spikes → immediate alarm to the commander. Key Refresh Cycles throughput allows secure keys to be refreshed in *seconds to minutes*, ensuring rolling one-time-pad availability. This is critical for sustaining secrecy in prolonged operations, even under bandwidth stress. Maritime and airborne QKD nodes are mounted on gimbal-stabilized platforms to compensate for movement, ensuring photon beam alignment. QKD nodes do not replace classical comms; they secure *key exchange*. Once keys are established, they plug into existing AES, OTP, or classified crypto modules.

Concept	Analogy	Operational Relevance
QKD nodes are deployed at military bases, command centres, and naval assets to establish a hardened cryptographic backbone. Each node consists of photon sources, quantum channels (fibre, satellite, or free-space), key management processors, and tamper-detection modules. These nodes enable continuous, intrusion-resistant key exchange and sustain rolling key refresh cycles even under bandwidth stress or electronic attack.	Like forward ammunition depots in logistics, each depot issues fresh, tamper-proof ammunition (encryption keys). Any attempt to sabotage the re-supply line is instantly detected, ensuring uninterrupted firepower.	Secures every echelon of the force: bases act as regional key hubs, command centres integrate QKD into C4ISR, and naval groups form a "floating QKD spine" anchored on carriers. Enables continuous, intrusion-proof encryption refresh for nuclear C2, expeditionary forces, SOF missions, and maritime strike groups even under GPS-denied or EW-contested conditions.

- **Continuous, Intrusion-Resistant Key Exchange:** QKD generates keys constantly rather than on-demand or through pre-distributed material. Keys can be refreshed every few milliseconds or seconds depending on channel quality. This creates a living stream of cryptographic material, where every new message can ride on a new, fresh key. Even if an adversary were hypothetically able to crack a key, its useful life would be so short that by the time they decipher it, the force is already operating with a new key. This provides battlefield-grade resilience under active cyber contest. Any eavesdropper trying to measure photons causes a disturbance, which manifests as a measurable rise in QBER. QKD continuously monitors this; if the error rate exceeds acceptable limits, the session is invalidated and fresh key generation re-starts. This means intrusion attempts are not silent, they raise alarms instantly. Just like radar detects a probing aircraft by its reflection, QKD detects a probing spy by the noise they introduce. Commanders are not blind to adversary surveillance; they get warning of attempted interception in real time.

QKD is not limited to one medium. It can operate via fibre optics, line-of-sight free-space optics, or satellites. Multi-channel deployment ensures redundancy: if fibre lines are cut, satellites take over; if satellites are jammed, ground-based optics sustain local keying. This maintains key continuity across the entire battle space from theatre HQ to forward patrols. Even under EW or physical sabotage, the network never loses its ability to refresh keys. Troops remain encrypted, leadership retains control, and adversaries cannot sever the "quantum lifeline." Classical systems risk man-in-the-middle (MITM) where adversaries replay or inject fake keys. In QKD, photons cannot be cloned (no-cloning theorem), and each exchange is tied to specific quantum states and time-stamps. This makes key injection or replay physically impossible. In a contested cyber battle space, adversaries cannot impersonate HQ or seed fake orders preventing catastrophic deception ops. QKD nodes at bases, ships, or satellites continuously share secure key streams. These keys are pushed into military Key Management Systems (KMS) that automatically synchronize across units. This allows coordinated force-wide refresh cycles ensuring all units are always aligned on uncompromised encryption.

Concept	*Analogy*	*Operational Relevance*
Deploys QKD nodes at bases, command centres, naval task groups, and space assets.	Like laying down secure "signal fortresses" at every major garrison or fleet hub, creating a backbone of trust.	Ensures every operational hub (land, sea, air, space) can originate and receive quantum-secured keys, anchoring the force-wide encryption grid.
Enables continuous, intrusion-resistant key exchange across the battle space.	Like a convoy constantly changing call-signs and passwords every few seconds any enemy trying to keep pace is left behind and exposed.	Keys are refreshed in near-real-time, QBER provides early-warning of eavesdropping, and multi-channel redundancy (fibre, free-space optics, satellites) keeps keys flowing under jamming or sabotage. Adversaries can neither inject fake keys nor replay old ones, ensuring uninterrupted secure comms even in contested EW zones.
Provides mission-ready encryption refresh cycles even under network strain.	Like frontline units receiving continuous ammo resupply despite ambushes or blockades, the logistics never stop.	Forward units, SOF detachments, drone swarms, and carrier strike groups all remain synchronized under fresh keys delivered through automated Key Management Systems. Even under cyber pressure, encryption never goes stale preserving command continuity and denying adversaries any time window for decryption.

Satellite and Airborne Quantum Communication Channels

Securing the Sky – From satellites to drones, ensuring encrypted command from above

Modern battlefields extend vertically into the skies and beyond Earth's orbit. Communications no longer travel solely across terrestrial fibre or radio, but through satellite constellations, airborne command relays, and swarming

UAVs. These channels are lifelines for expeditionary units operating at great distances, often beyond direct terrestrial infrastructure. Integrating quantum-secure links into satellite and airborne platforms ensures that this aerial backbone is hardened against interception, spoofing, or long-term quantum decryption threats.

Satellite and airborne QKD channels extend the protective shield of quantum security into the third dimension of warfare the aerial and space domains. By securing expeditionary comms, ISR feeds, and over-the-horizon coordination, QKD ensures that command from above is as invulnerable as the ground networks it supports.

Concept	Analogy	Operational Relevance
Space-Based QKD & UAV Relays – Satellites, drones, and airborne relays serve as "quantum routers," delivering intrusion-proof keys across vast distances.	Like re-supply aircraft dropping ammunition that tears if intercepted, making tampering impossible to conceal.	Ensures secure long-range command connectivity for carrier groups, deployed brigades, and coalition partners even in denied terrain.
Secure Expeditionary & OTH (Over-the-Horizon) Communications – QKD overlays protect satellite-ground and UAV links from jamming, spoofing, and interception.	Similar to encrypted carrier pigeons that self-destruct if intercepted, denying adversaries access.	Guarantees continuous, trusted comms for forward-deployed expeditionary forces, naval task groups, and SOF units beyond line-of-sight.
Quantum-Resilient ISR Feeds – QKD uplinks/downlinks shield satellites, HALE drones, and ISR UAVs from data theft or signal corruption.	ISR platforms are the "eyes of the force"; QKD is the eyelid shield protecting their vision.	Preserves integrity of reconnaissance, targeting, and early-warning data in hostile airspace, sustaining the commander's decision advantage in MDO.

QKD in air and space domains ensures that expeditionary and ISR communications remain inviolate, even when adversaries saturate the electromagnetic spectrum with jamming or attempt long-term decryption of intercepted data.

Quantum Key Distribution (QKD) can be embedded in both low-Earth orbit (LEO) satellites and airborne relays such as UAVs and high-altitude balloons. These platforms act as "quantum routers," distributing fresh encryption keys between distant ground forces, fleets, or command centres. Unlike classical key exchange, any interception attempt on a quantum channel disrupts the photons being transmitted, immediately exposing adversarial eavesdropping. Imagine re-supply aircraft delivering ammunition by parachute to forward-deployed troops. If an enemy tries to intercept mid-air, the parachute itself tears, and the drop is instantly compromised. QKD ensures that secure "crypto re-supply" through satellites and UAVs cannot be secretly intercepted.

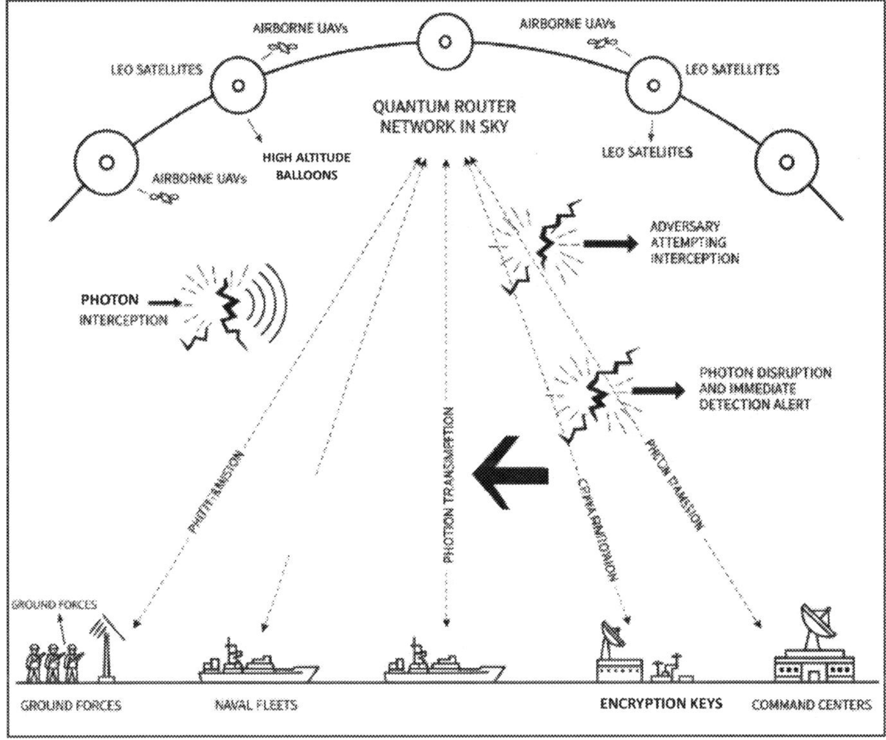

Fig. 3.19: Quantum Routers in the Sky: Securing Air and Space Communications

This architecture ensures commanders can securely direct operations across oceans, deserts, or denied regions where terrestrial fibre lines are absent. Expeditionary strike groups, forward-deployed special operations teams, and coalition partners benefit from uninterrupted key exchange regardless of geography.

Traditional over-the-horizon (OTH) links rely heavily on radio and satellite signals vulnerable to jamming or interception. By overlaying QKD onto satellite-ground and UAV-relay links, expeditionary forces gain continuous, intrusion-proof encryption, even when operating beyond visual range or in hostile EW environments. This is akin to sending orders by encrypted carrier pigeons that self-destruct if intercepted; adversaries cannot read or tamper with them without being detected.

Expeditionary forces, whether a naval carrier group projecting power overseas, or an airborne brigade deploying deep in land, retain secure communications even when adversaries attempt to deny access through EW or cyber measures. This resilience underwrites force mobility and operational independence.

Intelligence, Surveillance, and Reconnaissance (ISR) platforms such as satellites, HALE (High-Altitude Long Endurance) drones, and tactical UAVs carry some of the most sensitive mission data. If their communication links are compromised, adversaries could exfiltrate imagery, targeting coordinates, or sensor data in real time. Quantum-secured uplinks and downlinks ensure ISR data streams remain inviolate. Think of ISR systems as "eyes of the force." QKD serves as the eyelid shield, preventing adversaries from blinding or corrupting what those eyes see and transmit.

With QKD integration, ISR assets maintain data integrity in hostile air space, ensuring targeting data, early-warning feeds, and reconnaissance imagery are transmitted with uncompromised authenticity. This strengthens the commander's decision-making cycle in MDO (Multi-Domain Operations), where accurate ISR inputs are the difference between victory and mission failure.

Satellite and airborne QKD channels extend the protective shield of quantum security into the third dimension of warfare the aerial and space domains. By securing expeditionary comms, ISR feeds, and over-the-horizon coordination, QKD ensures that command from above is as invulnerable as the ground networks it supports.

- **Space-based QKD and Drone/UAV Relays:** Space-based QKD involves the use of dedicated quantum communication satellites that transmit entangled photons or single-photon streams to ground stations, enabling unbreakable key exchanges across continental and inter-theatre distances. This model bypasses the physical constraints of terrestrial fibre optic QKD, which suffers from photon loss over long distances. Complementing this, high-altitude airborne platforms such as HALE (High-Altitude Long-Endurance) UAVs, balloons, or ISR drones act as mid-tier quantum relays, extending secure quantum channels to forward-deployed units in austere environments where terrestrial or satellite connectivity is limited.

❏ **Strategic Reach:** Space-based QKD guarantees that command centres in the homeland can securely link with carrier strike groups, overseas bases, or deployed corps headquarters without the risk of interception or retroactive decryption. This supports strategic deterrence and crisis response.

❏ **Theatre Agility:** Drone/UAV relays extend secure channels to front-line SOF teams, expeditionary brigades, or naval task forces operating beyond line-of-sight or in denied electromagnetic environments.

❏ **Redundancy & Survivability:** Combining satellite QKD with UAV relays forms a layered quantum backbone, ensuring that if satellites are degraded by ASAT threats, airborne assets can bridge the gap locally.

❏ **ISR Protection:** Quantum uplinks secure the massive data feeds from space and aerial ISR platforms, satellites, HALE UAVs, and tactical drones ensuring reconnaissance imagery, targeting data, and early-warning alerts remain uncompromised, even under adversarial cyber-electronic warfare attempts.

Space-based QKD delivers the strategic layer of unbreakable comms across continents, while UAV relays supply the tactical layer, ensuring no unit from corps HQ to a forward SOF team is left unsecured in the battle space.

Concept	*Analogy*	*Operational Relevance*
Space-based QKD and airborne quantum relays create a multi-tiered secure backbone that transcends terrestrial limits. Satellites transmit entangled photons or single-photon QKD streams to ground stations, enabling long-range, unbreakable key exchange. Airborne platforms such as HALE UAVs, ISR drones, and balloons act as mid-tier quantum relays, distributing secure keys and channels to manoeuvre units operating in austere or denied environments. Together, they form a resilient layered architecture where satellites provide strategic reach and UAVs deliver tactical agility.	Strategic Arsenal Ship & Airborne Convoy. Satellites function like arsenal ships in orbit, continuously supplying secure "munitions" (keys) to friendly forces across continents. UAVs act as airborne supply trucks, distributing those munitions directly to forward-deployed brigades, naval assets, or SOF detachments. Even if one node is neutralized, the distributed nature of the network prevents loss of secure re-supply.	Strategic Reach: Ensures secure comms between homeland command centres and overseas task forces, supporting deterrence and rapid crisis response. Theatre Agility: UAV relays extend secure comms to SOF teams, expeditionary brigades, and carrier groups beyond line-of-sight or under EW pressure Redundancy & Survivability: If satellites are degraded by ASATs, UAV-based quantum relays provide a fallback for secure local links. ISR Protection: Encrypts uplinks from space and aerial ISR platforms, ensuring reconnaissance, targeting, and early-warning feeds remain uncompromised.

Space-based QKD delivers the strategic layer of secure communications across theatres, while UAV relays ensure that no forward-deployed unit is left unsecured, even in denied battle spaces.

- **Makes space and aerial ISR systems quantum-resilient:** Space and airborne ISR (Intelligence, Surveillance, Reconnaissance) systems transmit high-value data (radar sweeps, SIGINT captures, drone video feeds, targeting coordinates) over RF or laser links. These links, when encrypted with classical schemes, will eventually be decryptable by large-scale quantum computers (via Shor's Algorithm breaking RSA/ECC). Space and airborne ISR (Intelligence, Surveillance, Reconnaissance) systems transmit high-value data (radar sweeps, SIGINT captures, drone video feeds, targeting coordinates) over RF or laser links. This ensures that ISR data maintains end-to-end integrity and confidentiality from satellite or UAV sensor payloads, through relay platforms, to ground C2 nodes even decades after collection. Think of ISR platforms as scouts deep behind enemy lines. Without quantum protection, their reports could be intercepted, forged, or replayed by adversaries. QKD functions like issuing these scouts tamper-proof message scrolls, which self-destruct if intercepted, ensuring only the true commander receives the intelligence unaltered.

  ❏ **Preserves Strategic ISR Value:** Reconnaissance satellite imagery, synthetic aperture radar sweeps, and SIGINT intercepts can be secured against future quantum decryption even if adversaries stockpile encrypted ISR data today.

  ❏ **Protects Real-Time Feeds:** Live UAV video streams and SIGINT bursts delivered over contested airspace remain uncompromised even under EW or cyber pressure.

  ❏ **Hardens C4ISR Backbone:** Ensures that targeting, early warning, and BDA (Battle Damage Assessment) data maintain authenticity, preventing adversary misinformation or spoofed intelligence feeds.

  ❏ **Maintains Deterrence:** Adversaries cannot rely on delayed decryption to gain future insight into force postures or historical operational patterns, preserving long-term strategic surprise.

Quantum resilience ensures that ISR data, the eyes and ears of the force, remains uncompromised across decades, denying adversaries both immediate and delayed exploitation of reconnaissance feeds.

Concept	Analogy	Operational Relevance
Space-Based QKD & UAV Relays integrate quantum key distribution into satellite constellations and airborne assets (UAVs/drones, HAPS, manned ISR platforms), enabling quantum-secure links across continental and maritime theatres. QKD ensures any interception attempt collapses the photon state, instantly exposing hostile interference. Continuous key refresh cycles allow ISR and command data to be encrypted in real time, immune to both present and future quantum computer threats.	Like establishing airborne courier outposts and sky fortresses that deliver sealed, tamper-proof orders across vast distances. If an adversary attempts to intercept the courier mid-flight, the scroll disintegrates, alerting command instantly.	Expeditionary Reach: Enables secure comms for forces operating beyond line-of-sight, including carrier strike groups, forward-deployed SOF, and remote bases ISR Resilience: Quantum-protects space and aerial ISR platforms (recon satellites, SIGINT drones, AWACS) so their feeds cannot be intercepted, decrypted later, or spoofed by adversaries Hardens C4ISR Backbone: Preserves integrity of early warning, targeting, and BDA feeds under contested EW conditions Deterrence Value: Prevents adversaries from stockpiling today's encrypted ISR traffic for future decryption, denying them long-term operational insights into force posture and doctrine.

By embedding QKD into space and airborne relays, ISR becomes quantum-resilient ensuring the force's eyes and ears remain uncompromised even in decades of future threat environments.

Network Redundancy with Quantum Fail-Safes

Built to Survive – Quantum networks that adapt, re-route, and resist collapse

In modern warfare, the survivability of a military network is as important as its speed or reach. Quantum-secure networks are designed not only to encrypt data but to withstand disruption, degradation, or targeted attacks on their infrastructure. Unlike classical networks that can collapse if critical nodes or communication paths are destroyed, quantum networks integrate fail-safes at the cryptographic layer. Even under partial disruption, the system maintains encryption through alternate QKD routes, parallel nodes, and rapid re-keying cycles, ensuring that sensitive command traffic is never exposed, even momentarily.

A useful military analogy is the concept of defence-in-depth fortifications. Just as medieval strongholds built concentric walls so that breaching the first line did not guarantee enemy victory, quantum-secure networks create multiple fallback layers. If one QKD node or channel is disabled, the network immediately shifts to a parallel secure pathway, re-establishing trust keys without pausing or exposing traffic in the clear. In practice, this means adversaries cannot collapse a command network with a single strike or electronic attack; they face a system that adapts dynamically and preserves its protective shield.

Operationally, this resilience is decisive for C4ISR operations in contested zones. Whether in an air-sea battle with heavy jamming, a land campaign under cyber bombardment, or space operations facing satellite losses, quantum fail-safes ensure that encryption integrity survives even when the physical network is under duress. This not only protects real-time operational orders but also preserves confidence in intelligence flows, targeting data, and nuclear command-and-control systems. For expeditionary forces and forward-deployed commands, this translates into assured continuity of secure communications even in the face of aggressive EW, cyber, or kinetic assaults, strengthening both deterrence and operational superiority.

Concept	*Analogy*	*Operational Relevance*
Quantum-secure networks maintain encryption even if parts of the network are disrupted by cyber, electronic, or kinetic attacks. QKD keys can be re-established via alternate nodes and re-routed channels without exposing traffic.	Like fortresses with concentric defensive walls: Even if one wall is breached, inner defences hold strong. Adversaries cannot collapse the system with a single strike.	Ensures hardened C4ISR resilience in contested environments (EW zones, cyber-attacks, space denial). Guarantees continuity of encrypted orders, ISR feeds, and nuclear C2, even under partial network collapse.

- **Quantum-secure networks maintain encryption even during partial disruption:** Quantum-secure networks leverage *redundant QKD nodes and entangled photon channels* to ensure encryption keys can be continuously refreshed, even if some links are jammed, destroyed, or severed. Unlike classical networks, where a single-point failure in the key, exchange chain can collapse security. QKD architectures allow keys to be re-routed through alternate trusted nodes or via satellite relays. Encryption continuity is preserved by *proactive re-keying and multi-path entanglement distribution,* ensuring that at no point does a compromised or degraded link expose plain text or leave communications unencrypted.

For C4ISR systems operating under cyber and EW contestation, this ensures commanders retain access to encrypted situational reports, ISR feeds, and targeting orders without interruption. In nuclear command-and-control or *strategic deterrence architectures*, partial degradation from enemy attacks does not equate to loss of secure messaging. Adversaries cannot paralyze operations by targeting a few communication links, since the *quantum-secure backbone dynamically self-heals*, maintaining the confidentiality and integrity of mission-critical data flows.

Concept	*Analogy*	*Operational Relevance*
Quantum-secure networks maintain encryption even during partial disruption.	Like a convoy with multiple supply routes, if one is ambushed, supplies are immediately re-routed through alternate corridors.	Ensures encryption continuity through redundant QKD nodes, entangled photon multi-paths, and satellite relays. If a link is jammed, severed, or degraded, alternate paths sustain secure key refresh cycles, preventing any lapse into plain text exposure. Critical for C4ISR resilience under cyber/EW attack.
Supports hardened C4ISR operations in contested zones.	Comparable to a command bunker with blast-proof chambers: even if one room is compromised, the overall command function continues.	Enables field commanders to retain access to encrypted ISR feeds, targeting data, and orders even under heavy jamming or network attrition. Protects nuclear command-and-control and strategic deterrence architectures by ensuring adversaries cannot paralyze secure communications through partial network attacks.

- **Supports Hardened C4ISR Operations in Contested Zones:** C4ISR systems are the *nervous system of the modern battle space.* In contested zones where electronic warfare (EW), cyber intrusions, and kinetic attacks attempt to blind or paralyze command structures, continuity of secure communication becomes existential. Quantum-secure networks reinforce C4ISR by embedding *QKD-based key distribution at every critical link*, ensuring that even if a segment of the network is degraded, *encryption remains unbroken.* Quantum-secure networks reinforce C4ISR by embedding *QKD-based key distribution at every critical link*, ensuring that even if a segment of the network is degraded, *encryption remains unbroken.*

Comparable to a *hardened command bunker with multiple fallback chambers*: even if one entry is breached or collapsed, the chain of command inside remains intact. In the same way, quantum-hardened C4ISR ensures that even if *links are jammed, spoofed, or taken offline*, alternate paths (satellite, UAV relays, fibre backbones) maintain the secure key refresh cycle. The command's *"voice" is never silenced.*

In *near-peer conflict* scenarios, adversaries will attempt to saturate the electromagnetic spectrum with jamming, cyber-attacks, and GPS denial. Quantum-secure networks allow allied forces to maintain a *continuous encrypted C4ISR backbone*, sustaining operational tempo while the adversary experiences degradation. For *strategic deterrence*, hardened C4ISR ensures that nuclear command-and-control networks remain secure, eliminating any adversarial belief that communications can be disrupted or falsified. In *special operations* or expeditionary warfare, QKD-backed comms give forward teams *uninterrupted access to encrypted ISR and targeting feeds*, even in heavily jammed or GPS-denied theatres ensuring mission success in high-risk, high-value operations.

Concept	Analogy	Operational Relevance
Quantum-secure networks maintain encryption even during partial disruption.	Like a fortified bridge with multiple spans, if one span collapses, the others still carry the load.	Even under jamming, fibre cuts, or cyber-attack, encryption keys self-refresh and remain intact, ensuring adversaries cannot exploit weakened links. Commanders retain secure channels across multiple fallback pathways.
Supports hardened C4ISR operations in contested zones.	Comparable to a hardened command bunker with fallback chambers even if an entry point is hit, leadership stays protected and functional.	Ensures continuity of C4ISR in EW-heavy or GPS-denied battle spaces. Nuclear command, ISR relays, and special operations communications remain uncompromised, preserving operational tempo and deterrence credibility even under enemy saturation attacks.

Quantum Secure Field Units for Tactical Teams

Encryption at the Edge – Securing every squad, not just the HQ

Modern warfare demands not only that senior headquarters remain secure but also that tactical elements operating at the forward edge of battle are equally protected from interception, spoofing, and compromise. Quantum Secure Field Units (QSFUs) address this requirement by extending quantum-grade protection down to the level of special forces detachments, forward observers, and dispersed manoeuvre units, ensuring that encryption resilience is not confined to command echelons but carried into every tactical pocket of the battle space. Portable quantum encryption devices act as compact, ruggedized nodes capable of generating and receiving quantum-derived keys even in austere conditions. Much like a soldier carrying a personal encrypted radio rather than relying solely on divisional communications hubs, these units enable frontline teams to authenticate and secure their communications independently. By leveraging entangled photon-based or free-space optical key exchanges, the devices nullify the adversary's ability to eavesdrop or inject false signals even when tactical teams are isolated or operating behind enemy lines.

Crucially, these field units are designed for seamless interoperability with centralized QKD infrastructure, meaning that quantum keys generated at strategic nodes (such as command centres, satellites, or mobile HQs) can be securely extended to tactical endpoints. This ensures a tiered trust architecture, where keys are distributed from the national or theatre command level down to squads in the field, with each link safeguarded by the laws of quantum mechanics rather than reliance on vulnerable classical encryption.

For the operational commander, this capability translates into distributed security parity: tactical squads are no longer weak links in the chain of command but fortified extensions of the quantum-secure network. Whether calling for fire, transmitting ISR data, or relaying sensitive position updates, small teams can operate with the assurance that their communications are immune to quantum-enabled adversary decryption, even if compromised devices fall into hostile hands. In practice, QSFUs offer a decisive edge in special operations, irregular warfare, and dispersed manoeuvre campaigns, where the survivability of information flows often determines mission success.

Real-Time Monitoring and Quantum Threat Detection

Silent Guardians – Monitoring quantum links with quantum eyes

One of the defining strengths of quantum-secure networks is not only their ability to provide encryption based on physical laws, but also their ability to *self-monitor* for hostile interference. Quantum links are inherently sensitive; any unauthorized attempt to observe or tamper with the quantum states immediately introduces detectable anomalies. By pairing these physical properties with *dedicated quantum sensors and monitoring systems*, military networks achieve a new echelon of *real-time intrusion detection*. Unlike classical systems, where breaches may remain undetected for weeks or months, quantum-secure networks generate *instantaneous alerts* when adversaries attempt to eavesdrop, spoof, or degrade communications.

For the war fighter, this translates directly into battlefield survivability and command integrity. Quantum sensors embedded in communication channels instantly detect unauthorized access, fibre tapping, or photon interception, enabling defensive countermeasures before the adversary gains any intelligence value. The network provides actionable warnings if adversaries attempt signal manipulation, false traffic injection, or denial-of-service through degradation critical in EW-heavy operational theatres. Commanders gain a continuous operational picture of the communication battle space, integrated at the network layer, ensuring that quantum threats are identified and neutralized as part of broader C4ISR monitoring. Ultimately, real-time monitoring and quantum threat detection transform communications from a passive channel into an active security system, where the network itself becomes a living sensor grid against hostile interference. This closes one of the most dangerous gaps in classical cyber defence: the silent breach, which adversaries can exploit for months. In quantum-secure networks, there are no silent breaches, only instant detection, escalation, and response.

Concept	*Analogy*	*Operational Relevance*
Quantum networks integrate real-time monitoring through quantum sensors, instantly detecting unauthorized access, signal degradation, or spoofing attempts. Any intrusion disrupts the quantum state and triggers alerts.	Like sentries with night-vision goggles posted along a fortress perimeter able to spot even the faintest movements in the dark.	Provides continuous situational awareness of the communications battle space, ensuring commanders are warned of hostile activity in real time. Enables rapid countermeasures, hardening network integrity during high-intensity operations.

- **Real-time intrusion detection using quantum sensors and monitoring systems:** Quantum-secure networks employ quantum sensors that monitor entangled photon streams and single-photon states moving across fibre or free-space links. Because quantum states cannot be copied or altered without disturbance (per the no-cloning theorem and measurement disturbance principle), even the smallest attempt at interception introduces detectable errors in quantum bit error rates (QBER). Monitoring systems continuously measure QBER, phase shifts, and polarization changes, enabling the network to differentiate between natural noise and adversarial intrusion. Detection thresholds are set so that anomalies instantly raise an intrusion alert and trigger automatic key regeneration or rerouting protocols.

For commanders, this capability means early warning at the communications layer, even before hostile electronic warfare teams can exploit intercepted data. Real-time alerts ensure that command posts, mobile task forces, and expeditionary units are never operating on compromised keys. In practical terms, this allows automatic transition to fresh encryption keys and maintains command integrity during contested operations. It transforms communications from a reactive defence posture into a proactive security perimeter, reducing adversaries' window of exploitation to near zero.

Concept	*Analogy*	*Operational Relevance*
Quantum sensors continuously measure photon states (polarization, phase, and quantum bit error rates) across fibre and free-space channels. Because quantum states cannot be cloned or probed without disturbance, even minimal adversarial interception introduces measurable anomalies. Integrated monitoring systems distinguish between natural noise and hostile intrusion, triggering instant alerts, key regeneration, or re-routing.	Comparable to a trip wire and motion sensor network around a forward operating base: the slightest disturbance, whether vibration, shadow, or step, sets off an alarm. Here, the "trip wire" is the photon stream, which cannot be silently bypassed or spoofed.	Provides early warning at the communications layer, ensuring C2 nodes are never operating on compromised keys. Real-time alerts prevent enemy exploitation of data streams, enable automatic refresh of encryption keys, and preserve trust in mission communications even in contested EW zones. Transforms the communications backbone from a reactive defence to a proactive perimeter shield, closing the adversary's window of exploitation.

- **Alerts on unauthorized access, signal degradation, or enemy spoofing attempts:** Quantum monitoring systems leverage *Quantum Bit Error Rate (QBER) analysis* and *state-fidelity checks* to detect abnormal patterns. In normal operations, photon transmission exhibits predictable error thresholds caused by channel imperfections or environmental noise. Any deviation beyond those thresholds such as sudden spikes in QBER, abnormal polarization flips, or phase errors signals possible *unauthorized tapping, channel disturbance, or spoofing injection attempts.* Similarly, quantum signal degradation can be differentiated from natural attenuation by monitoring temporal consistency, ensuring anomalies linked to jamming or deliberate weakening attempts are flagged instantly. This functions like *an early-*

warning radar and IFF (Identification Friend or Foe) combined: radar detects the intrusion of hostile aircraft, while IFF verifies identity. If a spoofing attempt mimics a "friendly" communication packet, the quantum monitoring system still detects inconsistencies in the photon states just as radar sees flight vectors that don't match expected friendlies. It is a *watchtower and sentry line* that never sleeps, identifying both *direct enemy probes* and *subtle impersonation efforts*.

This capability means that *C2 and tactical nodes never unknowingly operate on compromised channels.* Unlike classical systems where an adversary may remain undetected for months, here intrusions are detected *in real time*. Network operators receive *automatic alerts* for immediate encryption refresh, re-routing to alternate QKD nodes, or raising higher-echelon cyber defence protocols. In contested battle spaces especially under *electronic warfare (EW) or cyber deception campaigns* this transforms the comms infrastructure from *fragile reliance on trust* to *constant validation and resilience*. It ensures that nuclear command links, expeditionary task force communications, and SOF missions are *immune to silent compromise or impersonation tactics.*

Concept	Analogy	Operational Relevance
Real-time intrusion detection using quantum sensors and monitoring systems.	Like an early-warning radar fused with a 24/7 sentry watch, continuously scanning for anomalies.	Quantum monitoring tracks Quantum Bit Error Rates (QBER) and photon state fidelity to instantly detect intrusion or tampering, ensuring defenders know of enemy actions as they occur.
Alerts on unauthorized access, signal degradation, or spoofing attempts.	Similar to a radar/IFF system exposing hostile aircraft pretending to be friendly, or a watchtower raising the alarm at suspicious movements.	Anomalous error spikes, polarization flips, or phase mismatches trigger instant alerts. Commanders are warned of tapping, jamming, or impersonation, enabling encryption refresh, re-routing, or escalation before compromise occurs.
Continuous quantum threat situational awareness at the network layer.	Functions like a battlefield common operating picture for communications health, always updating and highlighting risks.	Provides commanders with persistent visibility into comms integrity. No adversary can silently ride the network; every probe, jam, or deception attempt is exposed, preserving trust in C2 links under cyber and EW pressure.

- **Continuous Quantum Threat Situational Awareness at the Network Layer:** Quantum-secure networks operate with native telemetry. Every quantum bit (qubit) exchanged between nodes carries inherent diagnostic information about the link's integrity. By monitoring Quantum Bit Error Rate (QBER), photon arrival timings, and polarization/phase consistency across the entire network, commanders gain a live threat picture of comms health. Unlike classical systems, where eavesdropping or spoofing can go undetected until after compromise, quantum monitoring exposes intrusions the instant they occur, ensuring continuous situational awareness (SA) at the network layer. This is equivalent to having a persistent electronic perimeter defence around every communication channel like a forward operating base with motion sensors, IR trip wires, and drone over-watch. Just as no enemy patrol can approach without setting off layered alarms, no adversary, cyber or EW probe, can touch the quantum network without triggering error-based alerts that ripple up the command net.

Concept	*Analogy*	*Operational Relevance*
Quantum Intrusion Detection – Quantum sensors continuously monitor QKD links for anomalies, analysing photon statistics, timing, and phase/polarization states. Any deviation from expected norms indicates potential tampering, interception, or environmental disruption.	Like perimeter trip wires and IR sensors at a forward operating base, any disturbance instantly reveals an intruder's presence.	Provides real-time intrusion detection, ensuring no enemy probe can compromise a link without immediate detection. Supports immediate defensive action (key refresh, re-routing).
Automated Alerting on Unauthorized Activity – Quantum monitoring systems detect and flag unauthorized access attempts, signal degradation, or spoofing in real time, raising alerts to network command.	Similar to an AWACS or radar early-warning screen flagging hostile aircraft before they penetrate airspace.	Delivers actionable cyber-EW alerts directly to commanders, enabling proactive countermeasures and ensuring link continuity under contested conditions.
Continuous Situational Awareness (SA) – QKD networks provide persistent telemetry on link integrity (via QBER and channel diagnostics). This enables commanders to see a "live threat picture" across all network nodes.	Comparable to a battlefield COP (Common Operating Picture), where patrol reports and sensors provide commanders constant awareness of enemy movement.	Maintains persistent C2 trust across theatres. Ensures adversary quantum or classical cyber-attacks cannot silently corrupt communications, reinforcing C4ISR resilience in multi-domain contested battle spaces.

QKD networks provide a 24/7 "comms sentry" that constantly assesses the integrity of encryption keys and channels. When adversary interference is detected, commanders can refresh keys, re-route traffic, or escalate to backup pathways without waiting for a breach. Even in multi-domain operations, a single compromised link cannot silently corrupt the network. Every division satellite, airborne, naval, and field receives real-time status updates. This situational awareness ensures commanders never operate "blind" in the cyber domain, maintaining C2 dominance even under saturation cyber/EW attack.

Resilience Against Electronic Warfare and EMP Attacks

Shielded by Physics – When the battlefield turns dark, quantum keeps talking
Electronic Warfare (EW) and Electromagnetic Pulse (EMP) attacks represent some of the gravest threats to military command-and-control systems, capable of blinding sensors, silencing radios, and crippling entire communication grids. Quantum-secure networks, however, offer an unprecedented layer of resilience by leveraging the laws of physics rather than the vulnerabilities of classical electronics. Unlike conventional systems that rely on electromagnetic spectrum availability and electronic circuits prone to burnout, quantum communication often operates through optical fibres or satellite-based photon transmission mediums that are inherently less vulnerable to conventional jamming or high-energy EMP bursts. This physics-level immunity ensures that when adversaries attempt to "black out" the battle space, quantum channels continue to function as secure lifelines.

In a conventional battlefield, EW and EMP assaults are akin to artillery barrages designed to suppress communications wiping out radio chatter, frying circuits, and leaving units isolated. Quantum networks, by contrast, resemble hardened underground command bunkers. Just as those bunkers are engineered to withstand blast waves and continue coordinating operations under fire, quantum-secure communication pathways are engineered to survive hostile electromagnetic conditions and maintain a functioning command link. Where traditional comms are silenced, quantum acts as the *"last voice standing."*

The operational payoff is significant. Quantum-secure optical and satellite QKD links maintain integrity in conditions where traditional radios, microwaves, or digital comms collapse. This resilience enables hardened C4ISR continuity in the face of high-intensity electromagnetic disruption, ensuring that command directives, sensor feeds, and targeting data remain uncompromised. For deployed forces, this means forward brigades or naval task groups can continue to receive authenticated orders and relay ISR data even after an adversary executes EMP strikes or saturates the spectrum with jamming. In a future conflict where adversaries will almost certainly pair cyber offensives with EW saturation, quantum-secure networks represent a doctrinally indispensable safeguard ensuring that even in the darkest electromagnetic environments, commanders retain their most critical asset: *secure, trusted, and unbroken communications.*

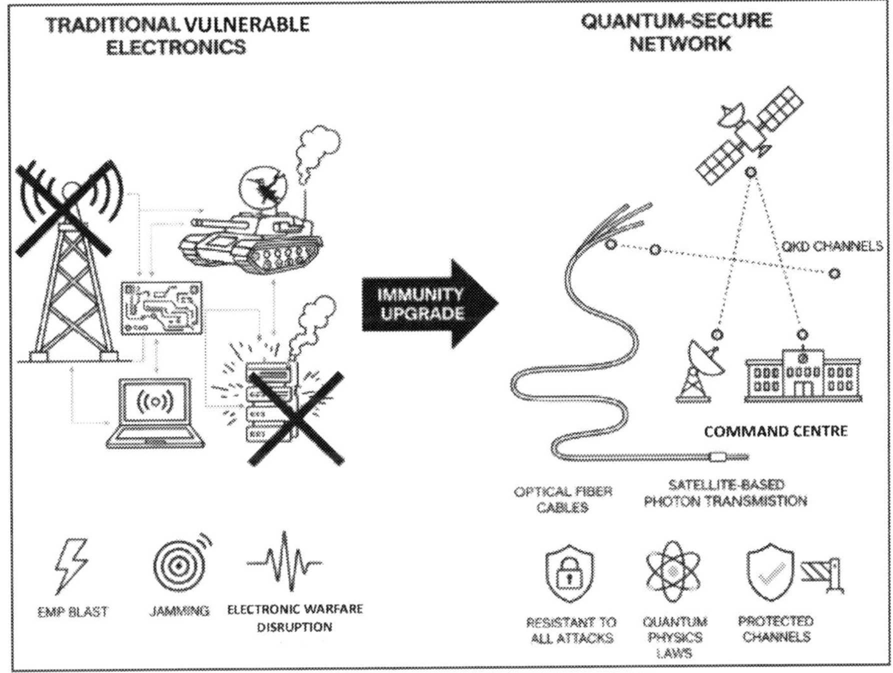

Fig. 3.20: Beyond Jamming: Photon-Based Networks That Can't Be Disrupted

Concept	Analogy	Operational Relevance
Quantum-secure networks leverage optical fibre and satellite-based QKD channels that are inherently more resistant to jamming, EMP, and high-intensity EW disruption. Instead of relying on vulnerable electronics and spectrum-dependent systems, these networks exploit photon-based channels secured by the laws of quantum physics.	Like a hardened underground bunker that continues to direct operations even under artillery bombardment, quantum communication channels endure where traditional radios and digital comms are disabled. They act as the "last voice standing" when the electromagnetic spectrum is saturated or neutralized.	Ensures continuity of C4ISR operations in the most hostile electromagnetic environments, preserving command directives, ISR flows, and targeting data even after EMP strikes or saturation jamming. Provides commanders with unbroken, trusted communication pathways, safeguarding operational integrity when adversaries attempt to black out the battle space.

Concept	Analogy	Operational Relevance
Resistance to Jamming – Quantum networks leverage photons in optical fibres or satellite channels that are outside conventional RF bands. Unlike radio or microwave comms, they are immune to noise flooding and spectrum saturation tactics.	Like issuing sealed written orders via a courier underground while enemy loudspeakers try to drown out radio orders on the surface. The jamming noise never reaches the protected channel.	Maintains C2 integrity in contested electromagnetic environments, ensuring uninterrupted orders to units when adversaries deploy wideband jamming.
Resilience to EMP – Quantum channels built on shielded fibre optics and hardened satellite relays are far less vulnerable to electromagnetic pulses, which primarily disrupt electronics and RF-dependent systems. Hardened infrastructure ensures quantum links survive EMP shocks.	Comparable to bunkered command posts that remain operational after artillery barrages, while exposed outposts collapse. Quantum links remain operational "under the bunker" of optical resilience.	Preserves nuclear C2, missile defence, and continuity of government comms after adversary EMP strikes intended to blind or paralyze decision-making.
Immunity to High-Intensity EW – Quantum communication does not depend on RF spectrum carriers, making it immune to barrage jamming, frequency saturation, and high-energy EW sweeps. Photons in optical or space channels bypass the spectrum battlefield entirely.	Like whispering orders through hidden tunnels while the enemy floods the battlefield with flares, noise, and decoys the underground path stays unaffected.	Ensures secure coordination of joint task forces, ISR feeds, and nuclear C2 even under adversary "maximum EW" conditions, providing assurance that secure comms will outlast spectrum-denial campaigns.

- **Quantum Networks' Resistance to Jamming, EMP, and High-Intensity Disruption:** Quantum networks differ fundamentally from traditional RF or digital communication systems, which are highly vulnerable to electronic warfare tactics such as jamming, spectrum flooding, and electromagnetic pulse (EMP) weapons. By operating primarily through photon transmission in optical fibres or via satellite-based quantum key distribution (QKD), these systems are not constrained by radio frequency spectrum congestion or dependent on electronics that can be fried by EMP surges.

  - ❑ **Resistance to Jamming:** Classical communications rely on radio frequencies (RF), which are inherently vulnerable to enemy jamming. A jammer only needs to flood the same frequency band with noise or power to render the signal unusable. Quantum-secure networks, however, rely on *photon transmission* via optical fibre or satellite free-space optics, where jamming is ineffective.

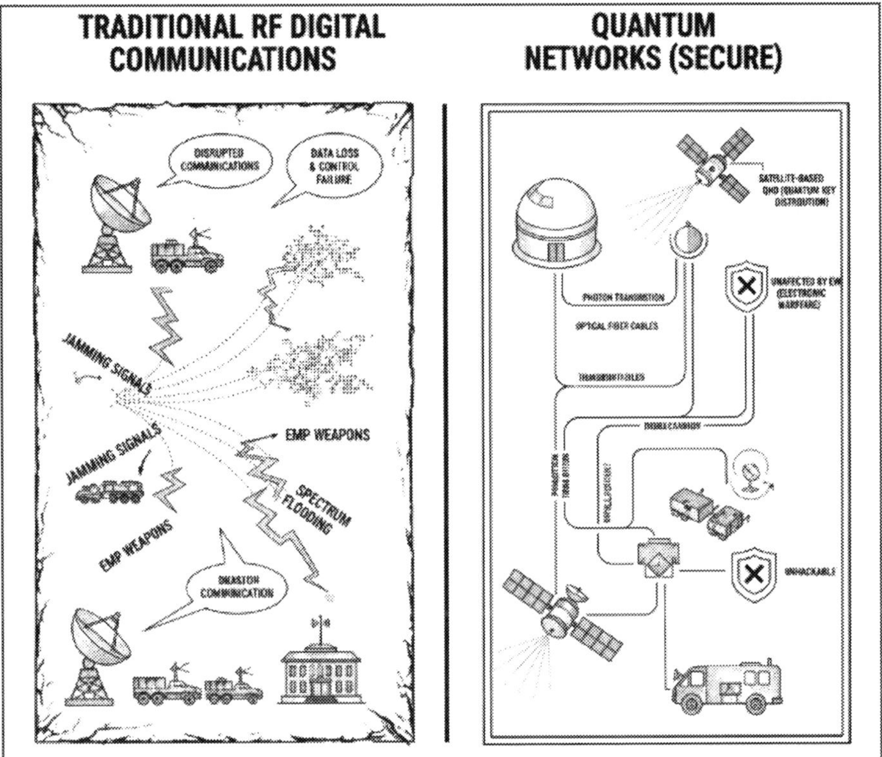

Fig. 3.21: Immune to Electronic Warfare: The Quantum Shield

Any interference attempt manifests as detectable quantum errors, allowing the system to discard compromised keys and reinitiate secure exchange rather than lose the channel entirely.

Photon-based QKD channels do not operate on conventional RF bands, so traditional EW jammers targeting spectrum bands have no impact. *Intrusion attempts* (like injecting fake photons or adding noise) cause measurable error rates in quantum states (quantum bit error rate, QBER). Once thresholds are exceeded, the network discards compromised keys, denying adversaries both denial-of-service and covert infiltration. In fibre links, adversaries must physically tamper with the medium to disrupt, which is difficult in hardened, buried, or shielded installations. In satellite-based quantum comms, jamming lasers must precisely intercept narrow optical beams in space a nearly impossible precision requirement compared to broad RF jamming.

This approach provides *assured comms in GPS-denied and EW-heavy zones*, where conventional radios or satellite uplinks would be saturated by hostile jamming. It also ensures *special operations teams, naval task groups, and forward observers* retain encrypted links even under heavy electronic assault while denying adversaries the ability to *silently degrade communications*. Any interference immediately flags the channel, prompting auto-refresh or re-routing without data leakage. It strengthens *C4ISR continuity* in high-intensity conflicts, ensuring encrypted situational awareness and targeting orders reach war fighters even when adversaries deploy massive EW assets.

Concept	Analogy	Operational Relevance
Quantum-secure networks are inherently more resistant to jamming, EMP, and high-intensity electronic disruption. Unlike RF-based comms, quantum channels transmit single photons via fibre or narrow satellite optical links. Traditional EW jammers, which flood frequency bands with noise, cannot target or overwhelm these optical quantum channels. Any intrusion or tampering only raises the quantum bit error rate (QBER), immediately detectable and automatically mitigated by discarding compromised keys. Fibre links are hardened against tampering, while satellite QKD beams are so narrow that intercept or jamming requires near-impossible precision.	Like stealth aircraft against radar jammers: conventional fighters are blinded by EW jamming, but stealth platforms simply give the jammer no radar surface to lock on to. Quantum channels offer the same advantage; they do not present a viable "frequency target" for hostile jammers, rendering the weapon ineffective.	Provides assured communications in GPS-denied and EW-heavy zones, where radios and satellite uplinks are easily saturated by hostile noise. Ensures special forces, naval groups, and forward observers maintain encrypted comms under heavy electronic assault. Denies adversaries the ability to covertly degrade or spoof communications, since any interference is flagged immediately. Strengthens C4ISR continuity in high-intensity conflicts, guaranteeing secure situational awareness and targeting even when adversaries employ massed EW and EMP assets.

- ❏ **Resilience to EMP:** Unlike traditional RF comms and copper-based lines that act as antennas for EMP surges, *optical fibre channels used in terrestrial QKD are immune to induced electromagnetic currents*. Light pulses (photons) in fibre are unaffected by EMP surges that typically fry electronic circuits. Quantum key distribution relies on *isolated photon detectors, single-photon sources, and dedicated optics*. These components can be shielded within *EMP-hardened enclosures* similar to those used in strategic command bunkers, reducing vulnerability to high-altitude nuclear EMP (HEMP) events. Satellite-based QKD links operate through *free-space optics* (laser-based comms). While the satellite bus electronics can be hardened like existing military satellites (EMP-resistant shielding, radiation-hardened processors), the actual *quantum communication channel (the laser link)* is unaffected by EMP, since electromagnetic pulses do not propagate in the vacuum of space. Even if ground-based electronics near an EMP event are temporarily disrupted, *quantum networks can re-establish fresh cryptographic keys almost instantly* once power and link stability return, ensuring continuity without long re-synchronization delays.

EMP acts like a shockwave that collapses entire radio grids in seconds, just as artillery suppresses unprotected infantry. Conventional comms "die in place," but quantum fibre and satellite links act like underground hardened bunkers; they absorb the blast overhead and continue relaying orders without collapse.

Concept	Analogy	Operational Relevance
Quantum-secure networks are inherently more resistant to EW tactics such as jamming, directed energy, and EMP (electromagnetic pulse). Unlike RF-based systems, QKD relies on photons in optical fibres or satellite links, which are less susceptible to electromagnetic surges or spectrum denial. Hardened QKD nodes with shielding and redundancy ensure continuity of secure key exchange and comms even in high-intensity disruption scenarios.	EMP and jamming are like artillery barrages designed to flatten communication lines. Conventional RF comms collapse under the barrage, leaving commanders "deaf and blind." Quantum-secure fibre and satellite channels, however, are like hardened bunkers and underground cables they absorb the shock, bypass the blast, and keep information flowing.	Maintains strategic deterrence messaging during EMP or high-intensity EW events, ensuring nuclear C2 continuity. Provides survivable C4ISR backbones for contested zones under deliberate blackout attempts. Protects ISR data flows and mission-critical targeting information when conventional RF nets collapse. Ensures force cohesion and operational tempo even in adversary attempts at total electromagnetic suppression.

❏ **Immunity to High-Intensity EW:** High-intensity electronic warfare (EW) encompasses full-spectrum assaults such as barrage jamming, directed energy weapons, and spectrum saturation designed to overwhelm classical communication channels. Traditional RF and microwave-based systems collapse because they depend on contested frequency bands and are vulnerable to electromagnetic interference. Quantum-secure networks, however, exploit the q*uantum states of photons* in optical fibres or free-space satellite links, which do not depend on traditional RF carriers. Since photons travel in dedicated, shielded optical channels or space vacuum paths, they are immune to frequency-based denial and less impacted by energy surges.

Operational Relevance:

- Ensures *unbroken secure C2 links* during adversary EW dominance attempts, preserving the chain of command.

- Allows *joint task force coordination* (naval, air, land, space) even when RF comms are neutralized.

- Protects *strategic assets* (e.g., nuclear C2, space-based ISR feeds) from collapse under large-scale spectrum denial campaigns.

- Provides *confidence in network resilience* during high-intensity war fighting, when adversaries escalate to maximum EW saturation.

Concept	Analogy	Operational Relevance
Quantum-secure networks inherently resist traditional EW tactics by operating through photon-based quantum key distribution (QKD) over optical fibres and satellites. Unlike RF-dependent comms, they are not susceptible to spectrum jamming. Attempts at interference manifest as measurable quantum disturbances rather than silent corruption. Hardened optical nodes and shielded repeaters provide resilience against EMP surges, while quantum verification protocols prevent adversaries from spoofing or injecting false signals.	Like a submarine's hardened hull designed to withstand pressure and shock, quantum networks are shielded by the very laws of physics. EMP bursts, RF jamming waves, or directed EW "floods" break harmlessly against their design, much like depth charges detonating outside a submarine's reinforced shell.	Ensures survivable command and control in the most hostile electro-magnetic environments. Even when radios are blinded by jamming, EMP disables legacy circuits, or broadband EW floods the spectrum and quantum-secure channels maintain mission-critical encryption flow. This guarantees continuity of hardened C4ISR operations, enabling orders, ISR data, and targeting directives to traverse contested battle spaces without degradation or compromise.

Chapter Four

Operational Security in a Quantum Battle Space

This chapter establishes how Quantum Secured Command and Control (C2) will protect the integrity, availability, and authenticity of orders and situational data across joint operations. It explains, in plain terms, how quantum derived keys and tamper evident channels enable real time authentication of orders and signals, preventing spoofing, replay, and man in the middle attacks— even under intense electronic warfare. Officers will see how these protections scale to Multi Domain Operations (MDO), preserving a common operational picture and ensuring that time critical decisions can be executed with confidence across land formations, carrier and submarine groups, and air task forces.

The chapter then details measures to prevent information leakage during strategic manoeuvres and to safeguard autonomous and AI controlled systems. It describes how quantum secure links compartmentalize access, enforce one time, forward secure keys, and provide compromise visibility at the physical layer, ensuring that sensitive intent, routes, and rules of engagement remain undisclosed. For autonomous platforms and swarms, the chapter sets out command link hardening, fail secure behaviour, cryptographic agility, and continuous attestation so that control cannot be hijacked, degraded, or silently influenced by an adversary.

A maritime focus follows, covering the embedding of Quantum Key Distribution (QKD) into naval communication backbones from ship to ship and ship to shore to satellite relays, along with integration into submarine communications where low probability of intercept and emission control are paramount. Practical patterns are provided for real time tactical key exchange between maritime battle groups, cross domain guards for coalition operations, and contingency modes that maintain secure coordination during jamming, deception, or link loss. The chapter also addresses resilience against interception and jamming through hybrid architectures that pair QKD with post quantum cryptography, frequency agile transports, and strict key lifecycle governance.

Finally, the chapter surveys global quantum communication initiatives and their implications for naval and joint security, including a focused review of rival efforts and the strategic consequences of early fielding. It outlines how China's leadership in large scale quantum communication infrastructure and space based experiments is shaping timelines, then offers a comparative assessment of strengths, gaps, and strategic focus areas across major powers. The section concludes with concrete guidance for Indian forces: priority use cases, technology readiness, interoperability and training pathways, test and evaluation in contested electromagnetic environments, and acquisition choices to achieve credible, scalable quantum secure C2 at the speed of operations.

4.1 QUANTUM-SECURED COMMAND AND CONTROL (C2) SYSTEMS

• **How quantum-secured C2 systems will enhance operational security in complex defence operations**

Quantum Command: Shielding Mission Integrity in the Fog of War

Ultra-Secure Communications between Decision Nodes

Quantum Mesh: Command certainty in a battlefield of uncertainty
In modern and future warfare, the most vulnerable link is not the soldier on the front line but the signal carrying the commander's intent. Classical cryptographic channels, while robust today, are finite in lifespan, eventually breakable either by quantum computing or sophisticated electronic warfare. QKD changes this equation by enabling unbreakable, physics-backed key exchange between decision nodes ranging from national command centres and theatre HQs to mobile field headquarters and allied C2 elements.

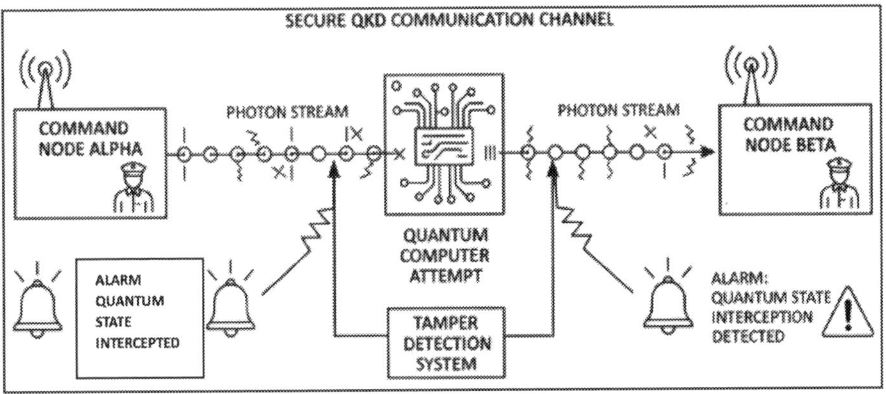

Fig. 4.1: Self-Alerting Security: Commands That Reveal Tampering Instantly

QKD ensures that every command passed between decision nodes is immune to eavesdropping, even by adversaries wielding quantum computers. Any interception attempt alters the quantum state of the photons, immediately alerting defenders to tampering. In military terms, this functions like a trip wire system, where any enemy approach disturbs the wire and signals intrusion. With QKD, the chain of command is not just encrypted, it is continuously monitored for breach attempts. Equally vital is the elimination of signal spoofing and false flag transmissions. In contested electromagnetic environments, adversaries may attempt to inject fake orders, confuse decision-making, or even simulate the voice or digital credentials of legitimate commanders. With quantum authentication, every command can be validated at the physical level, preventing adversaries from forging origin or destination. It is the equivalent of commanders issuing orders that carry a biological DNA marker, impossible to replicate by hostile forces.

Operationally, this secures the integrity of the command web in coalition and joint force operations. Decision nodes, whether a NATO HQ, an Indian naval fleet C2 centre, or a forward-deployed special operations task group, gain the assurance that all orders received and executed are genuine, untampered, and traceable. In practical terms, this quantum mesh acts as a battlefield immune system, detecting infiltration attempts and rejecting compromised signals before they can corrupt operations. In a battlespace clouded by electronic warfare, cyber intrusion, and information operations, ultra-secure decision-node communication ensures commanders retain clarity, continuity, and control. It preserves unity of effort, even under the most aggressive quantum-enabled adversary pressure, and keeps the operational chain of command unbroken.

- **QKD Securing Commands across Decision Nodes:** With QKD providing *unbreakable encryption for command transmissions* by exploiting the laws of quantum mechanics, specifically the no-cloning theorem and measurement disturbance principle ensures that any attempt to intercept or observe the encryption keys instantly changes their quantum state, triggering an alert and invalidating the compromised key. Applied to military C2, this means that orders exchanged between strategic headquarters, theatre-level command centres, deployed field units, and allied force nodes remain immune to eavesdropping even from adversaries armed with quantum-capable decryption systems. Unlike classical encryption, which can eventually be broken with brute computational force or stolen keys, QKD-generated keys are one-time-use and self-invalidating upon tampering, making them impervious to both present and future cryptanalytic threats.

  In a contested battle space, adversaries often target the "nervous system" of command by attempting to monitor or decrypt decision flows. By deploying QKD across satellites, airborne relays, and terrestrial fibre backbones, commanders can maintain uninterrupted and invulnerable C2 channels. This guarantees that critical decisions, whether nuclear deterrence alerts, joint task force manoeuvres, or time-sensitive targeting instructions cannot be silently observed, replayed, or pre-empted by hostile intelligence. It ensures trustworthy coordination across joint and coalition forces, preserving deterrence credibility and mission assurance even against quantum-enabled adversaries.

Concept	Analogy	Operational Relevance
QKD ensures that commands between command centres, field units, and allied forces are immune to eavesdropping even by quantum adversaries.	Like sealed military despatches that instantly self-destruct if intercepted, ensuring that only the intended recipient ever receives the order intact. Unlike classical envelopes that can be "steamed open," QKD makes interception impossible without destroying the message.	Guarantees invulnerable C2 channels across terrestrial, aerial, and space-based nodes, even in quantum-capable threat environments. Prevents adversaries from silently monitoring, replaying, or pre-empting command flows, ensuring trust, deterrence credibility, and seamless coalition coordination.

- **Eliminates risks of signal spoofing, interception, or false flag transmissions in contested environments:** Quantum cryptographic protocols bind each transmission with quantum-derived keys. Any attempt to forge or alter a signal even at the bit level collapses the entangled state, exposing the spoof attempt instantly. Unlike classical encryption, where encrypted data can be captured and stored for later brute-force decryption, QKD makes interception futile: observation irreversibly alters photon states, rendering intercepted data unusable and alerting defenders. Adversaries attempting to impersonate friendly units or inject counterfeit commands are thwarted, since quantum-verified keys are unique, non-replicable, and refreshed continuously. The system can instantly invalidate any command not authenticated through quantum exchange. In high-intensity cyber/EW battle spaces, quantum-secured signals preserve command authenticity even when conventional comms channels are degraded, ensuring decision-makers retain confidence in every transmission.

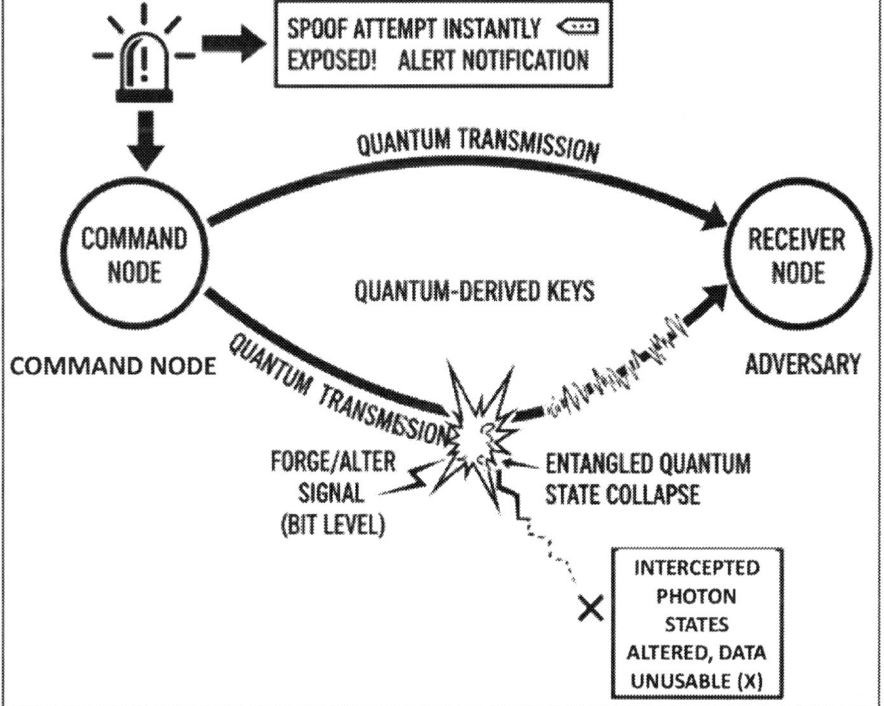

Fig. 4.2: Collapse on Contact: Quantum's Self-Destroying Defense Against Forgery

By eliminating spoofing, interception, and false flag risks, commanders preserve decision integrity. Units never act on compromised or counterfeit orders, reducing the possibility of ambushes, fratricide, or strategic misdirection. This resilience ensures coalition interoperability, preventing adversaries from driving wedges through misinformation in joint or multinational operations.

Concept	Analogy	Operational Relevance
Quantum-secured C2 eliminates risks of spoofing, interception, and false flag transmissions by ensuring every command is authenticated through QKD-derived keys. Any interception alters the quantum state, exposing the adversary instantly, while counterfeit orders cannot be validated.	Like a battlefield challenge-and-password system that changes with every breath, any intruder trying to imitate a friendly force is revealed immediately, as the recognition signal is impossible to copy or re-use.	Preserves decision integrity under contested conditions; prevents adversaries from injecting false commands, staging ambushes, or sowing mistrust. Secures coalition interoperability, ensuring allied forces act only on verified orders, even in cyber- and EW-heavy battlefields.

Real-Time Authentication of Orders and Signals

No More Doubt: Every command authenticated at quantum speed

One of the greatest vulnerabilities in modern C2 architectures is the risk of command injection where adversaries attempt to impersonate legitimate commanders, alter mission directives, or introduce false signals into the decision loop. Classical encryption, while robust, still relies on algorithmic protections that can be circumvented by advanced cyber exploitation or brute force quantum attacks in the future. Quantum cryptographic protocols fundamentally shift this paradigm by embedding authentication within the laws of quantum mechanics. Every digital command is secured by a quantum key that cannot be duplicated, altered, or replayed without immediately revealing adversarial interference. This ensures that orders transmitted between headquarters, forward operating units, and coalition partners retain absolute integrity, even under conditions of cyber saturation or electronic warfare (EW). In traditional warfare, the battlefield relied on challenge-and-password systems; a sentry demanding a code word to distinguish friend from foe. But passwords could be stolen, intercepted, or coerced from prisoners, creating avenues for infiltration. Quantum authentication is the equivalent of a dynamic challenge code that changes with every heartbeat, a system so fluid and incorruptible that even if an adversary intercepted a signal, it would become useless in the next instant. Like a secure chain of command reinforced at every echelon, quantum authentication ensures that no impostor can ever infiltrate the decision network, no matter how sophisticated the deception attempt.

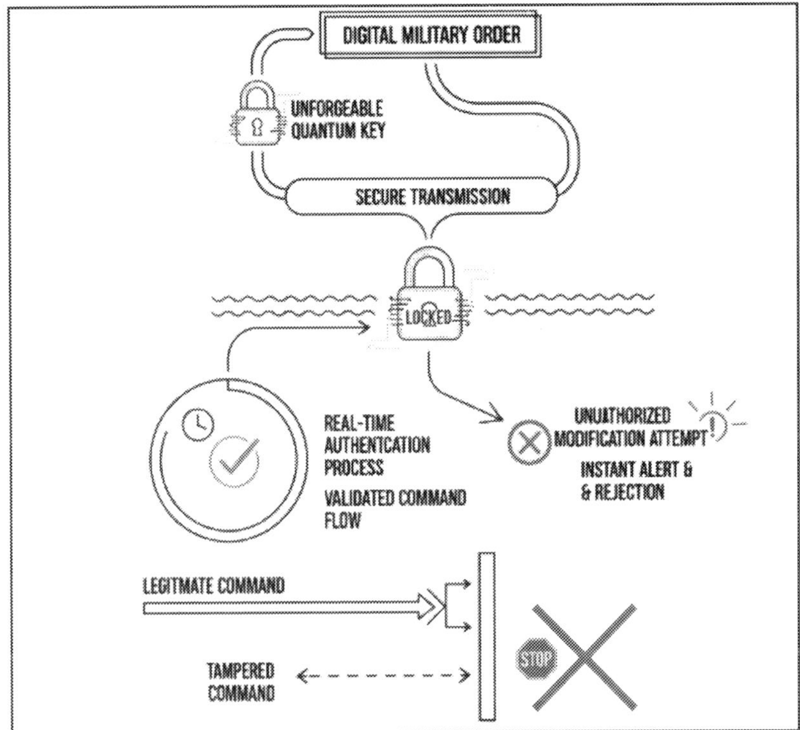

Fig. 4.3: Quantum-Sealed Orders: Authentication in Real Time

For military forces engaged in cyber-heavy and highly contested operational theatres, the authenticity of commands is as critical as their secrecy. A false order to re-position a brigade, to ground aircraft, or to disengage naval assets could create catastrophic consequences without a single shot fired. By implementing real-time quantum authentication, armed forces gain immunity against signal spoofing, false flag transmissions, and adversarial attempts to fracture trust within the command chain. This is especially vital in coalition and joint operations, where forces of multiple nations operate under a unified C2 umbrella. With quantum authentication, every participant can act with confidence that the directives received are not only secure but unquestionably genuine, reinforcing unity of effort and maintaining tempo in fast-moving combat situations.

Concept	Analogy	Operational Relevance
Quantum cryptographic protocols bind each digital order to an unforgeable quantum key, ensuring commands cannot be tampered with or counterfeited. Authentication occurs in real time, with any unauthorized modification detected instantly.	Like a battlefield challenge-and-password system where only those with the correct countersign are recognized as friendly, but executed at quantum speed with zero risk of an enemy stealing the countersign.	Prevents injection of false orders, signal spoofing, or adversary tampering in cyber-heavy conflicts. Provides absolute trust in command authenticity across the chain of command, ensuring unity of effort and preventing catastrophic misdirection in

- **Quantum cryptographic protocols guarantee authenticity of digital commands:** QKD can be extended into quantum digital signatures, which assign a unique, quantum-secure "stamp" to every command packet. Unlike classical signatures (RSA, ECC), these cannot be forged even by adversaries equipped with quantum computers. Each signed command is bound to a one-time-use quantum key. Even if an adversary intercepts the transmission, they cannot re-use it or substitute it without detection. Authentication does not require computationally heavy checks; it is enforced by the physical properties of entangled photons and quantum measurement. This ensures latency-free validation in time-critical operations.

Concept	Analogy	Operational Relevance
Quantum cryptographic protocols, including Quantum Digital Signatures (QDS), guarantee authenticity of all digital commands. Each order is bound to a one-time-use quantum key that cannot be forged, replayed, or substituted even by adversaries with quantum computers. Authentication occurs instantly at the physical layer of communication, and tampering attempts collapse the quantum state, exposing the intrusion.	Like a command seal pressed into wax, but unlike classical seals that can be stolen or forged, the quantum "seal" is self-destructive if tampered with. Only the legitimate commander's channel can issue an executable order.	Preserves chain of command under cyber and electronic warfare conditions by ensuring all orders are verified as authentic. Prevents false flag transmissions, spoofed drone commands, or injected cyber-orders from adversaries. Provides tactical edge by enabling soldiers, UAVs, and automated systems to execute without hesitation, knowing each signal originates from a trusted source.

Units can execute orders in high-tempo combat without hesitation, confident that every command has been authenticated at the quantum level. In cyber-heavy battlefields, where adversaries may attempt to inject false commands (e.g., redirect drones, spoof re-supply orders, or trigger premature strikes), quantum authentication prevents catastrophic deception. In cyber-heavy battlefields, where adversaries may attempt to inject false commands (e.g., redirect drones, spoof re-supply orders, or trigger premature strikes), quantum authentication prevents catastrophic deception.

Enhanced C2 in Multi-Domain Operations (MDO)

One Command, All Domains – Synchronizing land, air, sea, space, and cyber

In modern warfare, command and control (C2) is no longer confined to a single domain. Military operations increasingly unfold across land, air, sea, space, and cyber simultaneously, demanding a seamless and secure integration of all forces. Quantum-secured C2 systems address this challenge by enabling commanders to exercise authority with zero signal compromise, ensuring that communications channels remain impenetrable even against state-of-the-art adversary interception technologies. This ensures that the central principle of unity of command is preserved in multi-domain operations (MDO), where the ability to synchronize diverse capabilities is operationally decisive.

From a military analogy perspective, traditional C2 can be likened to commanding multiple divisions on a single battlefield, where orders flow in predictable and secure channels. However, in MDO, commanders must orchestrate a far more complex symphony akin to directing simultaneous campaigns across different terrains and theatres with enemy interference at every step. Without robust, trustworthy, and secure communications, the synchronization of air sorties with naval manoeuvres, cyber operations with electronic warfare, or space-based intelligence with ground-troop manoeuvring risks severe degradation. Quantum-secured C2 offers the equivalent of an unbreakable field telephone line linking every theatre, guaranteeing that instructions are delivered without delay, distortion, or compromise, regardless of adversarial attempts to jam, spoof, or decrypt signals.

Operationally, this capability underpins the very foundations of MDO strategies. Multi-domain battle spaces demand decentralized execution, where tactical units must operate with high levels of autonomy while remaining tethered to the commander's intent. Quantum-secured systems make high-speed and trustworthy command flows possible, empowering dispersed forces

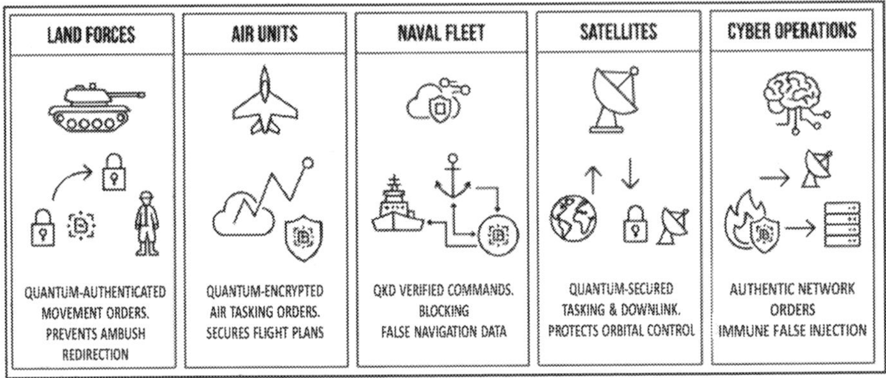

Fig. 4.4: Quantum Command Integrity Across All Five Domains

to act with confidence in the shared operational picture without fear of signal corruption or delay. In an environment where decision-making time is measured in milliseconds, such as missile defence or countering hypersonic threats, the capacity to transmit secure instructions and intelligence instantaneously across domains is a decisive enabler of mission success.

Ultimately, quantum-secured C2 does more than protect communications; it enhances operational tempo, safeguards decision superiority, and ensures that the principle of synchronized effects across domains is executed with precision. For commanders, it delivers the assurance that every manoeuvre, strike, and defensive action across land, air, sea, space, and cyber will be coordinated under a single, uncompromised command structure.

Aspect	Explanation
Concept	Quantum-secured C2 eliminates risks of spoofing, interception, and false flag transmissions by ensuring each command is authenticated through QKD-derived keys. Any attempt at interception alters the quantum state, immediately exposing the adversary, while counterfeit orders fail validation.
Analogy	Comparable to a battlefield challenge-and-password system, but instead of a static code, the "password" changes with every breath. Any intruder trying to imitate a friendly force is exposed instantly, since the authentication cue is impossible to copy or re-use.
Operational Relevance	Preserves decision integrity under contested conditions by preventing adversaries from injecting false commands, staging ambushes, or creating mistrust among units. Enhances coalition interoperability across domains, guaranteeing that allied forces act only on verified orders even in cyber- and EW-heavy battlefields, thereby sustaining cohesion and trust.

Domain	Concept Application	Analogy	Operational Relevance
Land	Secure, quantum-authenticated movement orders for ground forces ensure adversaries cannot redirect units into ambushes.	Like a sentry's challenge-password changing every second, ensuring infiltrators cannot impersonate friendly troops.	Prevents spoofed orders to brigades or battalions, preserving manoeuvre integrity and avoiding tactical deception in contested terrain.
Air	Air tasking orders are quantum-encrypted, ensuring every sortie receives authentic instructions.	Similar to an airborne refuelling signal that cannot be mimicked unless verified by a dynamic, uncopyable code.	Prevents adversaries from diverting aircraft, misdirecting strike packages, or grounding sorties through false flag transmissions.
Sea	Naval group commands are verified via QKD keys, preventing false navigation or targeting data from misleading fleets.	Like naval signal flags that instantly change design continuously, impossible for enemies to replicate.	Ensures safe convoy protection, synchronized naval fires, and prevents adversary attempts to scatter or deceive task forces at sea.
Space	Satellite tasking and downlink data commands remain tamperproof through quantum-secured access.	Like a telescope lens that clouds instantly if anyone else tries to look, revealing hostile tampering.	Guarantees operational continuity of ISR assets, secure satellite navigation cues, and uninterrupted command even under anti-satellite (ASAT) threats.
Cyber	Network operations orders remain authentic and immune to relay or false injection by hostile cyber actors.	Like a login system where the password regenerates before each keystroke, making intrusion futile.	Denies adversary attempts to corrupt decision loops, ensuring only valid authorized actions occur across coalition cyber nodes even in EW-heavy arenas.

- **Supporting MDO Strategies with Decentralized, High-Speed, and Trustworthy Command Flow:** Modern Multi-Domain Operations (MDO) demand a command structure that can thrive in a dispersed, contested, and fast-moving battle space. Traditional centralized command models, while secure in slow-moving conflicts, become brittle in environments where adversaries deliberately target the command structure with cyber intrusions, electronic warfare, and kinetic strikes. Quantum-secured C2 systems transform this paradigm by enabling decentralized, high-speed, and fully trustworthy command flows, ensuring operational resilience and tempo even under the most hostile conditions.

Quantum communication ensures that every transmission across the command network whether strategic orders from higher headquarters or immediate tactical updates from field units is authenticated and encrypted in real time. The trustworthiness of this flow stems from Quantum Key Distribution (QKD), which not only protects data but instantly signals if adversaries attempt to intercept or tamper with the command chain. This allows for decentralized execution, where commanders can delegate decision authority confidently to dispersed units, knowing command integrity cannot be compromised. At the same time, the high-speed nature of quantum-secured links enables decision cycles to accelerate rather than slow down under contested conditions.

Operational Relevance:

- **Decentralization**: In MDO, distributed units must act independently yet coherently. Quantum C2 assures subordinate commanders that their orders are valid and untampered, eliminating hesitation caused by fear of spoofing or corrupted instructions. This fosters mission command in its purest sense.

- **High-Speed Communications**: Time-sensitive missions like intercepting hypersonic weapons or neutralizing mobile missile launchers depend on sub-second data sharing. Quantum links provide near-instant authentication, preventing delays from verification processes or suspicion of data tampering.

- **Trustworthy Flow**: In adversary-contested electronic warfare environments, units are constantly exposed to deception attempts. Quantum authentication nullifies false flags and corrupted signals, allowing all elements, land manoeuvre, air strikes, naval fires, cyber disruptions, and space-based ISR, to operate with confidence in the reliability of tasking orders.

- **Coalition Interoperability**: Decentralized multinational operations, often complicated by differing standards and vulnerabilities, are streamlined under a quantum-secured framework. Every coalition partner accesses the same validated command picture, reducing friction and preventing adversary-driven divisions.

Preventing Information Leakage during Strategic Manoeuvres

Ghost Operations: When silence speaks louder than noise

One of the most acute vulnerabilities in conventional command and control networks arises not from intercepted content but from pattern-of-life leakage. Even when encrypted, adversaries can conduct traffic analysis to infer unit mobilization schedules, force concentrations, or command intentions simply by monitoring spikes in communications volume or directional flows. Quantum-secured C2 systems eliminate this risk by ensuring that movement orders remain untraceable and undecipherable. With QKD protecting every instruction, adversaries not only fail to decrypt the messages, but they cannot even reliably detect, trace, or analyse the communication patterns associated with strategic manoeuvres. This creates a communication environment in which the act of commanding itself becomes invisible, allowing for the execution of covert mobilizations and deception operations with unprecedented integrity.

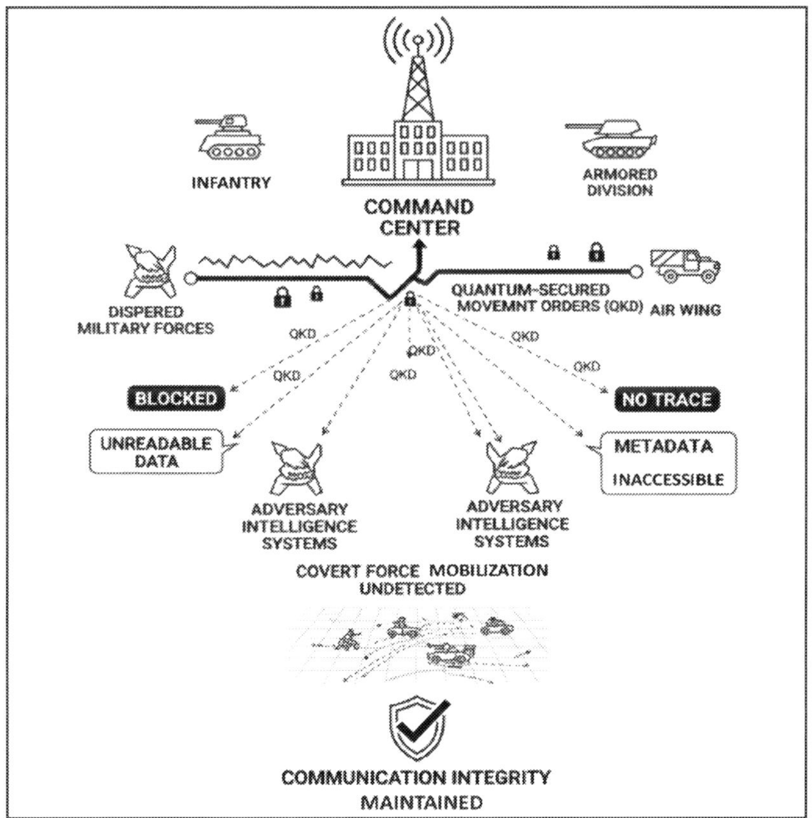

Fig. 4.5: Silent Orders: Quantum's Undetectable Command Flow

This can be likened to a silent night convoy moving under complete blackout conditions. In traditional operations, even if orders are hidden, the signals themselves, akin to the flicker of headlights, can betray activity to an attentive observer. Quantum-secured C2 turns off those "headlights," ensuring the convoy appears as though it does not exist. For adversaries attempting to intercept or track signals, the phenomenon is akin to pursuing shadows that vanish upon touch: no signal can be copied, re-played, or analysed without instantly betraying manipulation. Just as blackout movement in ground operations denies the enemy clues about troop or logistics flows, quantum security extends this invisibility to the command layer across all domains.

At the operational level, preventing information leakage during strategic manoeuvres translates directly into the preservation of surprise and deception, critical multipliers in warfare. Whether re-positioning armoured divisions, manoeuvring naval strike groups, or quietly mobilizing expeditionary forces, commanders can maintain total silence in the electromagnetic spectrum, denying adversaries intelligence on friendly intentions. This capability directly supports strategic deception plans, such as feints or diversionary manoeuvres, without the traditional risk of adversaries discerning patterns from encrypted but observable communications.

Element	Quick-Reference Guidance
Concept	Quantum-secured C2 denies adversaries the ability to detect, trace, or analyse command transmissions, even at the metadata level. By leveraging QKD, movement orders cannot be intercepted, decrypted, or patterned, ensuring covert mobilization and deception with uncompromised communication integrity.
Analogy	Comparable to blackout movement of a convoy where no headlight flicker reveals its presence. Quantum-secured channels are the digital equivalent, making command orders invisible; adversaries chasing signals encounter shadows that vanish upon touch.
Operational Relevance	Preserves operational surprise and deception by concealing order flows during mobilizations. Enables "ghost operations" where force deployments, diversions, or feints remain undetectable, denying adversaries situational awareness. Strengthens coalition coordination by ensuring interoperability without leaving signal trails, thus maintaining initiative under surveillance-heavy and contested environments.

In joint and coalition contexts, quantum-secured C2 enables multinational partners to execute synchronized covert operations while maintaining assured interoperability. The absence of detectable command traffic prevents adversaries from identifying seams in coordination or guessing at shared objectives. In high-intensity conflicts against technologically advanced adversaries, this contributes to strategic ambiguity, forcing opponents to waste resources on misdirected responses while friendly forces retain the initiative.

In essence, quantum-secured C2 transforms the conduct of manoeuvres into ghost operations where commands cannot be observed, traced, or compromised. By denying the adversary even the faintest signal of intent, it ensures that decisions remain concealed until revealed on the battlefield through decisive action.

- **Secure C2 and the Denial of Detection, Decryption, and Pattern Tracing:** Modern adversaries often do not require access to the content of military communications to piece together operational intent. Even if messages are strongly encrypted, hostile intelligence analysts can study signal patterns, transmission spikes, routing paths, and timing correlations, a process known as traffic analysis to predict unit mobilizations, concentrations, or staging for an offensive. In many cases, an adversary only needs to detect unusual signal volumes or directional flows to anticipate manoeuvres, even without decryption. This has historically been a critical vulnerability during strategic and tactical planning.

  Quantum-secured Command and Control (C2) overcomes this vulnerability by ensuring that movement orders cannot be detected, decrypted, or traced at the metadata level. Through Quantum Key Distribution (QKD) and quantum-resistant protocols, communications achieve three distinct layers of protection:

  - **Undetectability:** Communications do not generate identifiable patterns in the spectrum that can be harvested for traffic analysis, denying adversaries signal-based insights.

  - **Unbreakable Encryption:** Even if intercepted, the quantum-encrypted order is undecipherable and immune to retroactive decryption techniques, however advanced.

❑ **Pattern Masking**: Quantum-secured systems obscure timing and format cues, preventing adversaries from correlating communication surges with unit activity such as mobilization or redeployment.

This creates a communication silence effect, where the mere existence of manoeuvre orders is concealed, depriving adversaries of the ability to infer or anticipate friendly force dispositions.

This is akin to a commander transmitting orders through invisible courier routes that cannot be observed, intercepted, or mapped. In conventional radio traffic, even encrypted signals are like visible supply convoys; they may be locked, but the number of vehicles and their movement betrays direction and tempo. With quantum-secured C2, the "convoy" of orders is invisible; there are no tracks, no visible presence, and no rhythm to analyse. To the adversary, it appears as though no orders are being passed at all, even while entire formations mobilize.

Operational Relevance

❑ **Preservation of Surprise**: Adversaries cannot detect changes in command intensity that normally precede large-scale manoeuvres such as amphibious landings, airborne drops, or tank offensives.

❑ **Integrity of Deception Campaigns**: Strategic diversions or feints remain credible, since adversaries cannot distinguish between real and false order flows through signal analysis.

❑ **Protection in Surveillance-Heavy Environments**: Against peer competitors relying on signals intelligence (SIGINT), cyber espionage, and AI-driven traffic analysis, quantum-secured orders maintain an information blackout.

❑ **Coalition Advantage**: Multinational forces can issue synchronized movement orders without betraying coordination points or tipping off adversaries to joint operational timelines.

In effect, quantum-secured C2 transforms the command environment into signal invisibility, where movement orders are not only indecipherable but also undetectable and untraceable, leaving adversaries blind to the very existence of manoeuvre instructions until outcomes are physically revealed on the battlefield.

- **Supports covert mobilization and strategic deception plans with full communication integrity:** One of the timeless principles of warfare is that surprise magnifies combat power. Covert mobilization and deception plans are designed to deny adversaries foreknowledge of force movements and intentions, thereby shaping the battle space before the first shot is fired. Yet, in the modern era dominated by surveillance drones, satellites, SIGINT platforms, and cyber reconnaissance, the ability to mask mobilization has become increasingly difficult. C2 traffic itself often betrays operational patterns. Quantum-secured C2 directly solves this problem, giving commanders the ability to mobilize forces invisibly and stage complex deception operations with assured communication integrity.

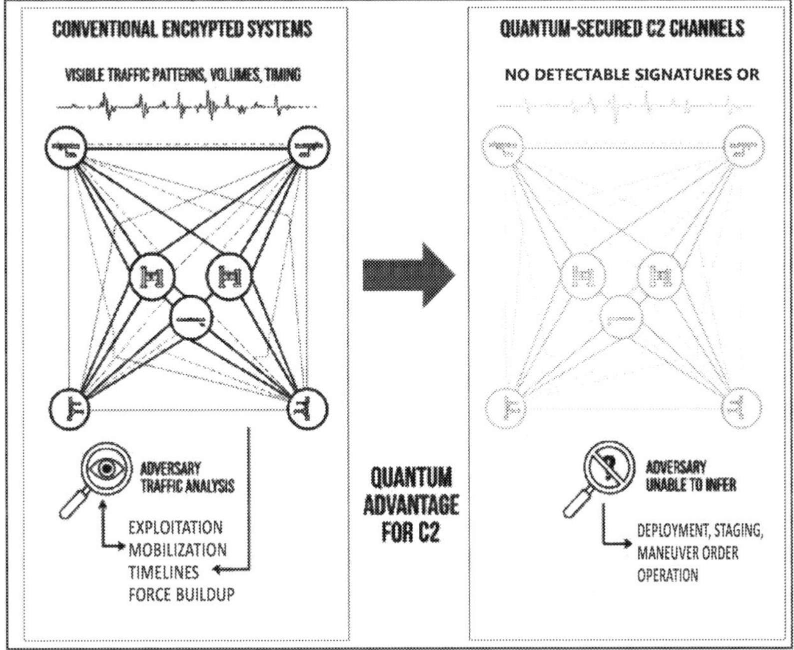

Fig. 4.6: Zero Signature Warfare: Mobilization Beyond Detection

Quantum-secured C2 systems reinforce covert mobilization by ensuring that all orders for deployment, staging, and manoeuvre are exchanged without leaving detectable or traceable signatures. Unlike conventional systems, where encrypted traffic volumes or timing can still be exploited through traffic analysis, quantum channels provide undetectable and tamper-proof order flow. This means that adversaries cannot infer mobilization timelines, force build-up, or deception vectors from communication activity.

In deception planning, this same integrity guarantees that real and false orders coexist without compromise. Authentic quantum-validated orders guide actual manoeuvres, while adversaries observing from outside remain blind, unable to distinguish operational reality from feints or diversionary steps.

Operational Relevance

❑ **Covert Mobilizations:** Armies can shift reserves, prepare amphibious or airborne operations, and re-position strategic strike assets without triggering enemy alert levels. This denies adversaries the ability to prepare defences in advance.

❑ **Strategic Deception:** False flag operations, feints, and misdirection's gain credibility because adversaries cannot validate or disprove communications activity. Commanders can draw enemy attention to one theatre while secretly massing forces elsewhere.

❑ **Survivability under Surveillance:** Even under persistent enemy ISR and SIGINT pressure, forces maintain communication without signalling intent. This frustrates adversaries, forcing them into reactive postures.

Element	Quick-Reference Guidance
Concept	Quantum-secured C2 ensures mobilization and manoeuvre orders are transmitted with complete communication integrity. Signals cannot be detected, traced, or decrypted, denying adversaries the ability to infer operational intent through traffic analysis. This enables covert force movements and deception operations without the risk of exposure.
Analogy	Like invisible couriers delivering sealed orders, no tracks left behind, no signs to follow. To the adversary, covert mobilizations unfold with the silence of a convoy under blackout, hidden until revealed in decisive action.
Operational Relevance	Supports covert redeployment of reserves, surprise offensives, and deception campaigns by shielding command flows from surveillance. Preserves coalition integrity by concealing signal patterns, frustrating adversary ISR and SIGINT, and ensuring strategic surprise is maintained in contested battlefields.

❏ **Coalition Integrity:** Multinational partners can plan covert manoeuvres without risking exposure of coalition signal flows, ensuring deception is executed without seams between allied forces.

Safeguarding Autonomous and AI-Controlled Systems

Quantum Autonomy: Securing machines that fight beside us

The increasing integration of autonomous and AI-driven systems into modern military operations brings unmatched speed, reach, and endurance to the battle space, but it also introduces new vulnerabilities. Drones, robotic combat units, autonomous ground re-supply vehicles, and unmanned surface or underwater assets, all depend on secure C2 channels to operate effectively. Adversaries capable of intercepting or falsifying these channels could attempt to hijack or re-program such platforms, turning them from force multipliers into liabilities. Quantum-secured C2 systems neutralize this threat by ensuring that these autonomous battlefield agents can only receive and execute commands encrypted and authenticated with quantum-derived keys. Any attempt at external interference alters the quantum state, exposing hostile activity instantly and preventing rogue control. The result is an unbreakable assurance that machines remain loyal to their operators, never to the adversary.

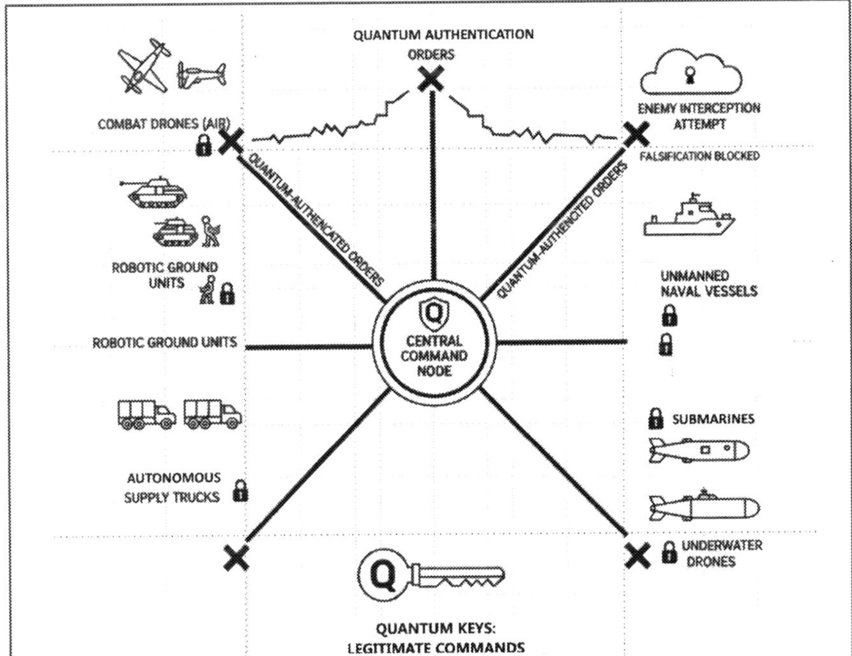

Fig. 4.7: Quantum Shield for Autonomous Forces: Unhackable Command and Control

This security can be compared to a trained guard dog that only heeds its handler's command phrase but, in this case, the phrase changes with every breath. Just as no intruder can guess or reuse the constantly shifting signal to redirect the dog, no adversary can replicate or forge commands to an AI-enabled drone or robotic unit secured by QKD. Attempts to issue false commands are immediately rejected as invalid, leaving the autonomous system deaf to enemy manipulation. In practice, this means unmanned platforms fight *only* for the commander who controls the cryptographic keys, never switching sides or becoming weaponized by hostile actors.

Operational Relevance

For commanders, the safeguarding of autonomous and AI-controlled systems with quantum-secured C2 provides both confidence in autonomy and assurance of operational integrity:

- **AI-Enabled Drones:** Intelligence, surveillance, and strike drones operate without fear of interception, degradation, or hostile re-tasking. This removes the risk of adversaries using stolen control links to disrupt missions or turn drones into weapons against friendly forces.

- **Unmanned Ground and Robotic Units:** From logistics convoys to robotic infantry support systems, unmanned ground platforms can execute tasks securely under contested cyber and EW environments, with no risk of ambush through diverted routing.

- **Naval and Aerial Unmanned Systems:** Autonomous wingmen aircraft or unmanned surface/underwater vessels maintain authentic linkage to command even in degraded communications zones, refusing any counterfeit instructions.

- **Strategic Deterrence:** Prevents adversaries from exploiting autonomy as a weakness. Instead of fearing compromised machines, commanders project confidence that autonomous assets strengthen deterrence by being immune to hijacking or reprogramming.

By bringing quantum assurance to autonomy, military leaders gain the ability to deploy unmanned and AI-enhanced systems as trusted teammates on the battlefield, secure in the knowledge that these platforms will never be turned against their own side. Quantum-secured C2 thus extends the principle of command loyalty from soldiers to machines—ensuring that every autonomous system remains a reliable extension of the commander's will under the most contested conditions of future warfare.

Element	Quick-Reference Guidance
Concept	Quantum-secured C2 ensures that AI-enabled drones, robotic units, and unmanned vehicles execute only commands authenticated with quantum-derived keys. This prevents hijacking, spoofing, or reprogramming of autonomous battlefield agents, guaranteeing their loyalty to the commander.
Analogy	Like a trained guard dog that only obeys its handler's secret command—which changes with every breath. No intruder can replicate or re-use the evolving signal, ensuring machines remain aligned solely with their operator's intent.
Operational Relevance	Protects unmanned aerial, ground, and naval platforms from hostile takeover in cyber- and EW-heavy battle spaces. Secures AI-driven logistics convoys, robotic combat systems, and drone swarms against adversary manipulation. Ensures autonomy and enhances combat power without introducing vulnerabilities, enabling machines to act as trusted teammates under joint and coalition command.

- **Ensuring AI-Enabled Systems Receive Encrypted, Authentic Commands Only:** AI-enabled drones, robotic units, and unmanned vehicles represent powerful force multipliers in modern military operations. However, their reliance on constant data exchange with command networks makes them particularly susceptible to interception, spoofing, or unauthorized re-tasking by skilled adversaries. Conventional encryption methods leave residual exposure since intercepted signals may eventually be decrypted, or false commands may be injected if channel authentication fails. Quantum-secured C2 closes this vulnerability by using QKD to generate encryption keys that are both unique and unbreakable. Every command sent to a platform is coupled with a quantum-authenticated key, meaning the system will only recognize orders that pass this validation. Any attempt to replace or forge a command instantly fails authentication, ensuring unmanned systems treat it as noise and refuse to act. This establishes an unassailable link of trust between commander and machine.

This can be compared to an exclusive command phrase given to a trained warhorse or working dog except that the phrase cannot be guessed, stolen, or re-used because it changes continuously with quantum verification. Even if an adversary attempts to whisper a similar cue, the animal refuses to respond. For AI-enabled units, quantum-secured C2 functions as that dynamic, unclonable signal: the machine obeys only its trusted handler, not an impostor.

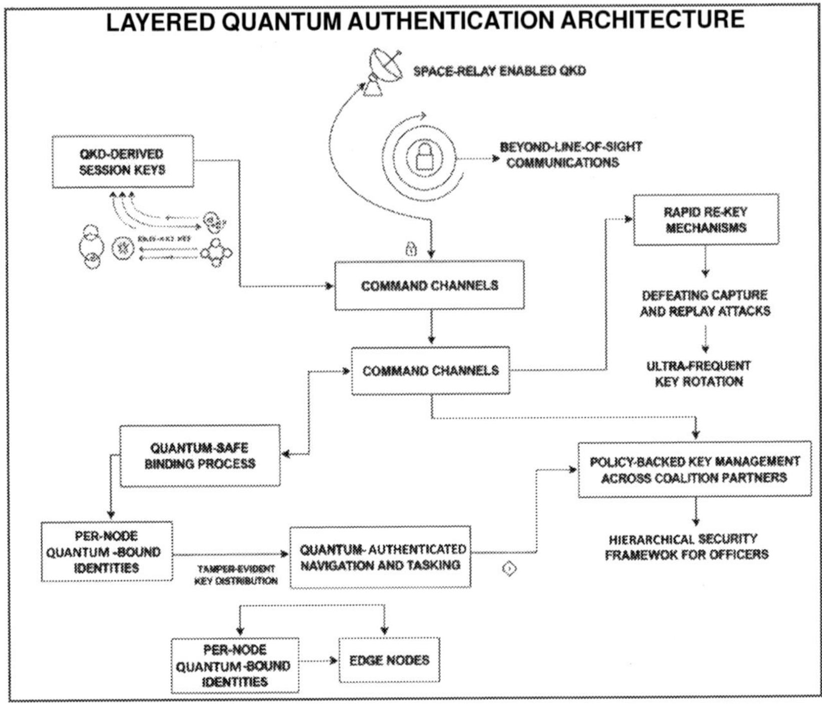

Fig. 4.8: Layered Defense: Quantum Authentication from Edge to Coalition

Operational Relevance

❑ **Combat Drone Operations:** ISR and strike drones remain immune to adversary hacking attempts; they strike only when directed by legitimate commanders, preventing enemy redirection or unauthorized targeting.

❑ **Unmanned Ground Vehicles & Combat Robots:** Robotic units entrusted with logistics, casualty evacuation, or fire support cannot be hijacked en route to compromise operational movement or redirect supplies.

❑ **Naval & Underwater Unmanned Systems:** UUVs or unmanned surface craft maintain constant authentic linkage with task force commanders, ensuring adversaries cannot insert false navigational data or override mission plans.

❑ **Integrity in Swarm Tactics:** Autonomous UAV or UGV swarms critical in overwhelming defences retain strict cohesion and follow legitimate tasking only, immune to disintegration from external interference.

❏ **Coalition Use:** In multi-national task groups, each unmanned system accepts commands solely from authenticated coalition authority, preventing adversaries from exploiting weak links between allies.

Quantum-secured authentication protects unmanned systems across air, land, sea, and swarm operations

Domain	Threats Addressed	Quantum-Secured Authentication Measures	Operational Effect
Air (UAS/loyal wingmen)	Command-link spoofing, GPS spoofing, false re-tasking, data-exfiltration of ISR feeds.	QKD-derived session keys for command/auth channels; quantum-safe identity binding for air vehicles; rapid re-key to defeat capture/replay; integration with post-quantum cryptography for compatibility where QKD links are intermittent.	Aircraft accept only authenticated tasking; hijack and re-task attempts fail; ISR and strike orders remain exclusive to authorized C2; resilience against traffic analysis and retroactive decryption.
Land (UGVs/robotic units)	Route diversion, mission override, malicious software updates, decoy beacons.	Quantum-authenticated control messages and updates; tamper-evident key distribution to edge nodes; quantum-safe re-key after contact loss; device identity attestation bound to quantum-derived secrets.	Robotic logistics and combat support refuse counterfeit routing or tasking; field updates are accepted only if quantum-authenticated; continuity of autonomous missions under EW/cyber pressure.
Sea (USVs/UUVs)	Navigation falsification, mission-plan overwrite, long-haul link interception.	Space/relay-enabled QKD for beyond-line-of-sight keys; quantum-authenticated navigation and tasking; one-time-pad at line rate where feasible; safe re-key for long-duration sorties.	Maritime unmanned assets reject forged waypoints and tasking; maintain assured control over extended ranges; mission plans cannot be replaced in transit.
Swarm operations (UAS/UGV swarms)	False leader injection, swarm fragmentation via counterfeit coordination, replay attacks.	Quantum-authenticated leader-election and coordination frames; per-node quantum-bound identity; ultra-frequent key rotation to null replay; PQC inter-op for mixed-networks.	Swarms preserve cohesion and follow only legitimate leaders; hostile nodes cannot split or re-task elements; cross-vendor, coalition swarms remain interoperable and trustworthy.
Cross-cutting (coalition C2)	Interoperability seams exploited for spoofing and identity fraud.	Policy-backed quantum key management, standards-driven accreditation of QKD link security, quantum-safe identity and authorization across partners.	Coalition unmanned assets accept commands only from authenticated authorities; reduced attack surface from mixed crypto baselines.

Mission-profile planning checklist, mapping quantum-secured authentication, and controls to risk reductions across domains

Mission profile	Core threats	Quantum controls to apply	Resulting risk reduction
ISR/strike (air UAS, loyal wingmen)	Command-link spoofing, GPS spoofing, replay attacks, false re-tasking, ISR feed theft.	QKD-derived session keys for C2 and payload; quantum-bound platform identity; rapid re-key on handoff; hybrid with NIST PQC for continuity when QKD intermittent; authenticated telemetry and tasking frames.	Orders accepted only from authorized C2, hijack attempts fail, ISR/strike tasking cannot be forged, resilience to traffic analysis and retroactive decryption.
Logistics re-supply (UGVs/ convoys/ robotics)	Route diversion, counterfeit tasking, malicious OTA updates, decoy beaconing.	Quantum-authenticated control and update channels; quantum-safe device attestation; per-mission one-time keys; automated re-key after link degradation.	Autonomous logistics refuse forged routes/updates, sustain missions under EW/cyber pressure, reduced compromise window if nodes are captured.
ASW patrol (USVs/UUVs with towed sensors)	Waypoint falsification, long-haul link interception, mission-plan overwrite during transit aerospace.	Space/relay-enabled QKD for BLOS keys; quantum-authenticated navigation/tasking; one-time-pad at feasible rates; secure mission vault with quantum-derived access aerospace.	Maritime unmanned assets reject forged waypoints, preserve mission plans over extended sorties, maintain assured control despite interception attempts aerospace.
SEAD swarms (air UAS swarms)	False-leader injection, swarm fragmentation via counterfeit coordination, re-play/flooding.	Quantum-authenticated leader-election and control frames; per-node quantum-bound ID; ultra-frequent key rotation; PQC inter-op for mixed-vendor/ally nets.	Cohesion preserved, hostile nodes cannot split or re-task elements, coordinated effects persist under EW and deceptive signalling.

- **Protects against hijacking or re-programming of autonomous battlefield agents:** Autonomous platforms UAS, UGVs, USVs/ UUVs, and robotic sensors/effectors depend on remote tasking, software updates, and inter-node coordination; each interface is a potential takeover point. Quantum-secured C2 anchors these links with quantum-derived session keys and authenticated control frames, so counterfeit orders, replayed packets, or malicious updates fail validation and are discarded. Even if an adversary records traffic for later exploitation, keys are forward-secure and rotated at high cadence, preventing retroactive decryption and long-dwell compromises. Think of each unmanned asset as a sentry dog that only obeys its handler's whistle pattern except that the pattern changes every instant in a way no outsider can predict or mimic. An impostor can shout commands or play recordings, but the dog hears only noise; any deviation from the authentic, moment-by-moment pattern is ignored, and attempts to interfere leave tell-tale traces.

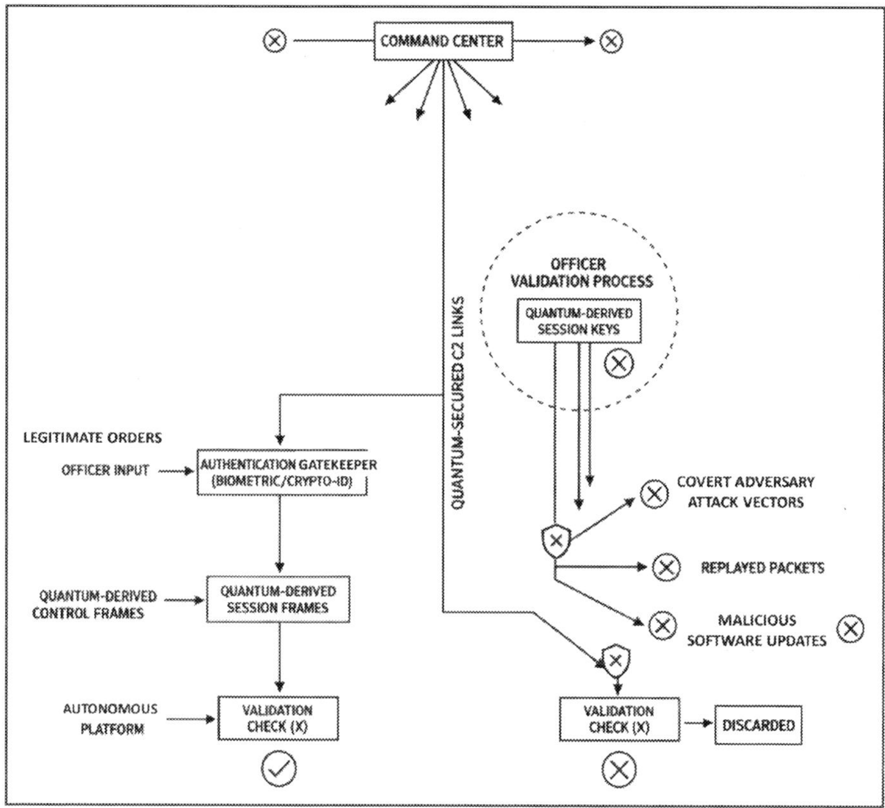

Fig. 4.9: The Quantum Gatekeeper: Automatic Rejection of False Commands

Operational relevance

❏ **Denies hijack-and-re-task**: Hostile attempts to redirect ISR drones, re-target loitering munitions, or alter waypoint stacks are rejected at the link layer, preserving mission integrity in EW-heavy airspace.

❏ **Blocks malicious re-programming**: Over-the-air updates and model weights for AI controllers are accepted only when quantum-authenticated end-to-end, preventing implanting of backdoors or behaviour toggles mid-mission.

- ❑ **Neutralizes leader spoofing in swarms:** Swarm coordination and leader election frames are quantum-authenticated, stopping false-leader injection and fragmentation attacks that aim to split or freeze formations.

- ❑ **Safeguards navigation-task coupling:** Authenticated tasking is cryptographically bound to platform identity and time, preventing adversaries from pairing forged PNT or sensor data with valid-looking commands.

- ❑ **Limits blast radius of capture:** If a node is physically seized, rapid key rotation and compartmented per-mission, keys constrain exposure; captured traffic cannot be decrypted and credentials cannot be re-used elsewhere.

- ❑ **Preserves coalition control:** Unmanned assets accept commands solely from pre-authorized authorities, closing interoperability seams that attackers exploit across mixed-vendor or allied networks.

- ❑ **Command-channel hardening:** Quantum-derived session keys for command, telemetry, and payload control; per-message authentication with high-frequency re-keying.

- ❑ **Identity and attestation:** Device identity bound to quantum-protected secrets; attested boot and signed autonomy policies to prevent rogue reconfiguration.

- ❑ **Update security:** Quantum-authenticated OTA pipelines for firmware, containers, and ML models; rollback protection and staged activation with multi-party release.

- ❑ **Swarm controls:** Authenticated synchronization/consensus messages; replay resistance and leader rotation tied to fresh keys to defeat flooding and capture.

- ❑ **Compromise response:** Automated key revocation and re-issuance; quarantine behaviours that degrade gracefully—safe loiter, return-to-base, or human-on-the-loop fallback.

Effects in operations

Mission set	*Benefits delivered*
Carrier air wing tasking	Authentic strike and recovery orders under heavy EW; no viable pathway for false re-tasking or delayed decrypts.
ASW barriers and pickets	Tamper-evident keys on sonobuoy nets/USV/UUV control prevent waypoint and plan overwrites.
Surface engagements	Inter-hull QKD-backed keys deny deception on fire-control and manoeuvre orders; fast re-key sustains tempo.
Amphibious/logistics groups	Secure replenishment and movement orders despite RF contention; key compromise on one leg does not expose the group.
Coalition maritime ops	Common quantum-anchored trust fabric reduces spoofing via interoperability seams; shared provenance for orders and data.

- ### Integration of QKD into naval communication architecture and its role in electronic warfare resilience

Quantum Shield at Sea: Securing Signals and Command amidst the Storm of Electromagnetic Warfare

Embedding QKD in Naval Communication Backbones

Quantum Backbone: Fortifying fleet-wide communication with unbreakable keys

Quantum key distribution embedded into naval communication backbones provides an unforgeable, tamper-evident key supply for ship-to-ship, ship-to-submarine, and ship-to-shore circuits, ensuring that fleet C2 and data links remain authenticated and confidential even under intense electronic warfare; practically, it hardens carrier strike groups and task forces against interception, deception, and retroactive decryption across RF, free-space optical, fibre, and underwater optical segments. Embedding QKD into naval backbones means integrating quantum key generation and refresh into the fleet's layered transport: shore fibre rings, shipboard optical trunks, line-of-sight free-space optical (FSO) links between hulls, and specialized channels for submarines and unmanned maritime systems. Keys produced via quantum states are information-theoretically secure and eavesdropper-detecting; any measurement attempt raises error rates, triggering automatic discard and re-key. The quantum layer feeds symmetric session keys to existing waveform stacks (HF/VHF/UHF, SATCOM alternatives, FSO, acoustic/blue-green underwater optics), so the data plane continues to use high throughput ciphers while the control plane gains tamper-evident keying. This preserves command integrity and confidentiality across the battle force without redesigning every radio; QKD becomes the backbone key factory for all enclaves.

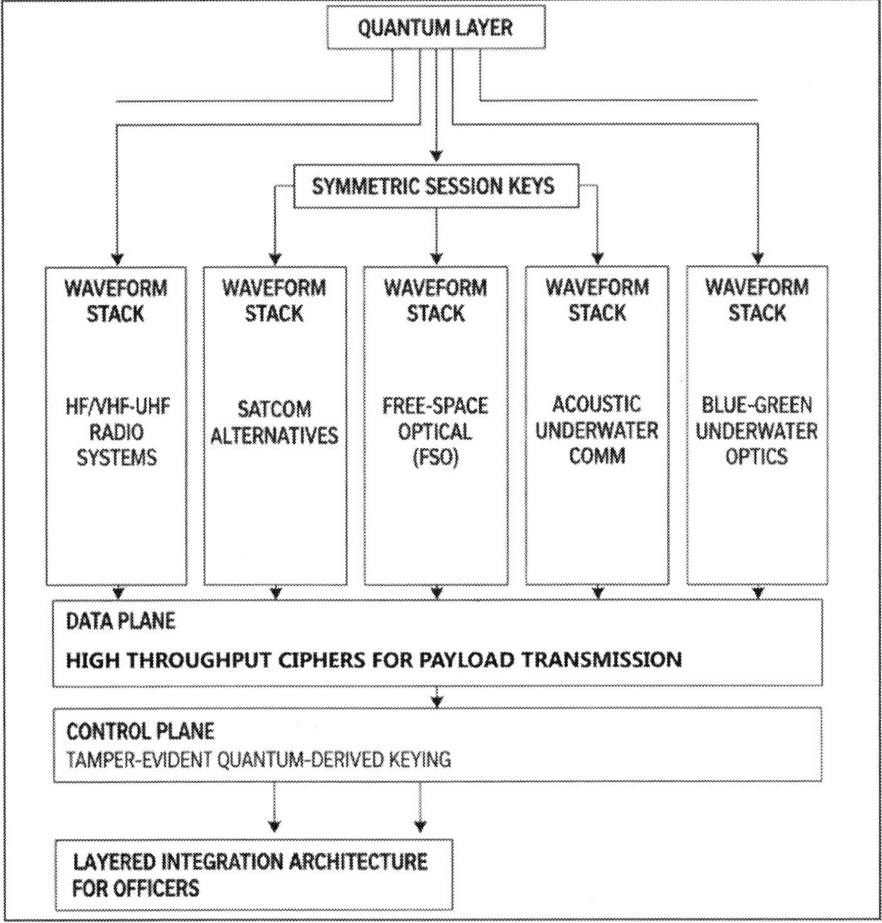

Fig. 4.10: One Quantum Layer, Every Waveform Secured: Universal Key Distribution

Treat the fleet's comms like a harbour protected by anti-tamper seals on every container: if any seal is touched, it shows immediately. QKD is that seal for cryptographic keys. The convoy (voice/data/video/tactical orders) still rides on familiar ships (existing links and waveforms), but their cargo manifests (keys) come from a bonded, incorruptible customs house. An adversary can watch the sea lanes or attempt a covert boarding, yet any interference with the seals is revealed on receipt, and the cargo is re-issued before fraud can propagate.

Operational relevance

- **Carrier strike group C2**: Ship-to-flag and ship-to-ship orders are keyed with quantum-derived material, blocking false tasking and preventing adversary collection from enabling later decryption; air wing tasking and strike packages retain authenticity during EMCON or EW pressure

- **Subsurface integration**: For SSNs/SSBNs/SSKs, QKD supports pre-deployment key loads via fibre ashore and opportunistic FSO/blue-green optical windows at periscope depth, enabling fresh keys without revealing content or permitting replay; patrol plans and mission updates cannot be forged or retro-decrypted

- **Amphibious and logistics groups**: Distributed groups maintain assured logistics and fire coordination with continuous re-key across LOS FSO relays and shipboard optical trunks; compromised or jammed RF legs no longer imply key compromise, reducing risk of route deception or cargo re-task

- **EW resilience in A2/AD**: Even when SATCOM is denied, shore fibre QKD rings and LOS quantum relays sustain key flow, allowing encrypted links to pivot transport layers without losing trust; attempts to tap pier, buoy, or relay nodes trigger discard and re-key, preserving tempo

- **Unmanned maritime systems**: USV/UUV tasking and telemetry accept only quantum-authenticated commands; forged waypoints, mission overwrites, or swarm leadership spoofing are rejected at the crypto boundary, keeping autonomous agents aligned with commander's intent

- **Enables fleet-wide** keying immune to interception by generating and refreshing keys from quantum states that reveal eavesdropping, then feeding these keys into existing radios and optical links so confidentiality and authenticity persist even under collection and jamming.

- **Facilitates highly secure data and command** exchanges between vessels, carriers, and maritime operations centres by providing continuous, compartmented re-key across shipboard optical trunks, inter-hull FSO bridges, and shore fibre rings; compromised segments re-key without exposing the rest of the force

Implementation cues for officers

- **Architecture:** Use shore-based fibre QKD rings as theatre key sources; distribute to task force via underway FSO relays or visiting units; cache per-mission key vaults aboard ships and subs for denied periods.

- **Integration:** Bind quantum-derived keys to platform/device identity; rotate session keys at tactical cadence tied to link state; reserve one-time pad only for narrow, highest-criticality circuits due to key-rate constraints.

- **TTPs:** Automate alarmed discard/re-key on elevated quantum bit-error rates; pre-plan denied-satcom key ladders; compartment keys by warfare area (CWC, SUW, ASW, AAW) and by unmanned control enclaves to limit blast radius.

Architecture by link type

Naval link	QKD integration approach	Key handling practice	EW resilience contribution
Shore ↔ Fleet (theatre backbone)	Fibre QKD rings at shore stations feed key material to embarkation points and relay nodes; underway replenishment via visiting units or FSO ship-to-ship bridges.	Compartment keys by warfare areas and task groups; schedule re-key tied to link health and mission phase.	Maintains theatre-level C2 keys when SATCOM is degraded; attempts to tap pier/shore fibre trigger discard/re-key alarms.
Ship ↔ Ship (surface action groups)	FSO QKD across masts/aerostats with beam tracking; hybrid paths (FSO+RF for control).	Rapid session key rotation at manoeuvre cadence; per-circuit key vaulting.	Blocks false tasking/inserts across surface links; preserves authenticity under jamming/spoofing pressure.
Ship ↔ Submarine	Pre-sail fibre QKD loads; opportunistic periscope-depth blue-green/FSO QKD refresh; port-call replenishment.	Time/mission-scoped key ladders; sealed "patrol vaults" with emergency re-key procedures.	Prevents forged patrol orders/waypoints; no RF exposure for key refresh, reducing detectability.
Shipboard internal (combat system/ backbone)	Shipboard optical trunks with QKD between enclaves (CIC, radio, CMS, EW suites).	Micro-segmentation; per-enclave keys; automatic quarantine on QBER spikes.	Limits blast radius of compromise; ensures intra-ship command authenticity under cyber-attack.
Fleet ↔ MOC (command ashore)	Shore fibre QKD to coastal relays; over water FSO QKD to picket ships; onward distribution within TF.	Hierarchical key management with role scoping; auditable key provenance.	Preserves flag-level C2 integrity; supports EMCON while retaining secure tasking.

Quantum-secured authentication and controls apply across unmanned domains

Domain	Primary threats	Quantum controls to apply	Implementation cues	Operational effect
Air (ISR/strike UAS, loyal wingmen)	Command-link spoofing, replay, GPS/PNT spoofing synergy, false re-tasking, ISR exfiltration.	QKD-derived session keys for command/telemetry; quantum-bound platform identity; authenticated tasking frames; hybrid QKD+PQC for continuity; rapid re-key at handoffs.	Use QKD to grow keys, then refresh AES session keys frequently; authenticate control plane with PQC signatures when QKD bootstrap not available; detect eavesdropping via QBER thresholds and re-route links.	Aircraft accept orders only from authorized C2; hijack attempts fail; ISR feeds remain exclusive; resilience to retroactive decryption and traffic analysis.
Land (UGVs/ robotic units)	Route diversion, counterfeit tasking, malicious OTA updates, decoy beacons.	Quantum-authenticated control and update channels; device identity attestation bound to quantum-derived secrets; per-mission compartmented keys.	Bind device identity to quantum-protected keys; stage OTA updates with authenticated channels and rollback protection; compartment keys per route/mission segment.	Logistics and combat robots reject forged routing or updates; missions continue under EW/cyber pressure; minimal blast radius if a node is captured.
Sea (USVs/ UUVs, long-haul patrol)	Waypoint falsification, long-range link interception, mission-plan overwrite.	Space/free-space QKD for BLOS key seeding; quantum-authenticated navigation/tasking; one-time-pad for critical bursts; periodic re-key on schedule etsi+1.	Use satellite/free-space QKD where fibre absent; combine with symmetric crypto for throughput; keep OTP for highest criticality segments due to key rate limits.	Maritime unmanned assets reject forged waypoints; mission plans preserved over extended sorties; assured control against interception.
Swarm operations (UAS/UGV swarms)	False-leader injection, swarm fragmentation, replay/flooding of coordination frames, kets-quantum.	Quantum-authenticated leader-election and coordination; per-node quantum-bound IDs; ultra-frequent key rotation to null replay.	Seed swarm with shared root via QKD, derive per-node sub keys; rotate keys each epoch; tie consensus to authenticated frames only.	Cohesion preserved; hostile nodes cannot split or re-task elements; effects persist under EW/ deception.
Coalition unmanned C2 (cross-cutting)	Interoperability seams, identity fraud across vendors/allies.	Policy-backed quantum key management; QKD for key growth + PQC for broad authentication; accredited link/device assurance.	Start from short shared secret or PQC bootstrap to authenticate QKD; save part of prior session keys to authenticate the next ("key growing"); align on NIST PQC suites for signatures.	Coalition assets accept commands only from authenticated authorities; reduced attack surface across mixed crypto baselines.

- **QKD can be integrated into ship-borne and submarine communication systems to generate encryption keys immune to interception:** Quantum key distribution can be embedded in ship-borne and submarine communications by adding quantum transmit/receive modules to optical segments (shipboard fibre trunks, line-of-sight free-space optical links, and periscope-depth blue-green/FSO windows) and coupling them to a key-management layer that feeds fresh, tamper-evident symmetric keys to existing radios and data links; any eavesdropping attempt introduces detectable errors, forcing automatic discard and re-key so encryption keys remain immune to interception and retroactive decryption. Integration centres on placing quantum endpoints at natural optical chokepoints within a ship's internal fibre backbone, on mast-mounted free-space terminals for ship-to-ship bridges, and in shore fibre rings that seed keys to the fleet before sortie, so keys are generated via quantum states and never exposed in the clear or derivable by computation. Each quantum session continuously tests channel integrity; elevated quantum bit-error rates indicate probing or tapping, causing the system to drop compromised bits and re-negotiate, ensuring only uncompromised keys reach link encryptors. The result is an information-theoretic key supply that remains secure even if adversaries record all traffic for future decryption attempts.

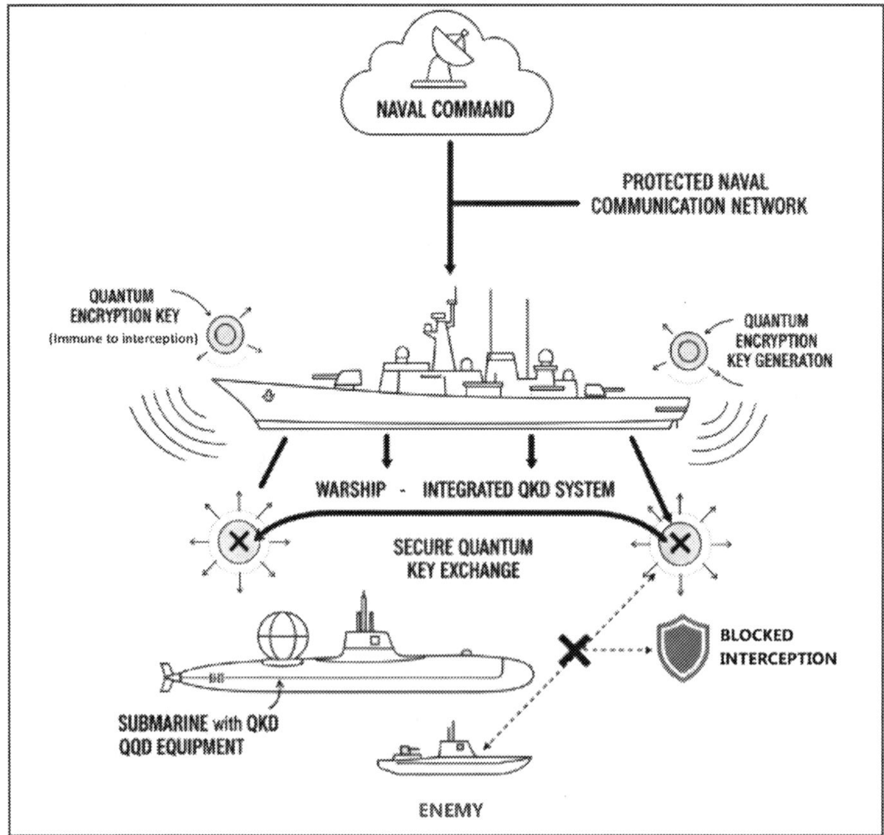

Fig. 4.11: Unbreakable at Sea: Quantum-Secured Naval Communications

Operational relevance

- **Ship-to-ship and ship-to-shore:** Inter-hull free-space quantum links and shore fibre rings furnish continuous re-key for C2 and tactical data nets, defeating link-layer spoofing and preventing later decryption of captured traffic under heavy EW.

- **Submarines:** Pre-deployment key loads via shore fibre and opportunistic periscope-depth optical refresh provide fresh keys without RF emissions, blocking forged patrol updates and waypoint overwrites while minimizing detectability.

- **Carrier air wing and escorts:** Quantum-derived keys feed existing link encryptors so strike tasking, air defence coordination, and combat-system data remain authenticated during SATCOM denial and jamming, preserving tempo.

- **Compartmentation and blast-radius control:** Keys are scoped per warfare area (AAW, ASW, SUW), circuit, and mission window; a compromise on any segment triggers alarmed re-key without exposing the wider task force.

Implementation cues

- **Placement:** Install quantum endpoints on shipboard optical trunks, mast-top FSO gimbals for LOS bridges, and shore fibre nodes; for subs, enable blue-green/FSO windows at PD and use pre-sail vaulting.

- **Key handling:** Use quantum-generated entropy to refresh symmetric session keys at tactical cadence; reserve one-time pad for narrow, highest-criticality circuits given key-rate constraints.

- **Assurance:** Automate drop/re-key on QBER anomalies; bind keys to platform identity and mission time; log key provenance for forensic trust.

- **Inter-op/fallbacks:** Pair quantum key supply with standardized quantum-safe authentication so control planes remain verifiable when optical paths are intermittent; pre-plan denied-satcom key ladders for A2/AD.

Mission effects crosswalk

Mission set	Risk	Quantum control	Outcome
Carrier air wing tasking	False re-task, delayed decrypt	Continuous QKD re-key to link encryptors	Authentic strike/recovery orders under EW pressure
ASW barriers/pickets	Waypoint/plan overwrite	QKD-authenticated tasking and nav-binding	Tamper-proof patrol plans and buoy/USV/UUV control
Surface engagements	Deception on fires/manoeuvre	Per-circuit rapid re-key and compartmentation	Tempo and trust preserved despite jamming
Amphib/logistics groups	Route deception, RF contention	Segment keys per leg; auto re-key on anomalies	Secure replenishment and movements; limited blast radius

- **Facilitates highly secure data and command exchanges between vessels, aircraft carriers, and command centres:** Quantum-secured keying creates a trusted, tamper-evident fabric that lets vessels, carrier air wings, and maritime command centres exchange data and orders with assured authenticity, confidentiality, and continuity even under jamming, interception, and satellite denial. It does so by generating fresh keys that expose eavesdropping attempts and then feeding those keys to existing link encryptors across fibre, free-space optical, and RF transports. Ground and afloat quantum endpoints continuously generate and refresh encryption keys whose integrity is physically verifiable; any interception raises detectable anomalies, prompting automatic discard and re-key. These quantum-derived keys are consumed by standard link encryptors and data links (C2 voice/data, tactical messaging, sensor feeds), preserving high throughput while upgrading trust in the key supply.

Operational relevance

- ❑ **Vessels to carrier:** Strike tasking, air-defence coordination, and deck cycle control remain authenticated during heavy EW; false re-tasking and delayed decryption attacks fail.

- ❑ **Vessels to command centres:** Theatre orders, COP updates, and fire coordination retain confidentiality and provenance across fibre ashore and over-water LOS bridges, sustaining tempo in A2/AD.

❑ **Submarines to surface:** Pre-sail vaulting and opportunistic optical refresh provide fresh keys without RF exposure, enabling mission updates that cannot be forged or replayed.

❑ **Coalition and unmanned integration:** Shared quantum-anchored identity and per mission key scoping prevent spoofing across mixed fleets and ensure USV/UUV/UAS tasking accepts only authorized commands

Below is a mission effects crosswalk table mapping key naval/joint mission sets to primary threats, quantum controls, and tactical outcomes, aligned with QKD-enabled, quantum-safe C2 practices.

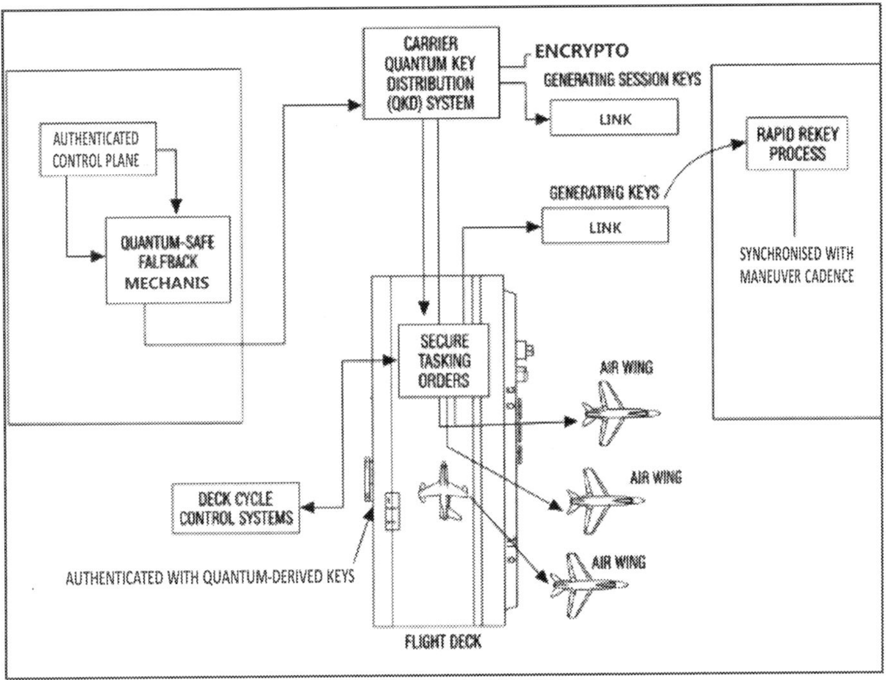

Fig. 4.12: Flight Deck Fortress: Quantum Keys at Maneuver Speed

Mission set	Primary threat	Quantum control	Tactical outcome
Carrier air wing	False re-tasking of sorties; interception and later decryption of strike/CAP tasking; coordination loss under EW and SATCOM denial.	QKD-derived session keys feeding link encryptors for tasking and deck cycle control; authenticated control plane with quantum-safe fallbacks; rapid rekey at manoeuvre cadence.	Authentic strike/recovery orders persist under jamming; hijack/spoof attempts fail; tempo maintained without revealing command patterns.
ASW barrier (pickets/USV/UUV)	Waypoint falsification; mission-plan overwrite; long-haul link interception during patrols.	QKD across shore fibre/FSO relays to seed keys; authenticated navigation-task coupling; compartmented keys per barrier sector.	Patrol plans remain tamper-proof; forged routes rejected; barrier coverage and cueing integrity preserved.
Logistics group (replenishment/ amphib support)	Route deception; counterfeit tasking; malicious OTA updates to unmanned logistics assets.	Quantum-authenticated control and update channels; per-leg key scoping; anomaly-triggered discard/re key.	Secure movements and replenishment despite EW; compromised segments isolated; no force-wide spill over.
Amphibious assault (JLOTS/ ship-to-shore)	De-confliction failures from spoofed orders; exposure during SATCOM degradation; signal-pattern leakage before H-hour.	Shore fibre QKD to marshalling nodes; LOS FSO QKD bridges afloat; tight key lifetimes and privacy amplification pre-assault.	Coherent ship-to-shore tasking under emissions control; surprise preserved; landing timeline integrity maintained.

Alarm and recovery TTPs

Alarm type	Immediate action	Role (primary/ supporting)	Decision time	Acceptable service impact
Eavesdrop detect (QBER above threshold or sudden rise)	Discard current raw/ sifted key, trigger automated rekey; increase privacy amplification; switch to alternate quantum path if available.	COMM-C leads; CIC/OPS informed.	1–5s for automated discard/re-key; 30–60s to stabilize.	Brief key outage acceptable; data plane continues on last good session key; no acceptance of new keys until clean.
Persistent elevated QBER (environ-mental degradation vs. probing)	Rate-limit key generation; lower symbol rate; adjust optics/pointing; if threshold persists, quarantine link and failover.	COMM C leads; EO/FSO tech team supports; OPS adjusts routing.	30–120s to tune; ≤2 min to failover.	Temporary reduc-tion in key rate acceptable; main-tain confidentiality by not relaxing thresholds.

Node compromise suspected (endpoint tamper alarm, access violation)	Quarantine node; revoke credentials; zero-size volatile keys; re-key affected enclaves; switch to pre-planned alternate nodes.	CIC authorizes; COMM C executes; OPS reconfigures circuits.	Immediate quarantine ≤10s; enclave re-key ≤5 min.	Short service interruption acceptable within compartment; no cross-enclave spill-over permitted.
Link degradation (FSO misalignment, weather loss)	Auto-reacquire beam; reduce modulation; shift to redundant path (RF/fibre) while preserving last good keys; schedule quantum refresh window.	COMM C leads; Bridge/Signal teams assist; OPS adjusts EMCON.	Auto re-acquire ≤30s; failover ≤60s.	Throughput reduction acceptable; integrity preserved; no acceptance of degraded keys.
Authentication failure on classical channel (MAC/signature mismatch)	Reject session; rotate to alternate authenticated control path (e.g., PQC signature set); audit logs; alert cyber defence.	COMM-C leads; Cyber cell supports; CIC informed.	Immediate reject; alternate path within 30-60s.	Brief control-plane disruption acceptable; no data-plane cutover until control plane authenticates.
Anomalous key provenance (audit hash mismatch, time/role scope violation)	Invalidate affected keys; roll back to last validated key epoch; initiate forensic capture; compartment re-issue.	COMM-C with Security Officer; CIC oversight.	≤5 min to rollback/re-issue.	Localized outage acceptable; force-wide continuity must not be impacted scholar.
Harvest-now indicators (SIGINT collection surge, targeted recording)	Increase re-key cadence; shorten key lifetimes; elevate privacy amplification; compartment keys per circuit.	COMM-C leads; OPS/CIC adjust comms posture.	Policy change ≤5 min.	No user-visible impact beyond slightly higher latency on keying.
Quantum endpoint health fault (temperature/clock/laser warnings)	Graceful key-rate reduction; maintenance switchover; schedule hot-swap; do not lower security thresholds.	COMM-C coordinates; Engineering maintains.	≤2–10 min, depending on spare availability.	Key rate reduction acceptable; security posture unchanged.
Multi-link anomaly (correlated QBER spikes across paths)	Assume coordinated probing; isolate affected region; increase thresholds; route keys via distant path; notify higher HQ.	CIC coordinates; COMM-C executes; OPS de-conflicts.	≤5 min to isolate/re-route.	Regional service degradation acceptable; no acceptance of suspect keys.

Securing Line-of-Sight and Optical Communication Links

Laser-Secure Links: Quantum keys riding light waves between warships

Line-of-sight quantum key distribution via free-space optics leverages polarized photons transmitted through laser beams between naval platforms to generate encryption keys that are information-theoretically secure. Unlike conventional FSO systems that rely on pre-shared keys or public-key cryptography, quantum-secured optical links embed key generation directly into the photon stream itself. Any attempt to intercept or measure these quantum states introduces detectable perturbations that immediately alert communicating parties to the presence of eavesdropping. The system operates by transmitting quantum-encoded photons across clear optical paths between ships, where successful photon receipt and measurement creates shared secret key material that feeds into existing link encryptors for high-throughput data exchange.

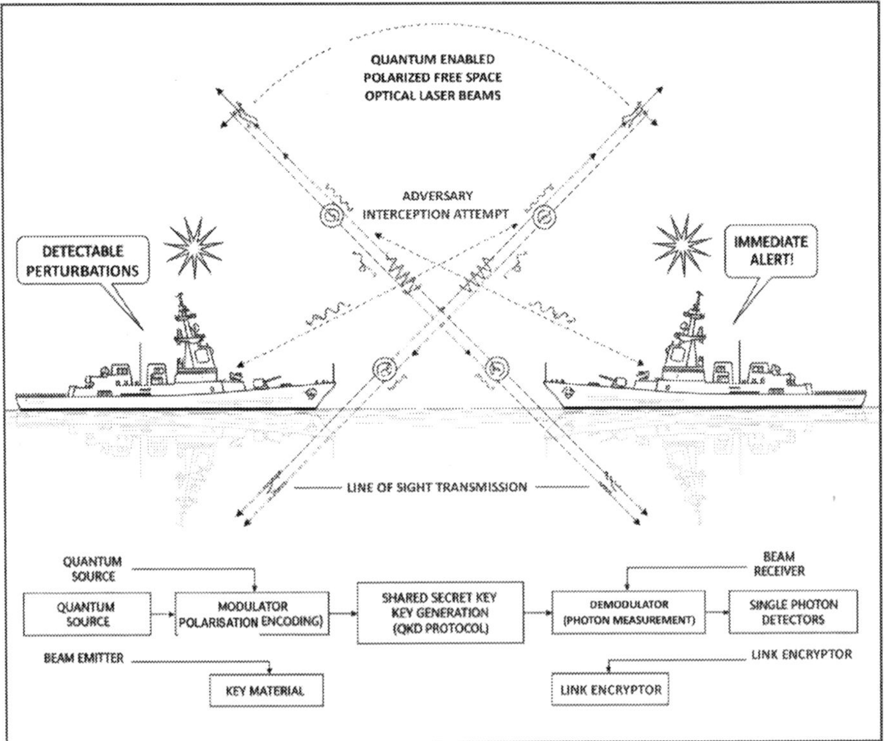

Fig. 4.13: Photonic Shield: Laser Keys That Detect Every Intrusion

This is comparable to semaphore signals between warships, but with an unbreakable twist: each flag position can only be read once before it changes permanently, and any spy attempting to observe the signal from a distance would cause the flags to flutter visibly, immediately revealing their presence to both sender and receiver. The ships continue using their normal communication methods for actual messages, but the "semaphore keys" that unlock those messages are impossible to steal without detection.

Free-space quantum optical links provide critical communication assurance for blue-water operations where satellite vulnerability or denial threatens traditional C2 continuity. Ship-to-ship QKD enables carrier strike groups to maintain authentic tasking and coordination across formation distances of 25-50 kilometres without satellite dependence. For amphibious operations, quantum-secured shore-to-ship optical bridges preserve command integrity during satellite-denied assault phases, ensuring landing coordination remains tamper-proof. The line-of-sight limitation becomes an operational advantage in contested environments, as quantum links cannot be intercepted beyond the optical horizon, creating natural compartmentation that limits adversary access to communication patterns.

Quantum optical links resist jamming attempts since they operate in narrow, directional laser beams rather than omnidirectional RF emissions. Attempts to disrupt the quantum channel through optical interference or spoofing fail because legitimate photon streams carry quantum signatures that cannot be replicated. This ensures continuous key refresh capability even under intense electronic warfare pressure, maintaining cryptographic strength when RF-based key management systems are compromised.

The combination of satellite-independence, tamper-evident key generation, and natural horizon-based compartmentation makes quantum-secured optical links essential infrastructure for maintaining C2 assurance in contested maritime environments where space-based assets cannot be relied upon.

FSO QKD vs. Satellite QKD Comparison for Blue-Water Operations

Aspect	*FSO QKD (Ship-to-Ship/Ship-to-Shore)*	*Satellite QKD (Ground/Ship-to-Space)*
Range/Coverage	Limited to line-of-sight: 25-50 km between ships; extended via relay chains through picket vessels.	Global coverage: LEO satellites provide intermittent coverage during overpasses; GEO satellites offer continuous but lower key rates.
Key Generation Rate	High rates possible due to short distances and minimal atmospheric interference; typically kbit/s to Mbit/s range.	LEO: kbit/s during overpass windows (few times daily); GEO: lower but constant rates due to 36,000 km distance and high channel losses (40-60 dB).
Availability	Continuous when ships are within LOS; weather-dependent (fog, heavy seas affect beam alignment).	LEO: Intermittent (flyover windows); GEO: Continuous but subject to atmospheric turbulence and pointing errors.
Vulnerability	Natural horizon-based compartmentation limits adversary access; difficult to intercept beyond optical horizon.	Satellites vulnerable to kinetic attack, jamming, dazzling; space-based assets are high-value targets in A2/AD scenarios by nature.
Implementation Complexity	Moderate: mast-mounted FSO terminals with auto-tracking; existing technology adaptable to naval platforms.	High: requires space segment, ground stations, precise pointing systems; significant infrastructure investment.
Weather Resilience	Affected by fog, precipitation, sea state (platform stability); backup RF links needed.	Atmospheric turbulence, scattering in last 20 km affects signal; beam broadening over long distances by nature.
Operational Security	No RF emissions; quantum-secured beam cannot be jammed conventionally; compartmented by nature.	Space-to-ground links create detectable signatures; satellites trackable; ground stations are fixed infrastructure nature.
Deployment Speed	Rapid: ships carry terminals; immediate availability upon formation.	Slow: requires satellite constellation deployment; ground station establishment.
Cost	Lower: shipboard terminals and relay nodes; leverages existing naval platforms.	Higher: satellite launches, ground infrastructure, space-qualified hardware.
Tactical Flexibility	High: formations can manoeuvre while maintaining quantum links; adaptive relay routing.	Limited: depends on satellite orbital mechanics; cannot be re-positioned tactically.

- **Quantum key distribution via free-space optics (FSO):** Naval FSO-QKD systems integrate multiple components: mast-mounted optical terminals with precision pointing and tracking systems to maintain beam alignment during platform motion; adaptive protocol switching between CV-QKD and DV-QKD based on distance and atmospheric conditions; and hybrid classical-quantum channels where quantum photons establish keys while classical laser pulses carry encrypted operational data. The system can form quantum-secured mobile ad-hoc networks (MANET) linking multiple naval platforms through relay chains, extending secure coverage beyond direct line-of-sight.

Operational advantages over satellites

❏ Immediate availability: No dependency on orbital mechanics or space asset availability; formations maintain quantum-secured communications regardless of satellite status.

❏ Natural EMCON compliance: Narrow laser beams produce no detectable RF emissions, preserving operational security during silent running.

❏ Compartmented security: Optical horizon naturally limits adversary access; intercept attempts require physical positioning within the beam path.

❏ High key rates: Short atmospheric path enables Mbps-class key generation compared to space-based systems limited by 36,000 km propagation losses.

❏ Weather adaptability: Multiple-input multiple-output (MIMO) FSO configurations with beam diversity overcome atmospheric turbulence effects.

FSO-QKD supports diverse naval mission sets through flexible deployment: carrier strike groups maintain air wing tasking security across formation distances; submarine operations use periscope-depth optical windows for secure mission updates without RF signature; amphibious forces establish quantum-secured ship-to-shore bridges during satellite-denied assault phases; and unmanned maritime systems receive authenticated control commands immune to spoofing.

The system incorporates multiple security layers: chaotic communication techniques can be layered with QKD for additional obfuscation of classical data channels; adaptive protocol selection maintains optimal key rates under varying atmospheric conditions; and MIMO configurations with spatial diversity overcome beam blockage from weather or platform orientation. Any detection of elevated QBER immediately triggers key discard and beam re-establishment, ensuring compromised keys never enter the encryption pipeline.

Enhancing Resilience against Electronic Signal Interception and Jamming

Silent Signals: Quantum communications that can't be jammed or decoded

QKD fundamentally transforms the security paradigm by leveraging the inviolable laws of quantum mechanics rather than computational complexity. When quantum states are transmitted between naval platforms, any attempt by an adversary to measure or intercept these states necessarily disturbs them creating detectable changes in the quantum bit error rate (QBER) that immediately alert both communicating parties to the presence of eavesdropping. This tamper-evident property is not dependent on the adversary's computational capabilities or technological sophistication; it is guaranteed by the fundamental nature of quantum measurement, making the system immune to both current and future cryptanalytic attacks. Unlike conventional encryption that can be intercepted and stored for later decryption when computational power advances, quantum keys that are compromised during transmission are automatically discarded ensuring no useful intelligence reaches hostile forces. This can be compared to a sealed message bottle that self-destructs if anyone other than the intended recipient touches it. Traditional encrypted communications like the strong protection of armoured couriers, can eventually be extracted if overwhelmed by a superior force or technology. Quantum communications are like messages written in disappearing ink that vanishes the moment an unauthorized person attempts to observe it, while simultaneously alerting both sender and receiver that interception was attempted. The bottle (quantum channel) cannot be opened without destroying its contents, making espionage not just difficult, but physically impossible.

Operational Relevance

In heavy electronic warfare environments where adversaries deploy sophisticated jamming, spoofing, and interception capabilities, quantum-secured communications provide naval forces with an unassailable communication backbone. Carrier strike groups can maintain authentic air wing coordination even when conventional SATCOM and UHF links are compromised, as quantum keys ensure that any false tasking or spoofed orders are immediately identified. Submarine operations benefit particularly from quantum security, as the system enables secure communication without electromagnetic emissions that could reveal position; periscope-depth optical quantum links allow mission updates and coordination while maintaining stealth. For amphibious operations in contested littoral environments, quantum-secured ship-to-shore communications preserve command integrity during the vulnerable approach and assault phases, when traditional RF communications face intense jamming and deception attempts.

The intrinsic detection capability of quantum systems also provides valuable intelligence regarding adversary intentions and capabilities. When quantum channels detect interception attempts, commanders gain immediate awareness of hostile surveillance efforts, enabling tactical adjustments and countermeasures. This transforms communications security from a passive defence into an active intelligence-gathering capability that reveals enemy electronic warfare activities in real-time.

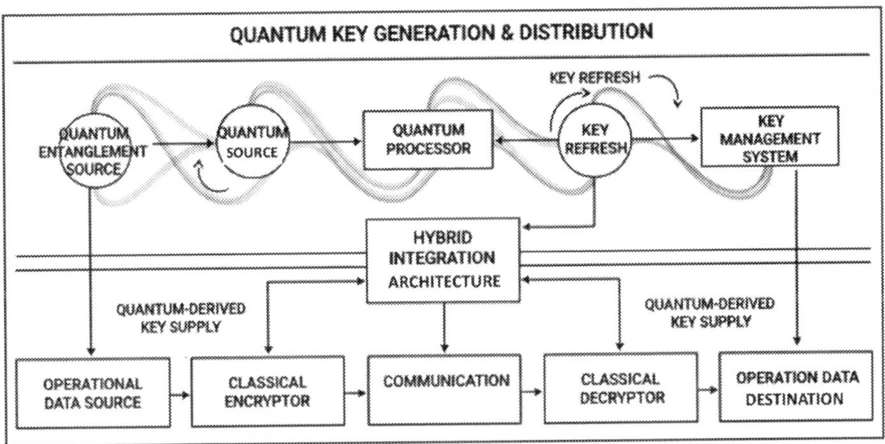

Fig. 4.14: Dual Channel Advantage: Quantum Security Meets Classical Speed

- **Implementation for naval forces**: QKD integrates with existing communication architectures through hybrid classical-quantum systems, where quantum channels generate and refresh encryption keys while conventional high-throughput links carry operational data encrypted with these quantum-derived keys. Mast-mounted free-space optical terminals enable ship-to-ship quantum links across formation distances, while fibre-optic connections at port facilities seed key material for extended operations. The system's resilience stems not from signal strength or frequency hopping, but from the fundamental impossibility of covert interception, ensuring that naval communications remain secure regardless of adversary electronic warfare capabilities.

This quantum-enhanced communication resilience enables naval forces to maintain operational tempo and coordination even in the most contested electromagnetic environments, where conventional communication security measures would fail under determined adversary pressure.

EW resilience comparison

Communication type	Vulnerability to jamming	Vulnerability to interception	Detection of compromise	Key recovery after attack
Conventional RF	High: RF jammers disrupt signals across frequencies.	High: signals can be recorded and analysed.	Limited: may go undetected for extended periods.	Manual re-key required; compromised data unrecoverable.
Frequency-hopping	Moderate: sophisticated jammers can track patterns.	Moderate: patterns eventually discoverable.	Moderate: unusual interference may indicate attack.	Re-key required; historical traffic at risk.
QKD-secured optical	Very Low: narrow laser beams resist conventional jamming.	Impossible without detection: quantum no-cloning theorem.	Immediate: QBER spikes reveal eavesdropping.	Automatic: compromised keys discarded instantly.

Mission-specific implementation

Mission set	QKD deployment method	Key tactical advantage	EW threat mitigation
Carrier air wing operations	Mast-mounted FSO terminals between ships in formation.	Authentic strike tasking immune to spoofing.	False re-task attempts fail; command integrity preserved.
Submarine patrol/ transit	Periscope-depth blue-green optical links.	Secure updates without RF emissions.	Position remains concealed; mission plans tamper-proof.
Amphibious assault	Ship-to-shore quantum bridges during approach.	Command integrity during contested landing.	Deception operations neutralized; timing preserved.
Surface action groups	Inter-hull FSO relay chains extending coverage.	Formation manoeuvres with continuous secure C2.	Jamming ineffective; coordination maintains tempo.

- **QKD links are inherently tamper-evident:** QKD links achieve tamper-evidence through fundamental quantum mechanical principles, specifically Heisenberg's uncertainty principle and the no-cloning theorem which make any measurement or interception attempt physically alter the quantum states being transmitted, creating detectable errors that immediately reveal eavesdropping to both communicating parties. The tamper-evident nature of QKD stems from two inviolable quantum mechanical principles that make covert interception physically impossible:

  - ❑ **Measurement Disturbance Principle:** When quantum states (typically photon polarizations) are measured, the act of measurement necessarily changes the system being observed. An eavesdropper attempting to intercept quantum key material must measure the photons to gain information, but this measurement inevitably alters the quantum states in detectable ways. Unlike classical signals that can be copied without affecting the original, quantum information cannot be perfectly duplicated due to the no-cloning theorem.

  - ❑ **Error Pattern Detection:** In the widely-used BB84 protocol, legitimate communication between Alice and Bob produces a baseline quantum bit error rate (QBER) of approximately 0-5% due to natural environmental factors. When an eavesdropper (conventionally called "Eve") attempts an intercept-and-resend attack, she must guess the measurement basis for each photon. With a 50% probability of choosing incorrectly and a 50% chance of Bob detecting the error when she guesses wrong, each intercepted photon has a 25% probability of creating a detectable error. This means even modest eavesdropping attempts create error rates of 12.5% or higher, far exceeding natural thresholds.

Detection Mechanisms Table

Detection method	Technical approach	Threshold parameters	Detection confidence	Response time	Military application
Statistical Analysis	Compare subset of received bits over public channel to calculate QBER.	BB84: >11% error rate indicates eavesdropping; Clean channel: 0-5% natural errors.	99.9999% confidence with 72 bits comparison; Detection probability approaches certainty.	Periodic (minutes); Post-session analysis.	Post-mission forensics; Pattern analysis of adversary EW capabilities.
Real-Time Monitoring	Continuous quantum channel quality assessment with automated key discard.	QBER thresholds: 11% (BB84), 15% (SARG04); Configurable per environment.	Immediate detection during active eavesdropping.	1-5 seconds automated response.	Tactical operations; live threat detection during mission.

QBER Threshold Analysis

Protocol	Clean channel QBER	Eavesdropping threshold	Detection accuracy	Bits required for 99.9% confidence
BB84	0-5% (environmental)	>11% indicates compromise	>99% with 100+ sample bits	72 bits minimum
SARG04	0-5% (environmental)	>15% indicates compromise	>99% with 200+ sample bits	96 bits minimum
E91/BBM92	5-6% (typical experimental)	>11% indicates compromise	Bell inequality violations provide additional detection	64 bits minimum

Attack Detection Performance

Attack type	Error rate induced	Detection probability	Sample size needed	Tactical implications
Full intercept-re-send	25% QBER (theoretical)	>99.9% detection	50-100 bits	Immediate compromise detection; Full adversary capability revealed
Partial intercept (50% packets)	12.5% QBER average	>95% detection	100-200 bits	Sophisticated adversary; Requires larger sample for confidence
Beam-splitting attack	Variable (2-10%)	80-95% detection	200-500 bits	Advanced threat; May approach detection threshold
Trojan horse attack	No QBER increase	0% via QBER	Requires side-channel detection	Sophisticated implementation attacks on detectors

Operational Response Matrix

QBER level	Threat assessment	Automated action	Manual intervention	Command notification
0-5%	Clean channel	Continue operations	Monitor trending	Routine status
6-10%	Environmental/equipment	Increase monitoring; Reduce key rate	Check alignment/ weather	Operations notice
11-20%	Probable eavesdropping	Discard keys; Switch backup path	Investigate source; Counter-EW measures	Immediate alert to C2
>20%	Active compromise	Halt transmission; Quarantine node	Security response; Forensic capture	Priority flash to higher HQ

- **Naval assets maintain encrypted communications even in heavy EW environments:** QKD enables naval assets to maintain encrypted communications in heavy EW environments by providing tamper-evident key generation that operates independently of RF systems vulnerable to jamming, spoofing, and interception, while delivering information-theoretic security that remains unbreakable even when adversaries employ sophisticated quantum computers and advanced cryptanalytic capabilities.

  - ❏ **Immunity to Traditional EW Attacks:** QKD-secured naval communications resist conventional electronic warfare through several fundamental advantages. Unlike RF-based systems that can be jammed across frequency spectrums, quantum optical links operate via narrow laser beams that are inherently resistant to broadband jamming. The directional nature of optical quantum channels makes them difficult to locate and target, as they do not emit omnidirectional signals that can be detected by electronic support measures (ESM) equipment. When adversaries attempt to disrupt quantum links through optical interference, the quantum states themselves detect this intrusion and trigger automatic countermeasures, transforming attempted attacks into intelligence opportunities about enemy capabilities.

  - ❏ **Operational Advantages in Contested Environments:** Naval operations in A2/AD scenarios benefit from quantum communications that maintain security even when conventional cryptographic systems are compromised. Carrier strike groups can coordinate air wing operations through quantum-secured tactical data links that resist spoofing attempts, ensuring authentic strike tasking and air defence coordination when RF communications are degraded. Submarine operations particularly benefit from quantum security, as boats can receive mission updates via blue-green optical links without surfacing or revealing their position through RF emissions, maintaining tactical advantage while staying connected to strategic command.

❏ **Multi-Platform Integration:** Quantum-secured communications enable seamless integration across diverse naval platforms under EW pressure. Surface action groups maintain formation coordination through ship-to-ship quantum links that operate at megabit data rates, supporting video teleconferencing and real-time tactical picture sharing even when SATCOM is denied. Amphibious forces use quantum-secured ship-to-shore bridges during contested landings, ensuring command integrity when conventional communications face intense jamming during the vulnerable approach phase. Coalition operations benefit from quantum key distribution that provides shared cryptographic foundation immune to exploitation, enabling multinational forces to maintain secure coordination despite different national cryptographic systems.

EW threat	Conventional vulnerability	QKD protection mechanism	Operational benefit
Broadband jamming	RF systems overwhelmed across spectrum	Narrow optical beams immune to RF interference; directional transmission resists jamming	Continuous secure communications maintained; enemy jamming ineffective
Spoofing/false injection	False commands accepted by legitimate receivers	Tamper-evident quantum states expose any interception; authentication coupled with QKD	False tasking attempts fail; command authenticity preserved under deception
SIGINT collection	Communications intercepted for later decryption	Quantum no-cloning theorem prevents copying; measurement disturbs states	Historical traffic remains secure; retroactive decryption impossible
ELINT/ESM targeting	RF emissions reveal platform locations	Optical quantum links produce no RF signature; natural EMCON compliance	Submarines maintain stealth; surface ships reduce electromagnetic signature
Man-in-the-middle attacks	Adversary inserts false relay nodes	Quantum channel integrity verification; authenticated classical channel required	Network topology remains trusted; inserted nodes detected immediately

Performance under EW Stress

Communication scenario	Data rate achieved	Detection distance	EW resilience factor
Ship-to-ship (25 km)	170 MB/s optical data; kB/s quantum keys	Optical horizon limited; no RF detection	High: immune to conventional jamming
Submarine-to-satellite	170 MB/s burst capability at periscope depth	No RF signature below surface	Very High: maintains stealth profile
Shore-to-fleet	Fibre QKD rates (GB/s) at port; FSO relay at sea	Fibre taps detectable; FSO horizon-limited	High: compartmented security
Coalition integration	Limited by lowest common cryptographic denominator	Shared quantum foundation enables high-grade inter-op	Moderate: requires policy alignment

Integration into Submarine Communication Systems

Deep Encryption: Taking quantum security below the surface

Special configurations extend QKD to submerged platforms by coupling blue-green laser apertures at periscope depth with deployable quantum buoys and tethered relays that surface briefly to close quantum links with ships, aircraft, or satellites; then pass fresh keys down fibre or optical tethers to the boat. This architecture decouples key exchange from vulnerable RF paths, leveraging quantum states whose measurement reveals intrusion; compromised bits are discarded and sessions re-keyed automatically, preserving confidentiality even against future decryption attempts. For trans-theatre reach, shore fibre QKD rings seed key vaults pre-sail, with opportunistic optical refresh during patrols to sustain high-assurance keys under emissions control.

Think of a submerged SSBN as a bank vault connected to a secure night-drop chute: the courier (quantum buoy or tethered relay) accepts deposits that are sealed with tamper-evident ink; if anyone touches the pouch en route, the seal shows, and the contents are invalidated and replaced before the vault accepts them. The vault never needs to open its doors to the street (no RF exposure), yet it continues to receive fresh, verified combinations that cannot be copied or replayed by an adversary without leaving evidence.

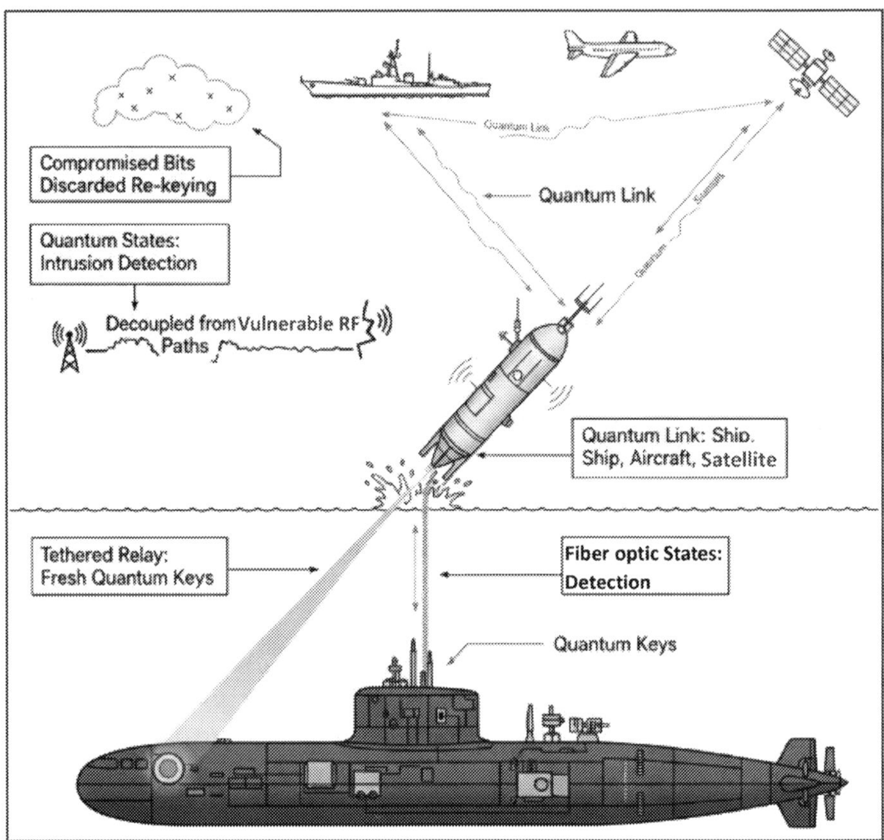

Fig. 4.15: Deep Quantum: Buoy-Relayed Keys for Submarine Stealth

Operational relevance

- **Nuclear command assurance:** Submerged ballistic missile submarines receive authenticated updates without surfacing or emitting RF, preserving stealth while ensuring launch authority and mission data remain immune to interception or harvest-now/decrypt-later threats in the quantum era.

- **Stealth under EW:** Blue-green optical QKD and tethered relays maintain key flow during heavy jamming and SIGINT collection; the quantum channel's tamper-evidence provides instant warning of adversary proximity or probing, informing manoeuvre and EMCON posture.

- **Tempo and reach:** Pre-sail fibre QKD vaulting combined with at-sea optical refresh supports higher-bandwidth secure exchanges than ELF/VLF, enabling timely intelligence and re-tasking while preserving emissions discipline across contested waters.

- **Nuclear command assurance:** Nuclear command assurance for submerged ballistic missile submarines is strengthened by integrating QKD into blue-green optical links, deployable buoys, and tethered relays that deliver tamper-evident, authenticated keys without RF emissions, ensuring launch authority and mission data remain immune to interception and harvest-now/decrypt-later threats in the quantum era.

  - **Optical quantum layer:** Blue-green wavelengths near 450–550 nm exploit favourable seawater transmission to exchange quantum states between a surfaced relay (expendable buoy, tethered mast, or periscope depth window) and airborne/surface/satellite endpoints, enabling key establishment while the submarine stays at tactical depth or exposes only minimal aperture area for brief windows.

  - **Relay modalities:** Expendable or recoverable quantum buoys close a satellite/air/ship quantum link topside, then pass sifted keys down a fibre tether; alternatively, a periscope integrated optical head uses short, opportunistic exposures to refresh keys directly, reducing temporal and spatial signature.

  - **Shore seeding and vaulting:** Prior to patrol, shore fibre QKD rings load per mission, per role "key vaults" with provenance metadata, enabling sustained operations when optical refresh windows are sparse and providing auditable continuity of command authority keys.

  - **Hybrid control plane:** QKD provides information-theoretic confidentiality for key growth, while quantum safe authentication on the classical channel prevents man in the middle, ensuring that both endpoints are legitimate even under EW conditions and adversary deception.

  - **Tamper evidence and alarms:** Elevated quantum bit error rate (QBER) during sifting signals probing or channel disturbance, triggering discard and re-key; this converts adversary interception attempts into immediate alerts without revealing sensitive payload content.

❑ **Compartmentation:** Keys are scoped to mission phase (pre deployment, deterrent patrol, crisis, EAM receipt), role (command, navigation, fire control), and time windows, limiting blast radius if a relay or node is compromised, and enabling selective zeroization and rapid re-issuance

Nuclear command assurance via QKD for SSBNs

Dimension	Technical detail	Assurance effect
Key transport path	Blue-green optical via buoy/tether/periscope to air/ship/satellite, with shore fibre QKD pre load	RF-silent key refresh; minimized detectability and signature exposure
Authentication	Quantum-safe signatures/MAC on classical channel binding identity, role, time	Prevents man-in-the-middle; assures legitimate endpoints before key use
Tamper evidence	QBER monitoring during sifting; automatic discard/re-key on threshold breach	Turns interception into alarms; no silent key compromise
Compart-mentation	Per-mission/role/time scoping; vaulting with auditable provenance	Limits blast radius; supports selective zeroization and re-issue
EW resilience	Directional optical links; immunity to RF jamming; brief exposures only	Maintains stealth and continuity under heavy EW pressure
Quantum-era security	Information-theoretic key growth; cipher text safe against future quantum decryption	Neutralizes harvest-now/decrypt-later risk to EAM and targeting data
Operations cadence	Pre-sail fibre seeding; opportunistic at-sea refresh; EAM key rotation	Sustains assured command without prolonged surfacing or emissions

• **Stealth under EW:** Blue-green (roughly 450–550 nm; often implemented around 480–532 nm) optical channels exploit the seawater transmission window to move quantum states between a surfaced relay (expendable buoy, tethered mast, or periscope head) and an airborne/surface endpoint while the submarine remains at depth or briefly presents a minimal aperture for key refresh. The quantum layer generates tamper-evident keys: measurement or optical interference changes the statistics of received states, raising QBER beyond thresholds and forcing automatic key discard and re-key, so compromised keys never enter the data plane. Directional beams and short exposure windows minimize intercept probability and emissions, while relay chains extend reach without resorting to wide-area RF. It is akin to passing a sealed, dye-trapped note through a narrow mail slot in a blackout room: the courier cannot be seen from the street, the slot is open only for seconds, and any attempt to pick or probe the seal stains the intruder's hands and voids the note. The household not only refuses the tampered message but also knows immediately that someone is on the doorstep allowing lights out, route changes, and guard shifts without revealing who is inside.

Operational relevance

❑ **Stealth under EW:** Because the quantum exchange rides in a tight optical beam and not in RF, broadband jamming is ineffective. SIGINT collection has nothing to record at range, and the platform's electromagnetic signature remains negligible. Attempts at optical dazzling or spoofing manifest as QBER anomalies, converting hostile proximity into actionable alerts that cue course changes, mast discipline, or timing shifts.

❑ **Assured command continuity:** Key vaults seeded pre-sail are opportunistically topped up via buoy/tether windows; session keys for EAM and targeting updates are rotated at tactical cadence without resorting to VLF/HF broadcasts, preserving both latency and concealment.

❑ **Compartmented risk:** Keys are scoped by time, role, and channel; if a relay is compromised or a link misbehaves, isolation and re-key are immediate and local, preventing cascade effects across mission systems and preserving force-wide trust.

§ **Tempo and reach:** Tempo and reach are enhanced by seeding submarines with large, role-scoped key vaults over shore fibre QKD pre-sail and then topping them up at sea via brief, directional blue-green optical refresh windows, which together support far higher encrypted data throughput than ELF/VLF while preserving strict emission discipline and stealth in contested waters.

❑ **Shore seeding and vaulting:** Before patrol, the boat receives per mission, per role key vaults over terrestrial fibre QKD, with provenance hashes, validity windows, and rollback checkpoints; this eliminates dependence on slow ELF/VLF during extended denied periods and ensures auditability of nuclear command paths.

Stealth under EW with blue-green optical QKD

Dimension	Technical detail	EW/stealth effect	Commander action
Optical band	Blue-green window (H+480–532 nm), wavelength tuned to local extinction	Minimizes attenuation; narrows intercept geometry	Approve wavelength plan per patrol area and season
Encoding/protocol	BB84/E91 with decoy states; adaptive phase/time-bin under turbulence	Low QBER in motion; resilience to polarization drift	Set protocol switching rules by sea-state/QBER trend
Pointing/tracking	Gimballed, low-divergence beam; auto-track with fine stabilization	Short exposures; low probability of intercept	Limit exposure to seconds; tie to mast discipline SOP
Filtering/gating	Narrowband optical filters; nanosecond-scale detector gates	Suppresses background; early anomaly detection	Enforce gating; monitor background baselines
Relay mode	Buoy/tether/periscope head; unmanned pickets for range	RF-silent key flow; extended reach in clutter	Task relay assets; pre-plan alt paths and timings
Alarm thresholds	QBER and photon-rate limits; provenance checks	Converts probing into alarms; no silent compromise	Auto-discard/re-key; quarantine; log and notify
Authentication	Quantum-safe signatures on classical control	Blocks MITM under deception/jam	Require auth pass before accepting new keys
Compartmentation	Keys scoped by role/time/channel; vaulting with rollback	Limits blast radius; rapid local recovery	Authorize rollback and selective zero-ize on breach
EMCON integration	Windowed exposures; terrain/weather masking	Preserves stealth while refreshing keys	Approve exposure windows; adjust mast policy
Manoeuvre cues	QBER/scintillation anomaly as proximity indicator	Inform route/tempo changes	Execute pre-briefed manoeuvre and relay shifts

❑ **At sea optical refresh:** During opportunistic windows, a tethered relay, expendable buoy, or periscope integrated head establishes a blue green optical QKD hop to a ship/airborne/satellite endpoint; fresh keys are sifted with QBER gating and fed into onboard KMS for rapid session rotation of C2, EAM, and targeting circuits without RF emissions.

❑ **Throughput decoupling:** QKD delivers keys, not payload; payload remains on high-rate channels (e.g., burst optical, low probability of intercept RF when authorized), so symmetric session keys can be refreshed at a cadence measured in seconds while encrypted data moves at Mbps–Gbps when the transport allows, compared to bits per minute ELF or sub kbps VLF.

❑ **Emission discipline:** Optical exchanges are narrow beam, seconds long, and horizon limited, minimizing intercept probability; no continuous RF beaconing is required for key management, preserving stealth and avoiding EW targeting.

Tempo and reach via pre-sail QKD vaulting + at-sea optical refresh

Dimension	Technical mechanism	Effect on tempo	Effect on reach	EMCON/ stealth impact	Commander action
Pre-sail key load	Fibre QKD seeding of role/time-scoped vaults with provenance and rollback	Immediate readiness; no wait on at-sea windows	Strategic duration sustained without external links	Zero RF/optical exposure in port	Approve vault scopes and retention policy
Optical refresh path	Blue-green QKD via tethered relay, buoy, or periscope head	Seconds scale re-key supports rapid re-tasking	Extends viable operations between shore intervals	Brief, directional exposure; horizon limited intercept	Authorize exposure windows and relay tasking
Session key cadence	QKD-fed symmetric key rotation tied to EAM/C2 SOP	Maintains low crypto latency for orders/intel	Keeps C2 responsive across oceanic distances	No continuous RF keying; minimal signature	Set rotation targets and thresholds
Payload transport	Encrypted data on burst optical or authorized LPI/LPD RF	Mbps–Gbps data feasible when transport allows	Beyond ELF/ VLF constraints for mission data	RF used only when cleared; optical preferred	Select transport per threat/weather
Anomaly handling	QBER/provenance alarms trigger discard, rollback, reissue	Avoids tempo loss from silent compromise	Ensures continuity with last good epoch	No extra emissions; internal recovery	Execute rollback playbook; log and proceed
Compartmentation	Keys partitioned by role/channel/time	Localized impact of incidents	Reissue limited scopes; no force-wide reset	Preserves stealth by avoiding broad re-key	Approve selective zero-ize/reissue
Authentication	Quantum-safe signatures on classical control	Fast, assured key acceptance	Trusted refresh from any approved node	Prevents MITM without RF chatter	Enforce auth pass/fail gating
Relay architecture	Unmanned pickets extend optical arcs	Flexible rendezvous for refresh	Wider oceanic coverage without SAT reliance	Low signature footprint for relays	Task pickets; define alternates
Planning integration	Pre-planned refresh arcs and alternates	Predictable, brief re-key events	Maintains reach when weather/ EW shifts	Minimizes exposure time and angles	Align refresh with manoeuvre/EMCON

❑ **Resilience and continuity:** If refresh windows are missed due to sea state or threat posture, the pre seeded vault's time scoped ladders sustain operations; upon QBER anomalies or provenance mismatches, keys are auto discarded, the event is logged, and the boat rolls back to the last validated epoch and attempts a new refresh on an alternate relay vector.

❑ **Compartmentation and blast radius control:** Keys are partitioned by role (command, navigation, fire control), channel, and time; compromise of any relay or enclave triggers localized quarantine and re-issue without affecting other mission functions.

❑ **Hybrid authentication:** The classical control plane employs quantum safe signatures/MAC to authenticate endpoints before accepting new quantum derived keys, closing man in the middle avenues under deception and jamming.

❑ **Planning and manoeuvre:** Patrol plans include pre approved refresh arcs, mast exposure ceilings, and alternate relay tasking (e.g., unmanned pickets) so the CO can time refresh with environmental cover and threat lay-downs while keeping latency low for re-tasking

Real-Time Tactical Key Exchange Between Maritime Battle Groups

Fleet Sync: Quantum trust across the battle group in motion

Real-time tactical key exchange between maritime battle groups uses quantum key distribution to refresh symmetric session keys continuously across moving formations, preserving authenticated, confidential coordination during high-mobility manoeuvres and sudden operational pivots while denying adversaries any covert path to harvest or replay keys. Real-time tactical key exchange equips each battle group element carriers, escorts, logistics vessels, unmanned pickets with quantum endpoints that generate fresh keys on the move via short, directional optical hops within the formation and beyond-line-of-sight relays to command nodes, then feed those keys into existing link encryptors at a cadence matched to manoeuvre tempo. The quantum layer provides tamper-evidence: any interception attempt perturbs the quantum states, driving the quantum bit error rate above thresholds, which triggers automatic discard and re-key without exposing the data plane. In practice, keys rotate on the order of seconds to minutes, preventing accumulation of risk during dynamic re-tasking, air defence re-allocations, or composite warfare role hand-overs. Think of the battle group as a convoy whose vehicles swap one-time road ciphers every few intersections using a silent hand signal that cannot be copied: a pickpocket watching from the pavement is forced to step into the lane to see the gesture, which instantly changes the sign and alerts the drivers. The convoy stays synchronized even as vehicles change lanes or routes, and any suspicious approach prompts an immediate cipher refresh without stopping traffic.

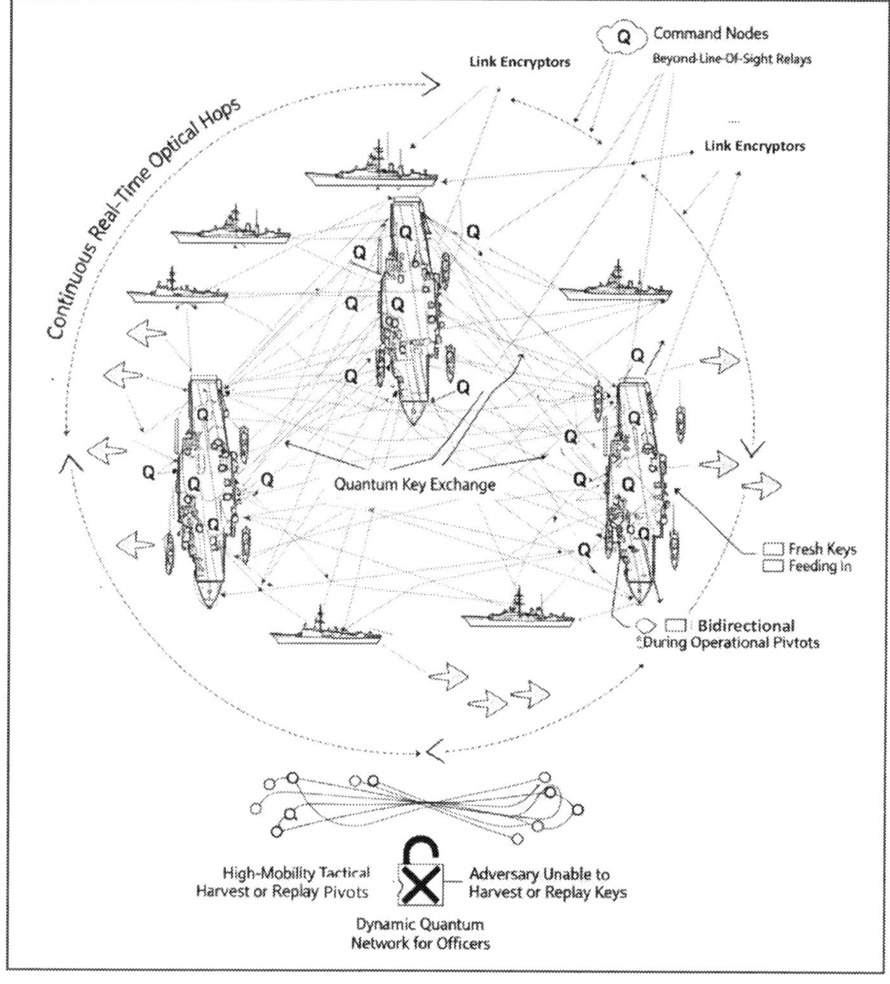

Fig. 4.16: Formation in Motion: Real-Time Quantum Keys at Battle Tempo

Operational relevance

- **Dynamic dispersion:** As the group widens or contracts for threat sectors and sea state, inter-ship quantum refresh maintains a single trust fabric, so composite warfare commanders can rapidly re-assign sensors and shooters without crypto lag or re-key pauses.

- **Pivot speed under EW pressure:** High-frequency re-keying mitigates key exposure during jamming and deception; even if a leg is stressed, keys never age long enough to be useful, and alarms translate probing into actionable cues for formation geometry and emission control.

- **Coalition agility:** Per-link, per-role scoping keeps cross-national enclaves interoperable without forcing force-wide re-key; new entrants to the group establish authenticated quantum roots quickly, while departing units revoke cleanly.

Real-time tactical key exchange across a battle group

Dimension	How it works	Commander's benefit	Risk	Mitigation
Key refresh cadence	Seconds–minutes via intra-group optical QKD; faster during pivots	Orders and sensor/weapon tasking stay ahead of EW/deception	Missed window due to sea state	Bank keys in good conditions; alternate relays
Trust fabric scope	Per-role, per-link, per-time scoping with hierarchical derivation	Rapid re-role without force-wide re-key	n-squared key growth at scale	Hierarchical keys; local sub-key derivation
Tamper evidence	QBER thresholds auto-trigger discard/re-key	Probing becomes an alarm; no silent compromise	False positives near thresholds	Environment-tuned thresholds; provenance checks
Mobility support	Auto-tracked, narrow-beam optical links via gimballed terminals	Maintains coherence while manoeuvring	Beam loss in turns	Predictive pointing; picket relays on arcs
Coalition inter-op	Quantum roots per nation; quantum-safe auth on control plane	Plug-and-fight partners without key seams	MITM on classical control	PQC signatures; authenticated channels
Unmanned relays	USVs/UAVs extend optical arcs, cache keys	Coverage continuity in complex geometries	Relay capture or loss	Compartmented keys; rapid revocation
EMCON compliance	Directional optics; minimal exposure windows	Low signature; reduced ESM targeting	Optical detection attempt	Short windows; deception beams
Harvest-now defence	Keys never in computable form; frequent rotation	No future decryption of recorded traffic	Complacency on cadence	SOP-driven cadence tied to threat state

- **Swarm and unmanned control:** Picket USVs/UAVs serve as optical relays that extend quantum coverage and provide manoeuvrable apertures; their own control channels accept orders only if authenticated with freshly derived keys, preventing leadership spoofing in the presence of EW.

- **Harvest-now mitigation:** Because keys are born from quantum exchanges and never transmitted in computable form, recorded cipher text has no future exploitation value, preserving operational secrecy against quantum-era decryption.

- **Dynamic key refresh via QKD**

  Dynamic key refresh via QKD lets dispersed maritime battle groups keep a single, trusted cryptographic fabric in motion by rotating symmetric session keys at manoeuvre cadence, detecting any interception through quantum-induced error spikes, and re-keying automatically without pausing tactical coordination. This turns key management from a slow, centralized activity into a continuous, tamper-evident background function that scales with formation changes, preserves coalition interoperability, and denies adversaries any harvest-now/decrypt-later advantage.

  Each platform (carrier, escorts, auxiliaries, picket USVs/UAVs) hosts quantum endpoints that generate fresh key material via line-of-sight optical exchanges and, when available, satellite or shore-seeded quantum networks; the quantum layer provides keys, while existing data links carry encrypted traffic at operational rates. Quantum measurements are inherently disturbance revealing, so any probing manifests as elevated QBER; affected keys are discarded instantly, and the group autorotates to new keys while logging provenance for forensic assurance and command oversight. Keys are scoped by role (AAW, ASW, SUW, Strike), link, and time window, enabling local re-key during dispersion, re-composition, or pivots, with hierarchical derivation to avoid n-squared key explosion as units join/leave the net. Imagine a convoy exchanging a silent, uncopiable hand-signal at every intersection to update route ciphers; if anyone tries to crowd close enough to see it, the gesture changes on the spot and drivers receive a visible cue to alter spacing, lanes, or pace. The convoy never stops, the signal cannot be recorded for later use, and intruders convert themselves into warnings rather than threats.

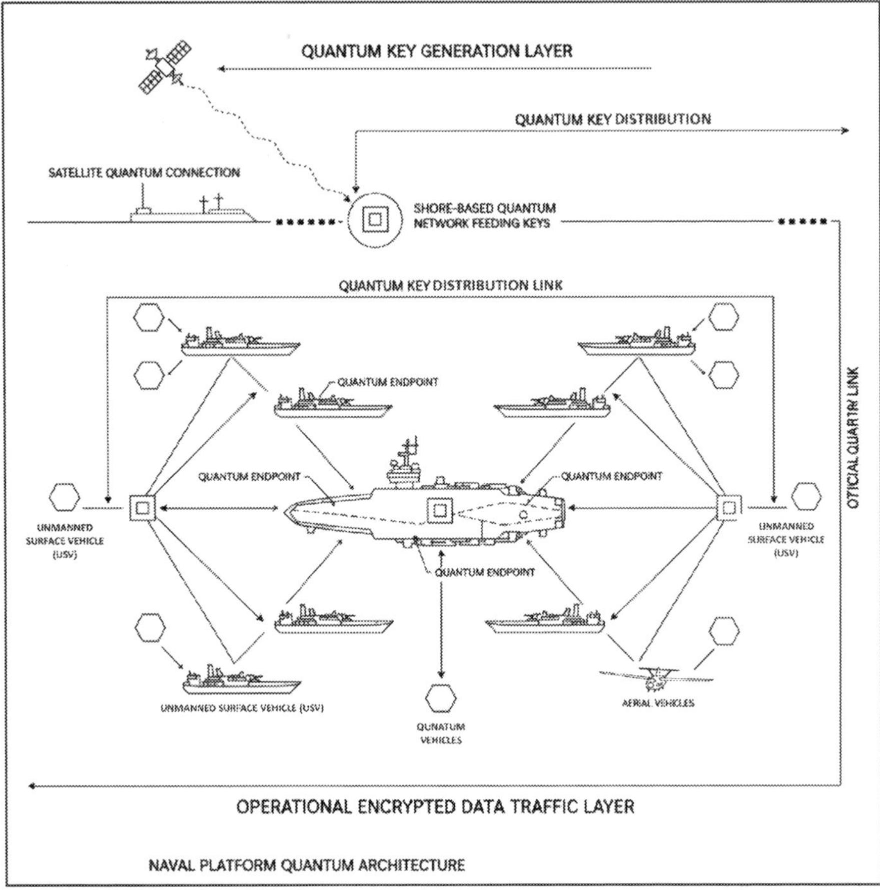

Fig. 4.17: Every Platform Secured: Fleet-Wide Quantum Integration

Operational relevance

❑ **Dispersed operations:** As the battle group spreads for sensor baselines, missile kinematics, or weather, per-link quantum refresh maintains a coherent trust fabric without forcing force-wide re-key, keeping composite warfare command agile.

❑ **High-mobility pivots:** During rapid course/speed changes, strike launches, or AAW surges, second-to-minute re-key cadences prevent key aging, so even stressed or partially degraded links never carry long-lived secrets.

❑ **Coalition fluidity:** New entrants obtain authenticated quantum roots rapidly and derive local sub-keys; departing units can be cleanly revoked by scope, preserving interoperable security without exposing national crypto internals.

❑ **EW resilience:** Directional optical exchanges resist RF jamming; attempted optical interference appears as QBER anomalies, which not only blocks key acceptance but also cues changes in formation geometry and EMCON posture.

❑ **Harvest-now mitigation:** Because keys are created from quantum exchanges and never exist as computable artefacts in transit, recorded cipher text remains useless, even if future quantum computers outpace classical cryptography.

Dynamic QKD key refresh for dispersed battle groups

Dimension	How it works	Commander's benefit	Principal risk	Mitigation
Refresh cadence	Seconds–minutes via intra-group optical QKD; accelerated during pivots	Crypto stays ahead of manoeuvre and EW	Missed windows in heavy seas	Bank keys in good weather; alternate relays/wavelengths
Trust scoping	Role/link/time-scoped keys with hierarchical derivation	Local re-key without force-wide disruption	n-squared key growth	Hierarchical keys; local sub-key derivation
Tamper evidence	QBER thresholds auto-discard compromised keys	Probing becomes an alarm, not a breach	Threshold false alarms	Environment-tuned thresholds; provenance checks
Mobility support	Gimballed, narrow-beam optical tracking	Maintains trust during dispersion/re-composition	Beam drop on hard turns	Predictive pointing; picket arcs for continuity
Coalition agility	Rapid root establishment; scoped revocation	Plug-and-fight partners, clean detach	Inter-op seams	Quantum-safe auth policies and joint SOPs
EW resilience	Optical directionality; RF-silent keying	Jamming/ESM largely ineffective	Optical dazzling attempts	Narrowband filters; fast re-key; manoeuvre cues
Harvest-now defence	Keys never computable in transit	Future decryption of captures denied	Complacency on rotation	SOP-driven cadence tied to threat state
Command oversight	Provenance logging; rollback to last good epoch	Assured audit trail and controlled recovery	Cascading revocations	Scoped zeroization; enclave compartmentation

- **Prevents key compromise during high-mobility operations or sudden strategic pivots:** Dynamic key refresh through QKD prevents key compromise during high mobility operations or abrupt strategic pivots by turning key management into a continuous, tamper evident background process that tracks manoeuvre tempo. As platforms disperse, recombine, or execute sharp course/speed changes, directional optical exchanges generate fresh symmetric keys in seconds-to-minutes, so any single key's exposure window is too short to be operationally exploitable. If an adversary probes or attempts interception, the quantum bit error rate rises beyond thresholds, the in progress key is discarded, and a new exchange is initiated without pausing tactical coordination. This eliminates the "stale key" problem common to static re-key schedules and denies attackers the chance to harvest keys for later decryption as battle geometry shifts. Technically, resilience comes from three design choices. First, per link, per role, time boxed scoping confines each key to a specific sender receiver pair, warfare function (AAW/ASW/SUW/Strike), and validity window, so joins/leaves and role handovers don't force force wide re key or create n squared exposure. Second, hierarchical key derivation seed's role roots to edge nodes and derives ephemeral subkeys locally at the cadence of manoeuvre, keeping distribution scalable while preserving rapid revocation. Third, quantum safe authentication on the classical control plane prevents man in the middle during deception or jamming, ensuring only authenticated endpoints can accept fresh quantum derived keys. Together, these measures keep trust synchronized with movement: the faster the formation pivots, the faster keys rotate, and the narrower the window for any compromise.

Operationally, dynamic refresh protects crypto continuity precisely when forces are most vulnerable during dispersal for sensor baselines, massing for strikes, or reorienting to new threats. Inter ship optical paths with auto tracking gimbals sustain refresh during turns; unmanned pickets extend optical arcs to preserve coverage while elements re-position. Alarm doctrine ties re-key cadence to threat state: suspected "harvest now" collection, jamming spikes, or QBER drift automatically increase rotation rates, turning adversary pressure into cues for faster, safer cryptographic turnover. Because keys are never transmitted in computable form and never age long enough to be valuable, recorded traffic remains cryptanalytically sterile even if enemy capabilities improve later.

Risk is contained by design. Weather or scintillation that shortens refresh windows is offset by banking surplus keys during favourable periods, using alternate wavelengths, or momentarily bridging with low probability of intercept RF for payload while the quantum layer catches up. Beam loss in sharp manoeuvres is mitigated with predictive pointing and relay geometry on turn arcs. Implementation side channels are addressed with optical isolation, nanosecond detector gating, and continuous device attestation, with provenance logs enabling instant rollback to the last known good key epoch. The result is a manoeuvre matched trust fabric that prevents key compromise not by assuming a quiet spectrum, but by shrinking and validating key lifetimes in lockstep with tactical motion.

Dimension	Mechanism	Compromise pathway blocked	Commander action
Key lifetime	Seconds-minutes rotation tied to manoeuvre tempo	Keys never age enough to be exploited mid-pivot	Set cadence by threat state and phase
Tamper evidence	QBER thresholds discard in-progress keys	Interception turns into alarms, not silent leaks	Enforce auto-discard/re-key; quarantine context
Scoping	Per-link, per-role, time-boxed keys	Joins/leaves don't expose force-wide secrets	Approve scopes; revoke by role/time
Distribution	Hierarchical roots with local ephemeral subkeys	Avoids n-squared exposure and stale shared keys	Delegate derivation at edge; audit derivations
Mobility sustainment	Auto-tracked optical links; picket relays on arcs	Beam loss doesn't force crypto pauses	Task relays; plan turn geometry
EW/deception guard	Quantum-safe auth on control plane	Blocks MITM during jamming/spoofing	Require auth pass before key acceptance
Weather/ scintillation	Bank keys in clear windows; alternate wavelengths	Reduced refresh windows don't stall trust	Pre-brief alternates; bank margin
Rollback/ provenance	Logged epochs; instant rollback on anomaly	Stops cascade from a single bad exchange	Drill rollback; isolate affected scopes

Mitigating Risks from Adversarial Quantum Eavesdropping

Quantum vs. Quantum: Outpacing the adversary's listening post

Dynamic quantum key distribution turns eavesdropping into a detectable event and keeps keys so short lived and tightly scoped that even rapid dispersal, re-composition, or abrupt operational pivots never expose a reusable secret, thereby preventing key compromise while preserving tempo and coalition interoperability. Battle group elements exchange quantum states over line-of-sight optical links (and, when available, via airborne or satellite relays) to derive fresh symmetric keys at a cadence matched to manoeuvre, rather than on fixed schedules tied to communications windows. Any adversary measurement perturbs the quantum states, spiking the quantum bit error rate and causing instant discard and re-issue, so an in-progress key never enters service if touched. Keys are scoped to link, role, and short time windows, and derived hierarchically from role roots to avoid n-squared key explosions as units join and leave. Consider a convoy where drivers swap a new, uncopiable hand signal every few intersections. If someone leans in to watch, the signal changes on the spot and the drivers receive a cue to widen spacing or change lanes. The convoy keeps moving; the signal cannot be recorded for later use; and intruders reveal themselves instead of stealing the route plan.

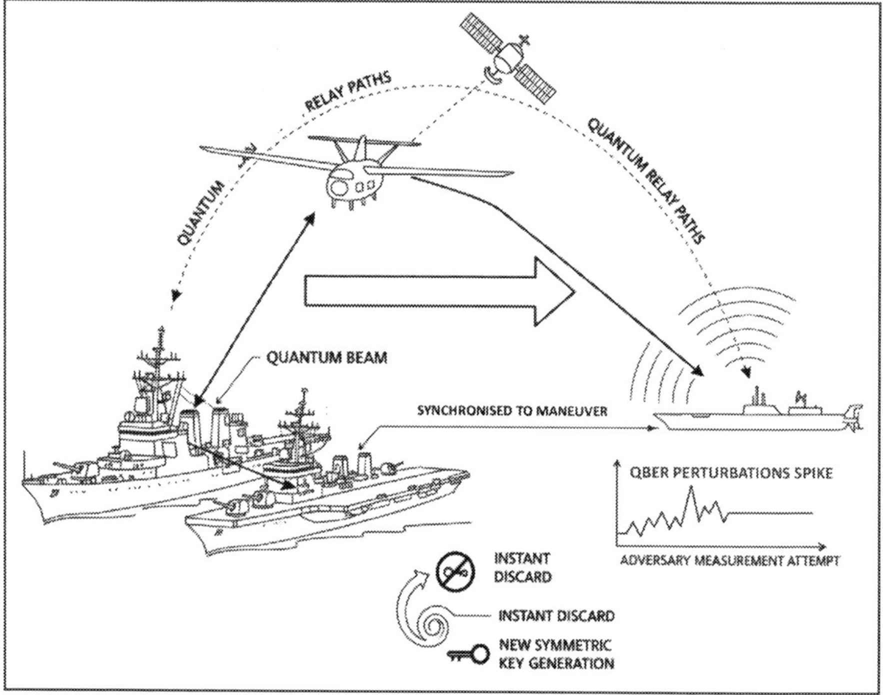

Fig. 4.18: Keys That Move With You: Maneuver-Tempo Quantum Security

Operational relevance

- **High-mobility manoeuvres:** Keys refresh in seconds to minutes during turns, speed changes, and sensor/weapon reassignments, eliminating "stale key" exposure and denying adversaries a window to harvest credentials.

- **EW and deception pressure:** Directional optical exchanges resist RF jamming; attempted optical probing manifests as QBER anomalies, automatically rejecting suspect keys and cueing formation geometry and EMCON adjustments.

- **Coalition fluidity:** Per role, per link scoping lets new participants establish authenticated quantum roots quickly while departing units are cleanly revoked without force-wide re-key, preserving tempo during reorganizations.

- **Harvest-now defence:** Because keys are born from quantum exchanges and never exist as computable artefacts in transit, recorded traffic remains cryptanalytically sterile even if future quantum computers arrive.

Dimension	*Mechanism*	*Compromise prevented*	*Commander's action*
Re-key cadence	Seconds-minutes tied to manoeuvre and threat state	Keys never age enough to be exploitable mid-pivot	Set SOP thresholds; accelerate on EW cues
Tamper-evidence	QBER spikes auto-discard in-progress keys	Probing converts to alarms, not silent leaks	Enforce auto-discard/re-key and quarantine
Scoping	Per link, per role, time-boxed ephemeral subkeys	Joins/leaves don't expose force-wide secrets	Approve scopes; revoke by role/time
Distribution	Hierarchical role roots; local derivation at edge	Avoids n-squared exposure and shared-key re-use	Delegate derivation; audit provenance
Mobility sustainment	Gimballed narrow-beam optics; picket relays on arcs	Maintains refresh through turns and dispersion	Task relays; plan geometry for turns
EW/deception guard	Quantum-safe auth on classical control	Blocks MITM during jamming/spoofing	Require auth pass before key acceptance
Weather resilience	Key banking; alternate wavelengths; brief LPI/LPD bridges	Missed optical windows don't stall trust	Pre-brief alternates; bank margin
Rollback	Logged epochs; revert on anomaly	Stops cascade from a single bad exchange	Drill rollback; isolate affected scopes

Cadence control ties re-key intervals directly to the threat state and manoeuvre tempo, accelerating key rotation during suspected collection, jamming spikes, strike launches, or AAW surges to ensure secrets never age into liabilities. Hierarchical distribution seeds role roots for AAW, ASW, SUW, and Strike at the edge and derives ephemeral sub-keys locally per link and time slice, avoiding n-squared scaling while constraining blast radius if any node is stressed. Authentication hardening employs quantum-safe signatures or MAC on the classical control plane so only authenticated endpoints can accept fresh quantum-derived keys despite deception attempts. Optical resilience is achieved with gimballed, auto-tracking terminals using narrow divergence, spectral filtering, and tight temporal gating to maintain low QBER through sea motion, while unmanned pickets extend optical arcs and preserve refresh continuity during tight turns. Alarms and rollback procedures mandate that on any QBER or provenance anomaly, the system discards in progress material, quarantines the link context, and reverts to the last validated epoch never admitting degraded keys for new sessions thereby preserving command integrity

4.2 GLOBAL QUANTUM COMMUNICATION INITIATIVES AND NAVAL SECURITY

- **Review of global quantum communication efforts by rival nations**

Quantum Giants at War: Unveiling the Global Race to Command Secure Communication Frontiers

USA: Leading Quantum Innovation with Military Applications

Forging the Quantum Shield: America's Quest for Unbreakable Defence Networks

R&D, standards from NIST, and public-private-academic partnerships to field satellite and terrestrial QKD prototypes, deploy quantum-safe cryptography across defence networks, and harden command-and-control against harvest-now/decrypt-later threats under joint, coalition operations. The U.S. approach integrates physics-based key generation (QKD) with standardized quantum-safe cryptography (post-quantum cryptography, PQC) to secure both near-term and future force networks. Department of Defence programs fund quantum networking test beds, free-space optical demonstrations, and space-to-ground segments, while NIST's PQC standards provide immediate, deployable authentication and key-establishment mechanisms resilient to future quantum computers. This dual track QKD where feasible, PQC broadly creates layered assurance for high-value C2, ISR distribution, and nuclear command support in contested EM environments.

The DoD has systematically integrated quantum technologies across multiple research portfolios, focusing on networking, sensing, timing, and secure communications aligned with joint all-domain command and control requirements, characterized by multi-service integration where the Army, Navy, Air Force, and Space Force have established dedicated quantum research programs, with DARPA coordinating cross-service efforts and advancing

Fig. 4.19: Dual-Track Defense: QKD Innovation Plus PQC Standards

quantum sensing for navigation and timing applications independent of GPS, while the Defence Innovation Unit (DIU) accelerates quantum technology transition from laboratory to operational deployment through programs advancing quantum sensing capabilities from lab environments to orbital platforms within months. This coordinated approach emphasizes hybrid architecture demonstrations where quantum key distribution provides tamper-evident entropy generation while post-quantum cryptography handles identity assurance and authentication, enabling immediate deployment of quantum-resistant security measures while building towards physics-based key distribution capabilities. Current operational applications target resilient position, navigation, and timing (PNT) systems that maintain accuracy when GPS is denied or degraded, quantum-enhanced sensing for submarine detection and navigation, and secure communications networks that resist both classical and quantum-enabled attacks, with the Pentagon's fiscal year 2026 budget allocating over $ 179 billion for research, development, test, and evaluation activities that position quantum technologies as foundational elements alongside artificial intelligence and space systems.

The National Quantum Initiative (NQI) coordinates quantum research and development across civilian and defence agencies, creating a unified national strategy for quantum technology advancement through comprehensive agency coordination that aligns efforts across NIST, NSF, Department of Energy, DOD, NASA, and NSA, ensuring that civilian quantum research advances complement defence applications while avoiding duplication of efforts and maximizing synergistic benefits. The initiative encompasses strategic infrastructure development through the establishment of quantum test beds, research centres such as the NSF Quantum Leap Challenge Institutes and DOE National Quantum Information Science Research Centres, and industry consortia including the Quantum Economic Development Consortium (QED-C) that provide shared experimental environments for advancing quantum networking, sensing, and computing technologies while fostering public-private partnerships. Central to the NQI's mission is comprehensive workforce development that includes educational programs, training centres, and quantum-literate workforce initiatives necessary for both defence applications and broader national competitiveness in quantum technologies, with the National Quantum Coordination Office (NQCO) providing technical and administrative support to coordinate interagency activities and serve as the primary point of contact for federal civilian quantum information science activities.

U.S. government initiatives also encompass a comprehensive multi-agency approach led by the DoD quantum portfolio, which includes multi-year programs in quantum networking, resilient PNT, sensing, and secure communications aligned to joint all-domain command and control use cases, with demonstrations emphasizing hybrid architectures where QKD supplies entropy and PQC assures identity. The NQI provides government-wide coordination to scale quantum information science research and development, test beds, workforce development, and industry consortia, aligning civilian, defence, and intelligence community funding for quantum networking and metrology efforts across agencies including NIST, NSF, DOE, DOD, NASA, and NSA. NIST leadership drives the selection and standardization of PQC algorithms, provides implementation guidance for migration to quantum-safe systems, and advances measurement science for quantum networking components such as integrated photonics and timing systems, establishing the cryptographic foundation that enables immediate deployment of quantum-resistant security while QKD capabilities mature.

USA: Quantum Innovation at a Glance

What	*Key points*	*Why it matters to commanders*	*Near-term actions*
Strategic approach	Dual-track security: QKD for physics-based keying; PQC for standardized quantum-safe crypto	Layered assurance against harvest-now/decrypt-later; continuity under EW and deception	Prioritize PQC migration now; pilot QKD where LOS/space links permit
DoD portfolio	Networking, secure comms, resilient PNT, sensing; multi-service with DARPA/DIU acceleration	Joint C2 aligned; faster lab-to-fleet transition and TTP maturation	Task units for FSO QKD trials; integrate crypto agility in CONOPS
QKD prototypes	Satellite-to-ground, air/sea free-space optical, fibre backbones	Seconds–minutes re-key cadence; EMCON-compliant intra-group trust	Field auto-tracking optical terminals; define QBER alarm thresholds
NIST standards (PQC)	Standardized KEMs/signatures; migration playbooks and testing	Immediate hardening of identity, gateways, software supply chain	Enforce PQC at identity/access points; update crypto inventories
Orchestration	Role/link/time-scoped key management with provenance and rollback	Limits blast radius; rapid recovery under attack	Implement scoped key policies; drill rollback to last good epoch
Partnerships	Primes + start-ups + academia + labs; shared test beds and red-teaming	Ruggedized gear, faster innovation, detector/side-channel hardening	Join inter-op test beds; adopt hardening base-lines (isolation, gating)
Operational use	Carrier C2, ASW barriers, logistics/JLOTS, coalition gateways	Spoof-resistant tasking; secure patrol plans; interoperable C2	Map QKD to high-value links; set re-key cadence by threat state
EW resilience	Directional optics resist RF jamming; tamper-evident keys	Probing becomes an alarm, not a breach; maintains tempo	Tie manoeuvres to alarm doctrine; pre-brief alternate relay geometry
Risk controls	Weather/scintillation, scale complexity, side-channels, MITM	Mission assurance without force-wide crypto resets	Bank keys in clear windows; hierarchical keys; PQC-auth control plane
Outcome	Layered, manoeuvre-ready quantum security	Preserves C2 integrity and coalition trust under quantum-era threats	Execute PQC-first, QKD-where-feasible roadmap; measure and iterate

Strategic partnerships with private sector and academia form a comprehensive ecosystem where industry delivers ruggedized quantum terminals, integrated photonics, quantum random number generators (QRNGs), and orchestration stacks, with start-ups accelerating component innovation while prime contractors integrate these technologies into platform architectures and military systems. University and national laboratory consortia advance critical protocols including decoy-state and continuous/discrete-variable QKD, device physics research, and networking control planes, with cooperative agreements and research centres shortening the timeline from laboratory innovations to fleet-ready systems. Open standards and shared test beds create collaborative experimentation environments that de-risk interoperability challenges, while red-team quantum hacking exercises inform detector hardening, optical isolation, and timing-gate defences, ensuring that deployed quantum systems can withstand sophisticated adversarial attacks.

China: Pioneering Large-Scale Quantum Communication Infrastructure

The Quantum Great Wall: China's Ambition to Secure the Digital Silk Road

China is pioneering a vertically integrated quantum communication architecture that pairs space- based QKD (anchored by the Micius satellite) with continental-scale fibre backbones and trusted-node hubs, creating a state-directed, defence-relevant network aimed at secure command-and-control, intercontinental liaison, and strategic signalling across the Digital Silk Road. China's model fuses quantum satellites with expansive terrestrial quantum backbones to deliver end-to-end keying at national and intercontinental scales. Space nodes like Micius provide long-range, line-of-sight quantum links for cross-theatre key distribution, while ground networks built along trunk lines linking political, military, and industrial hubs propagate keys across trusted nodes into operational enclaves.

Construction of the Micius satellite and military implications

Micius, China's first quantum communication satellite, validated satellite-to-ground QKD, intercontinental key exchange, and secure video links, proving precision pointing, acquisition, and tracking across thousands of kilometres. For military planners, this demonstrates:

- Beyond-horizon re-key for dispersed task forces and overseas facilities, providing synchronized key refresh between central command and forward elements.

- Interoperable ground-station architecture (telescope arrays, narrowband filters, timing gates) that can be proliferated across military districts, enabling predictable space windows for re-key during crises.

- A template for medium/high-orbit quantum satellites to increase coverage dwell and reduce reliance on any single platform key to survivability against anti-satellite risks.

Deployment of extensive ground-based quantum networks

China's trunk-line QKD networks connect major urban and industrial centres, extending to financial and governmental institutions and, by design, having reach to military districts and coastal bases. Trusted-node architectures move sifted keys segment by segment, with metro-scale QKD rings feeding data centres, signals brigades, and naval shore commands. For operations, this yields:

- Persistent, low-latency key availability at bases and ports for pre-sail vaulting and rapid turnaround.

- Role/time/channel scoping at scale, enabling per-mission compartmentation and clean revocation without force-wide re-key.

- A pathway to integrate fibre QKD with free-space optical bridges across straits and littorals, shaping maritime C2 resiliency.

State-driven national programs and quantum supremacy in communications

A centralized funding and governance model concentrates resources on quantum communications milestones with military utility, incentivizing rapid prototyping (e.g., micro/nanosat payloads), integrated photonics, and country-wide standardization of key formats and orchestration. The strategic effect is a compressed lab-to-network timeline, unified doctrine for QKD employment, and sovereign supply chains for terminals, detectors, and QRNGs reducing exposure to export controls and giving freedom to optimize for military use cases. Emphasis on QKD for secure command and control

China's doctrine highlights QKD as a physics-anchored hedge against future decryption, positioning it for high-consequence links: nuclear command support, strategic messaging, and hardened fleet coordination. Operational TTPs likely include QBER-based alarm doctrine for automatic discard/re-key, scoped key hierarchies to limit blast radius, and hybrid posture (QKD for key growth, national algorithms for authentication) to prevent man-in-the-middle during deception and heavy EW.

Result: China's "Quantum Great Wall" is not a single system but an integrated, state-driven space–ground complex aimed at strategic C2 assurance, diplomatic signalling, and allied service provision. For naval operations, it promises persistent key availability from shore to fleet, intercontinental re-key opportunities via satellite windows, and doctrine that treats eavesdropping attempts as triggers for automatic re-key and posture shifts raising the bar for maritime electronic warfare and strategic communications security.

Dimension	What China has built	How it works end-to-end	Military payoff	Key risks	Mitigations
Architecture vision	Vertically integrated space–ground quantum network	Space-based QKD (Micius-class) fused with continental fibre QKD trunks and trusted-node hubs	Strategic C2 assurance, inter-continental liaison, Digital Silk Road signalling	Trusted-node exposure; constellation single points	Compartmentation; proliferated satellites; alternate pass plans
Space segment (Micius)	First quantum comms satellite; validated Sat-ground QKD, intercontinental keys, secure video	Precision PAT, narrowband filtering, time-tagging; decoy-state and entanglement modes	Beyond-horizon re-key for dispersed task forces and overseas facilities	Weather/line-of-sight limits; counter-space threats	Orbit diversity; revisit-rate growth; ground diversity
Ground backbone	Beijing–Shanghai trunk and national metro rings	Segment-by-segment key relay via trusted nodes into operational enclaves	Persistent, low-latency keys at bases/ports for pre-sail vaulting and rapid turnaround	Insider/ physical node compromise	Layered physical/ cyber controls; tamper-evident enclosures
Naval extension	Coastal FSO bridges; ship-shore apertures under EMCON	Directional optical refresh aligned to sea state and exposure windows	Stealth-preserving re-key for fleets during dispersal/ massing/pivots	Scintillation; pointing loss in manoeuvres	Predictive pointing; relay geometry; key banking windows
Ground-station design	Interoperable apertures (telescope arrays, filters, timing gates)	Standardized interfaces across military districts for predictable space windows	Theatre-wide synchronized re-key during crises	Aperture bottlenecks; schedule contention	Pre-assigned pass priorities and alternates

Dimension	What China has built	How it works end-to-end	Military payoff	Key risks	Mitigations
Key orchestration	Centralized, doctrine-driven key management	Mission/theatre/role/time metadata; provenance logs; rollback to last good epoch	Limits blast radius; rapid recovery without force-wide re-key	Policy rigidity	Delegated authorities; drilled branches/sequels
C2 doctrine	QKD for key growth; national algorithms for authentication	QBER-gated acceptance; auto-discard/re-key on anomaly	Eavesdropping becomes an alarm, not a breach	Classical MITM on control plane	Strong auth, device attestation, signed updates
Operational TTPs	Scoped keys and cadenced rotation	Selective zero-ize/re-issue; cadence tied to threat state	Maintains tempo under EW; preserves authenticity	Cadence drift; stale keys	SOP thresholds; telemetry-driven cadence boosts
Export posture	Cross-border satellite QKD demos	Compatible stations/formats for BRICS/Digital Silk Road	Secure liaison; geopolitical leverage via "quantum services"	Dependency backlash	Interface control; selective scope for partners
Scaling strategy	State-driven standards and certification	Conformance on devices, protocols, and orchestration	Faster lab-to-network timelines; inter-operable deployments	Vendor lock-in; supply bottlenecks	Tiered vendors; domestic photonics/QRNG capacity
Strategic effect	"Quantum Great Wall" space-ground complex	Intercontinental reach plus terrestrial persistence	Raises bar for maritime EW and strategic comms security	Targetable ground and space nodes	Redundancy; dispersal; rapid reconstitution playbooks

Construction of the world's first quantum communications satellite (Micius) and its military implications

Micius, launched by China in 2016, is the first dedicated quantum communications satellite and has demonstrated satellite-to-ground QKD, intercontinental key exchange, and secure video links, proving the feasibility of long-range, tamper-evident keying with direct military implications for beyond-horizon command and control, crisis re-keying, and coalition signalling. Micius (Quantum Experiments at Space Scale, QUESS) validated space-based quantum links using entangled photons and decoy-state BB84 to distribute keys between satellite and widely separated ground stations, overcoming atmospheric loss and pointing challenges at thousands of kilometres. These experiments culminated in the first intercontinental quantum-secured video conference and multi-thousand-kilometre key exchanges, establishing a technical baseline for strategic communications that are immune to harvest-now/decrypt-later attacks.

Micius achieved high-fidelity entanglement distribution and decoy-state QKD with precision pointing, acquisition, and tracking (PAT), narrowband filtering, and time-tagging to maintain low quantum bit error rates over long paths. The program demonstrated satellite-to-ground key rates sufficient to seed symmetric encryption on conventional links, proving the hybrid model where QKD provides keys and classical channels carry mission data. For military planners, Micius shows that beyond-horizon re-key is possible during orbital passes, enabling synchronized key refresh across theatres without reliance on vulnerable RF SATCOM, thereby preserving authenticity and confidentiality under electronic warfare pressure. The space-to-ground architecture allows dispersed forces and overseas facilities to receive tamper-evident keys aligned to pass schedules, supporting crisis posture changes and rapid reconstitution of cryptographic trust

Micius links into China's terrestrial QKD backbone via trusted nodes, allowing satellite-derived keys to be injected into city-to-city fibre routes that connect political, industrial, and military hubs, thus extending space-sourced entropy across operational networks. While trusted nodes introduce physical security dependencies, the integrated space-ground design provides persistent coverage and theatre-level compartmentation that can be tuned to mission roles and time windows. China has used satellite QKD to conduct international demonstrations, including reported links with foreign partners, signalling technological leadership and offering a prospective "quantum services" layer to Digital Silk Road participants that could align partner infrastructure with Chinese standards. Such demonstrations serve both diplomatic and military messaging by showcasing secure channels that are claimed to be immune to interception and future decryption, shaping perceptions of communication superiority.

Micius provides a template for proliferated LEO and potentially higher-orbit quantum satellites that increase dwell time and revisit rates, improving resilience against kinetic or reversible counter space threats through constellation design. Operational doctrine can exploit QBER-based alarm thresholds to auto-discard suspect keys and re-key on subsequent passes, converting eavesdropping or dazzling attempts into triggers for posture adjustment rather than silent compromise.

Limitations include dependence on clear weather windows, strict PAT requirements, and the need to secure ground stations and trusted nodes, which remain potential targets for physical or cyber intrusion. Adversaries assessing the system should prioritize endpoint hardening assessments, ground-segment disruption, and attacks on classical authentication channels, which remain essential alongside QKD for man-in-the-middle resistance.

Micius proves that space-sourced quantum keys can be delivered globally to shore nodes that support fleets, enabling pre-sail vaulting and scheduled at-sea re-key opportunities tied to orbital passes for dispersed naval forces. This creates a pathway to strategic-grade, tamper-evident key distribution that enhances command continuity in contested environments, while integrating with terrestrial QKD trunks for theatre persistence.

State-driven national programs aiming for quantum supremacy in communications

In China, state-driven national programs pursue quantum supremacy in communications by tightly integrating space-based QKD (exemplified by Micius), nationwide fibre QKD backbones with trusted nodes, and centrally coordinated funding, standards, and industrial policy to field a sovereign, scalable, and defence-relevant quantum communications ecosystem.

Centralized strategy and governance

China's centralized strategy and governance for military quantum communications concentrates policy, funding, standards, acquisition, and operations under a unified state architecture, integrating space-based QKD, nationwide fibre QKD trunks, and trusted-node ground segments into a doctrine-driven network that can deliver tamper-evident keys to strategic and theatre C2 on demand. This model compresses the lab-to-network timeline, aligns deployment with operational doctrine, and sustains a sovereign supply chain for terminals, detectors, photonics, and orchestration software, ensuring that communications security remains both scalable and under national control.

At the top level, a national lead body sets end-to-end objectives for secure command and control, strategic signalling, and critical infrastructure protection, translating these into synchronized roadmaps for satellites, ground stations, trunk-line QKD backbones, and metro rings. Multi-year budget lines are tied to concrete milestones, satellite batches, ground aperture rollouts, urban trunk completion reducing fragmentation across ministries and forces.

Doctrinal integration codifies key acceptance criteria, re-key cadence, and incident response so that navy, air, rocket, and strategic support forces execute a common playbook under EW pressure, preserving tempo and unity of effort when contested. Standards, compliance, and orchestration are mandated centrally to guarantee interoperability from satellite downlinks to base enclaves. Common key formats, metadata schemas for mission/theatre/role/time, sifting protocols, and device conformance baselines are enforced alongside a national orchestration layer that scopes keys by role and time windows, maintains provenance and audit trails, and institutionalizes rollback to the last validated epoch on anomalies. A certification regime at central laboratories qualifies terminals, detectors, QRNGs, and photonic subsystems against defined QBER floors, timing jitter, isolation levels, and environmental hardening thresholds, ensuring only compliant equipment enters operational networks.

Space-ground fusion is governed by a single authority that arbitrates capacity across satellite pass schedules, ground-station windows, and trunk bandwidth to inject space-derived keys into terrestrial backbones. Pre-planned crisis playbooks assign alternate ground stations, relay vectors, and metro rings to guarantee synchronized re-key despite weather, scintillation, or counter-space disruption. The layered design leverages space windows for inter-theatre reach, fibre trunks for persistence, and coastal free space optical bridges and ship-shore apertures to extend keys to naval formations while respecting EMCON, thus preserving stealth without sacrificing re-key cadence. Industrial mobilization and a sovereign supply chain are driven through directed procurement that aggregates national demand for integrated photonics, detectors, cryogenics, gimbals, and narrowband filters, lowering per link costs and reducing import exposure. Vendor tiering assigns prime integrators to deliver ruggedized terminals while second-tier firms supply QRNGs and modulators; research institutes transfer IP under state guidance to sustain lifecycle support. State-run red teams conduct quantum hacking trials, detector blinding, timing/after pulsing, and trojan horse probes to harden fielded gear and update acceptance tests, continuously feeding improvements back into standards and certification.

Command, control, and operational governance emphasize role-based scoping and alarm doctrine. Keys are provisioned by mission phase (peacetime, crisis, combat), echelon (national, theatre, task force), and warfare area (AAW/ASW/SUW/Strike), enabling selective zeroization and rapid re-issue without force-wide re-key. Environment-tuned QBER thresholds gate acceptance; breaches trigger automatic discard, quarantine of the link context, and immediate re-key via pre-planned alternates, with alerts pushed to joint C2 for posture adjustment. Tactics, techniques, and procedures integrate EMCON and manoeuvre guidance with quantum alarms, using persistent anomalies to cue geometry changes, aperture discipline, and wavelength plan shifts that preserve stealth and continuity under pressure.

Security, counterintelligence, and legal authorities underpin trusted-node protection and endpoint integrity. Ground nodes and metro hubs are designated critical defence infrastructure with layered physical, cyber, and insider-threat controls; access and maintenance adhere to strict military security rules. Continuous device attestation and provenance logging enable rapid isolation of compromised enclaves and generate evidentiary trails for forensic analysis and command review. National policies define key retention, zeroization, and audit periods; any deviation from cadence or acceptance policy requires senior authorization, maintaining tight executive control over cryptographic posture.

Externally, the centralized approach supports a calibrated coalition and export posture. Compatible ground stations and key interfaces are offered to selected partners, aligning them with Chinese orchestration formats and enabling cross-border QKD demonstrations for political signalling. Dual-use leverage ensures civil telecom quantum backbones are pre-emptiable in crisis, with priority routing and scoping rules shifting capacity to defence circuits on order, thereby converting national infrastructure into a mobilizable military asset.

Risks are addressed within the governance model. Single-point policy rigidity is mitigated through pre-approved alternate re-key paths, delegated execution authorities, and simulation-driven war gaming that validates branches and sequels before crises occur. Trusted node dependencies are contained by compartmentation, tamper-evident enclosures, split knowledge procedures, and rapid re-key from alternate segments. Counter-space and weather vulnerabilities are offset by constellation proliferation, diverse orbits, and widely distributed ground sites, while fibre trunks and coastal FSO bridges assure continuity when passes or atmospherics degrade links. The result is a vertically integrated lever to field, standardize, and operationalize quantum-secured communications at national scale ensuring naval and joint C2 can be re-keyed on demand, that eavesdropping attempts become actionable alarms, and that cryptographic trust can be rapidly reconstituted under electronic warfare and crisis conditions.

Space–ground fusion as a force multiplier

In China's military context, space-ground fusion uses quantum satellites to inject tamper-evident keys into national fibre QKD trunks and metro rings, creating a continuous, doctrine-driven key supply that enables synchronized re-key for strategic and theatre C2, resilient operations under EW, and exportable "quantum services" to aligned partners for geopolitical leverage. Space-ground fusion aligns quantum satellites with terrestrial QKD backbones so that orbital passes deliver long—range key material, while trusted node trunks distribute those keys to military districts, naval bases, and theatre headquarters on persistent schedules designed for crisis operations and EMCON compliance. This architecture turns space windows into force-level re-key events and uses ground networks for endurance, ensuring that eavesdropping attempts produce QBER alarms and trigger automatic discard and re-issue without exposing operational traffic. By centralizing orchestration, China synchronizes key rotation across echelons and theatres, preserving command authenticity even when RF SATCOM is contested or subject to harvest-now/decrypt-later risks.

China's Micius satellite, validated to ground QKD, intercontinental key exchanges, and secure video links, has proved precision pointing, acquisition, and tracking required to sustain low QBER over thousands of kilometres for defence-relevant tasks. Military strategy builds on these demonstrations to pursue proliferated quantum satellites that increase revisit rates and reduce single-point failure risks, enabling predictable pass -based re-key for dispersed forces and overseas facilities. Space-derived keys seed symmetric ciphers on classical high throughput links, allowing mission data to flow while the quantum layer focuses on tamper-evident key growth and alarm doctrine.

China's city-to-city QKD trunks and metro rings extend satellite-injected keys through trusted nodes to political, industrial, and military hubs, providing persistent availability for pre-sail key vaulting and theatre rotations. Trusted nodes segment the network into controllable hops with physical security and audit trails, enabling role/time scoping and rapid revocation to limit blast radius if a node is compromised. For naval missions, shore-based trunks supply bases and fleet HQs, while coastal free-space optical bridges can push keys across littorals to formations under EMCON, preserving stealth and tempo. Centralized orchestration defines key formats, provenance logging, and rollback rules, tying acceptance to QBER thresholds that convert probing into immediate alerts for manoeuvre and posture changes across commands. Doctrine scopes keys by mission phase, echelon, and warfare area, institutionalizing selective zeroization and re-issue so that incidents do not trigger force- wide re-key or operational pauses. Crisis playbooks pre-assign alternate ground stations, relay vectors, and pass priorities to guarantee synchronized re-key even under weather, scintillation, or counterspace pressure, preserving strategic signalling and C2 continuity.

Space-ground fusion reduces dependence on RF keying by using directional optical channels from space and fibre trunks on the ground, both of which are less susceptible to broadband RF jamming and SIGINT collection than conventional links. Any optical interference that perturbs quantum states manifests as elevated QBER, preventing suspect key material from entering service and alerting commanders to hostile proximity or dazzling attempts without revealing payload content. For fleets, scheduled pass-based re-key plus shore-fed trunks allow EMCON-compliant operations while maintaining rapid re-key cadence during dispersal, massing, or pivot manoeuvres. China has showcased satellite-enabled QKD with foreign partners, signalling intent to export compatible ground stations and formats that align external networks to Chinese orchestration practices and standards. This creates a "quantum services" posture for BRICS and Digital Silk Road participants, enabling secure liaison channels while shaping cryptographic dependencies advantageous to Chinese strategic interests. Such demonstrations reinforce claims of communications immune to retrospective decryption, enhancing diplomatic signalling and coalition coordination narratives.

The strategy anticipates constellation growth (LEO emphasis with potential higher-orbit augmentation) for coverage and survivability, mitigating single-satellite risks and allowing more frequent re-key windows that match operational tempo. Ground diversity multiple apertures and trunk redundancies along with standardized trusted node procedures supports continuity when segments are degraded, while certification regimes harden detectors and timing systems against known quantum hacking vectors. Combined, constellation and ground proliferation support sustained operations despite weather, scintillation, or selective disruption.

Military operations use cases

- National C2: Strategic leadership circuits receive synchronized keys during passes with injection into trunks, preserving the authenticity of orders in A2/AD contexts.

- Naval C2: Bases obtain persistent keys via trunks for pre-sail vaulting; at sea, formations exploit pass windows and coastal FSO bridges for re-key aligned to EMCON.

- Theatre logistics and ISR: Trusted node distribution provides scoped keys for logistics corridors and intelligence hubs without force-wide re-key during incidents, maintaining operational tempo

Space-ground fusion relies on trusted nodes that must be physically secured, and on clear atmospheric conditions for optimal optical performance; these are mitigated by compartmentation, redundancy, and crisis alternates embedded in doctrine. Satellites face kinetic and reversible threats; constellation proliferation and pass planning reduce vulnerability windows while trunks ensure persistence when passes are unavailable. Adversaries evaluating this posture should focus on ground station security, classical authentication channels, and relay timing rather than attempting undetected interception of the quantum channel itself.

Bottom line: China's space-ground fusion transforms quantum communications into a force multiplier by fusing intercontinental reach with terrestrial persistence, orchestrating synchronized re-key across echelons, and embedding alarm-driven doctrine, yielding a military communications posture that treats eavesdropping as a detectable trigger and maintains C2 integrity under electronic and strategic pressure.

Dimension	What China is doing	How it works (space–ground fusion)	Operational effect for PLA C2	Commander actions
Strategic concept	Treats quantum comms as a state-directed, defence-relevant utility	Satellites inject tamper-evident keys into national fibre QKD trunks and metro rings	Continuous, doctrine-driven re-key across strategic and theatre echelons	Schedule pass-based re-key cycles; enforce theatre key scopes
Space segment	Proliferating LEO quantum satellites following Micius-class demos	Precision PAT, decoy state/entanglement QKD; pass windows seed symmetric keys	Beyond horizon synchronized re-key without RF reliance	Align orbital passes to alert states and joint readiness levels

Dimension	What China is doing	How it works (space–ground fusion)	Operational effect for PLA C2	Commander actions
Ground backbone	Beijing-Shanghai trunk lines extended to major hubs and coastal bases	Trusted nodes segment keys city-to-city; metro rings serve bases/HQs	Persistent theatre key availability; rapid revocation/rollback	Pre-sail vaulting; assign alternate trunk/ring routes
Maritime extension	Coastal free-space optical bridges and ship-shore apertures	Directional optical refresh aligned to EMCON windows	Stealth-preserving key continuity for fleets at sea	Tie mast/aperture discipline to re-key SOPs and sea state
Orchestration and doctrine	Centralized national orchestration and acceptance policy	Common formats, mission/theatre/role/time metadata; QBER-gated acceptance	Converts probing into alarms; prevents silent key compromise	Enforce auto-discard/re-key; quarantine and provenance logging
EW resilience	Optical links and fibre trunks minimize RF exposure	Any interference disturbs quantum states, elevating QBER instantly	Jamming yields alerts, not breaches; tempo preserved	Cue manoeuvre/posture on alarm persistence and severity
Crisis playbooks	Pre-assigned alternates for passes, stations, and trunks	Automatic failover to alternate apertures, relay vectors, and metro rings	Synchronized re-key under counterspace, weather, or scintillation	Drill branches/sequels; validate cadence under degraded conditions
Compart-mentation	Keys scoped by echelon, warfare area, and time window	Role/time scoping limits blast radius; selective zeroization and re-issue	Incidents stay local; no force-wide re-key pauses	Approve scopes; set zeroization triggers and re-issue priorities
Security and assurance	Certification, red team quantum hacking, device attestation	Hardens detectors (blinding, after pulsing), timing, and isolation	Reduces side channel exposure; raises acceptance confidence	Mandate certified terminals; periodic red team validation
Coalition and influence	Exportable "quantum services" for aligned partners	Compatible stations/interfaces align partners to Chinese orchestration	Secure liaison channels; geopolitical leverage	Define partner key scopes and cross-border re-key windows
Survivability and scale	Constellation growth; ground diversity and redundancy	More frequent pass windows; redundant trunks/metro rings	Continuity despite localized disruption or counterspace effects	Resource alternate orbits and aperture sites; stock spares
Constraints	Trusted-node security; weather/pointing limits	Compartmentation, redundancy, and pass planning mitigate	Maintains service under degradation; known attack surface	Prioritize node hardening; strengthen classical auth channels

Defence-oriented standardization and scaling

China's defence-oriented standardization and scaling drive for quantum communications focuses on enforcing common technical baselines, centralized orchestration, and certification across space and terrestrial segments so military C2 can re-key predictably at national scale, with tight blast-radius control, rapid rollback, and interoperable deployment patterns from strategic echelons to naval formations.

China's approach mandates end-to-end standards spanning key formats, protocol choices, device interfaces, and orchestration metadata so that satellite downlinks, trusted-node fibre trunks, and base enclaves operate as a single, interchangeable system. Key material carries mission/theatre/role/time tags; acceptance is gated by environment-tuned QBER thresholds; and provenance is logged at every hop. These standards codify how keys are generated (e.g., decoy-state BB84 or entanglement-based modes), sifted, authenticated on classical channels, compartmented by scope, and fused into platform key management systems ensuring consistent behaviour under electronic warfare stress. A national orchestration layer enforces policy for scoping, rotation cadence, acceptance/reject criteria, and incident recovery. Re-key schedules are synchronized with orbital pass plans and theatre trunk capacities so strategic circuits, fleet headquarters, and task forces receive coordinated refresh without contention. Automatic rollback to the last validated epoch is institutionalized: on anomaly, affected links are quarantined, scoped keys are zeroized, and pre-assigned alternates ground stations, relay vectors, metro rings take over. This creates predictable, drillable behaviour across services and echelons.

China drives scaling through a certification pipeline that tests terminals, detectors, QRNGs, integrated photonics, gimballed optics, and timing subsystems against defined thresholds (QBER floors, timing jitter, optical isolation, polarization/phase stability, environmental hardening). Only certified gear is admitted to operational nets; field acceptance includes red team evaluations for detector blinding, after pulsing, Trojan horse injection, and timing side channels. Conformance profiles ensure that equipment from different state-aligned vendors remains interoperable, reducing integration frictions as deployments expand.

Standardized topologies city-to-city trunks, metro rings, edge key vaults enable repeatable builds that scale across provinces and military districts. Trusted nodes are specified with layered physical, cyber, and insider-threat controls, tamper-evident enclosures, dual-person access, and split knowledge handling. Templates define how keys are staged at shore bases (pre-sail vaulting), refreshed at theatre HQs, and extended over coastal free space optical bridges to fleets under EMCON, allowing rapid replication of proven patterns wherever operational demand increases. Scaling to force level relies on hierarchical roots (national/theatre/service) from which ephemeral per link subkeys are derived locally for short time windows. Scoping by warfare area

(AAW/ASW/SUW/Strike), platform class, and mission phase allows selective zeroization and re-issue, preventing force-wide re-key in incidents. As more nodes are added, the number of keys under any root is constrained by policy, and rotations are staggered to balance capacity with security, keeping compute and logistics overheads tractable.

Defence standards bind satellite pass schedules to ground trunk orchestration: pass windows are pre-allocated by priority; ground stations have pre-briefed alternates; and metropolitan rings receive quota-based injections. As constellations grow, revisit rates increase and re-key windows multiply, enabling higher cadence across more units without degrading acceptance criteria. Ground diversity (redundant apertures, diverse fibres) is standardized to maintain service continuity under localized disruption.

Acceptance policies embed alarm thresholds that translate perturbations into immediate operational cues. Persistent QBER drift, scintillation anomalies, or photon-rate deviations trigger discard/re-key and cross-domain actions: aperture discipline, course/formation adjustments, wavelength plan changes, or shifts to alternate relays. Training and exercises normalize these responses so that re-key and manoeuvre are synchronized behaviours rather than ad hoc reactions.

Scaling is supported by planned spares for certified terminals, modular upgrades (detector modules, filters, QRNG boards), and periodic re-certification. Firmware and orchestration software follow signed update regimes with quantum-safe authentication to protect the classical control plane. Mean-time-to-restore targets and field replaceable unit standards reduce downtime at trusted nodes and ship-shore apertures, keeping re-key cadence stable as the footprint expands.

Standardization gives industry clear conformance targets, enabling tiered production: primes assemble ruggedized terminals and gimballed optics; subtler vendors supply chips, photonics, QRNGs, and control electronics. State-guided roadmaps sequence capacity expansions with deployment phases, lowering unit costs and shortening lead times. Shared test beds and interoperability trials push updates back into standards, reducing field integration risk as new vendors are onboarded.

Defence-oriented standards are packaged for selective partner export: compatible ground stations, orchestration interfaces, and acceptance policies are offered to aligned states. This external scaling binds partner networks to Chinese formats and procedures, enabling cross-border re-key while maintaining Chinese control of interfaces and updates extending influence without exposing core cryptographic internals.

By hard-coding standards, certification, and orchestration into the buildout, China can scale quantum-secured communications from strategic core to fleet edge without losing interoperability or response discipline, ensuring synchronized, tamper-evident re-key remains routine, even as the network grows in size, complexity, and operational load.

Comparative Assessment: Strengths, Gaps, and Strategic Focus

Measuring Quantum Might: Who Leads, Who Lags, and What It Means for Future Conflicts

The balance of quantum communication power reflects differing national bets. China leads in space-ground QKD scale and pace, the USA leads in standards and layered PQC and QKD integration, and Russia pursues selective QKD deployment and cryptanalytic hedging, each shaping naval C2 resilience and escalation dynamics in future conflicts.

China has demonstrated end-to-end space–ground quantum links with Micius and is building continental QKD backbones, indicating high maturity in intercontinental keying and national-scale terrestrial rollout under a state-directed model. The USA emphasizes layered adoption near-term PQC standardization via NIST and targeted QKD pilots prioritizing interoperability, crypto agility, and rapid retrofit across defence networks. Russia is accelerating through selective QKD corridors, investment in devices and terminals, and a hedge posture focused on PQC migration and cryptanalytic capability, indicating measured scale but clear strategic intent. China's centralized orchestration, trusted-node templates, and doctrine for QBER-gated acceptance enable synchronized re-key across echelons, aligning space pass windows with ground trunk distribution for naval and theatre C2. The USA aligns DoD test beds, satellite/FSO trials, and PQC migration with joint C2, producing a hybrid posture where QKD feeds symmetric keys and PQC secures identity and control planes. Russia's approach integrates QKD on strategic terrestrial routes, prepares space-ground interfaces, and embeds alarm-to-manoeuvre cues in EW/INT doctrine for continuity under pressure.

China's scale is national by design, with continental trunk lines, metropolitan QKD rings, and interoperable ground stations linking political, industrial, and military hubs into a unified quantum-secure backbone. Space-based demonstrations signal an ambition to proliferate satellites and export "quantum services" to partners, extending secure, cross-border key distribution aligned with Digital Silk Road objectives. This space–ground fusion enables end-to-end keying at both national and intercontinental scales, with trusted-node hubs propagating keys into operational enclaves for persistent command-and-control.

In the USA, scale is driven by a standards-first approach that rapidly retrofits post-quantum cryptography across vast legacy networks for identity, gateways, and software assurance. Quantum key distribution expands selectively through defence pilots prioritized for high-value maritime, space, air, and fibre links, with a strong emphasis on coalition interoperability and crypto agility. This creates immediate hardening against harvest-now/decrypt-later risks while hybrid QKD+PQC deployments mature where mission geometry and payoff are highest.

Russia's scale remains selective and corridor-based, concentrating on leadership nodes and key national routes secured by terrestrial QKD and prepared ground segments for eventual satellite uptake. Growth pathways include partner demonstrations that provide contingency links beyond Western-controlled infrastructures, while broader state systems pivot to quantum-safe cryptography and cryptanalytic readiness. The result is purposeful coverage on critical circuits, with doctrine emphasizing scoped keys, rapid rollback, and alarm-to-manoeuvre transitions under electronic warfare pressure

China's investments serve strategic C2 assurance, Digital Silk Road signalling, and exportable "quantum services," binding partners to Chinese orchestration and formats. The USA prioritizes coalition interoperability, zero-trust migration, and harvest-now/decrypt-later mitigation across vast legacy networks, emphasizing rapid standardization and hybrid fielding. Russia aims to deny Western interception and retrospective decryption while sustaining C2 under sanctions and contested infrastructure, pairing selective QKD with PQC migration and indigenous supply chains.

China's operational posture fuses space-pass synchronized re-key with terrestrial quantum backbone distribution to preserve theatre-wide command authenticity under electronic warfare, demonstrate deterrence through visible space achievements, and cultivate partner dependencies via standard-aligned interfaces that enable cross-border interoperability. In the USA, a standards-first, PQC-centric hardening immediately lowers decryption risk across allied networks, while hybrid pilots deliver manoeuvre-matched re-key on priority maritime and joint links sustaining operational tempo without waiting for universal QKD coverage. Russia's corridor-focused QKD and system-wide PQC migration harden leadership communications and transform adversary probing into alarm-driven re-key and routing adjustments, maintaining C2 continuity despite sanctions pressure and contested infrastructure.

Operational implications for naval security

- **China:** Space-pass re-key fused with fibre trunks supports EMCON-compliant fleet operations and synchronized theatre re-key, raising the bar for EW disruption during dispersal or massing.

- **USA:** PQC-first plus targeted QKD creates immediate hardening of identity and gateways while piloting manoeuvre-matched re-key on high-value maritime links.

- **Russia:** Terrestrial QKD corridors for strategic posts and shore-fed key vaults support continuity, with doctrine converting probing into re-key and routing changes rather than compromise

Country	Strategic tagline	Technology maturity & scale	Military integration & readiness	Competitive advantages	Key vulnerabilities	Strategic focus for naval C2
USA	"Forging the Quantum Shield"	Layered posture: PQC standardized now; targeted QKD prototypes in space/air/sea/fibre; strong crypto agility and retrofit pathways	Joint DoD test beds; PQC-first migration across identity/gateways; QKD piloted on high-value links; zero-trust-aligned orchestration	Standards leadership (PQC), coalition interoperability, hybrid QKD+PQC architectures, robust public–private–academic ecosystem	Slower space-QKD scaling; supply-chain dependencies; integration across large legacy fleets	Immediate PQC hardening, selective QKD where LOS/space allows, manoeuvre-matched re-key for carrier groups and and coalition gateways
China	"The Quantum Great Wall"	Highest scale in QComms: Micius-class space QKD plus continental fibre trunks and trusted nodes; inter-continental demos	Centralized orchestration; QBER-gated acceptance; doctrine-driven re-key tied to satellite passes and ground trunks	Vertical integration and pace, space–ground fusion, standardized deployment templates, exportable "quantum services"	Trusted-node exposure; counterspace risks; weather/pointing limits; classical control-plane attack surface	Pass-based synchronized re-key for theatres, persistent shore-to-fleet key supply, partner alignment via compatible ground stations
Russia	"Quantum Bear Awakens"	Selective QKD corridors on strategic routes; ground readiness for future space QKD; PQC migration as hedge	Focus on strategic command posts and terrestrial corridors; alarm-to-manoeuvre EW doctrine; provenance logging	Sovereign cipher posture, targeted deployments, crypt-analytic emphasis, roll-back-and-re-key playbooks	Coverage limits; sanctions/supply constraints; node security; atmospheric constraints on FSO	QKD for leadership links and shore vaults, PQC-first for identity/software, EMCON-aligned refresh and preplanned alternates

Chapter Five

Quantum Warfare; Future Fight Scenarios

"Quantum Warfare: Future Fight Scenarios" provides a concise operational picture of how quantum communications, computing, and sensing will redefine the conduct of war across land, sea, air, space, and cyber. The chapter translates core capabilities into mission effects: low probability of intercept quantum links for secure coordination; quantum accelerated planning that compresses decision cycles; and quantum sensors that expose stealth, spoofing, and GPS denial tactics. Through well structured vignettes—littoral blockades, contested air corridors, and cross domain strike packages—it demonstrates how quantum technologies harden force posture, improve target custody across the kill chain, and preserve a coherent common operational picture under electronic attack and cyber pressure.

A dedicated section addresses Quantum Enabled Autonomous Systems and the employment of quantum enhanced AI in autonomous drones, unmanned submarines, and intelligent weapons. It explains how quantum accelerated optimization and perception sharpen swarm routing, multi sensor fusion, and dynamic re tasking, enabling time critical prosecution while maintaining resilience to jamming, deception, and comms disruption. The section emphasizes command assurance and safety: cryptographic agility, continuous integrity attestation, human on the loop authority, mission bound constraints, and fail secure behaviors to ensure lawful, predictable performance in dense, cluttered battlespaces.

Naval warfare receives focused treatment, detailing the integration of quantum technologies into defensive and offensive maritime platforms. The chapter outlines how embedding Quantum Key Distribution within fleet backbones, pairing quantum aided navigation with passive sensing, and conducting real time tactical key exchange between carrier, surface, and submarine groups can sustain command coherence from blue water to littoral operations. Submarine communications and emissions control are addressed with concepts for covert, tamper evident coordination; carrier and expeditionary groups are shown employing quantum guided strike planning, deconfliction, and anti submarine warfare with extended detection ranges and reduced targeting latency.

The concluding section frames Quantum Cybersecurity in Defense—securing C2, critical infrastructure, and weapon systems against quantum enabled cyber-attacks—alongside a forward looking review of global initiatives. It presents a comparative view of rival investments, highlighting China's pursuit of large scale quantum communication infrastructure and space based links, and distills strengths, gaps, and strategic focus areas for India and partners. The chapter closes with an actionable roadmap: prioritize quantum secure communications for joint C2, pilot quantum enhanced planning and wargaming, field quantum assisted sensing in high payoff missions, and institutionalize training, red teaming, and interoperability standards to ensure readiness as quantum capabilities transition from laboratories to line units.

5.1 QUANTUM TECHNOLOGIES IN NAVAL WARFARE

How quantum communications, computing, and sensing will change future battles

Quantum Edge Warfare: Unhackable links, unstoppable compute, unseen sensors – AI at the helm, deterrence by design

Quantum Communications: Enabling Unhackable, Real-Time Military Networks

Locking down the battlefield with zero-leak, zero-lag communications

Quantum communications use the physics of light single photons and entanglement to distribute encryption keys that reveal any eavesdropper the instant they probe the channel. This is called quantum key distribution (QKD). If an adversary tries to "listen," the quantum state changes, error rates spike, and the key is discarded before it ever protects traffic. Practically, QKD feeds fresh symmetric keys into own existing radios, SATCOM, fibre trunks, and optical links, hardening the network without changing how operators pass orders or data. Imagine every unit's comms vault uses glass-sealed keys that shatter if touched by anyone but the intended operator. The moment an intruder tries to copy a key, it breaks, alarms flash, and a new key is issued automatically without any pause in the fight. QKD turns adversary interception into an alarm, not a breach, while own data rides conventional links encrypted with pristine, short-lived keys.

Operational relevance for naval and joint forces

- **Assured C2 under EW:** Directional optical links (ship-to-ship, ship-to-shore, satellite-to-ground) and fibre QKD trunks provide tamper-evident keys that preserve order authenticity even under heavy jamming. Elevated error rates (QBER) trigger instant discard and re-key, converting probing into manoeuvre cues and posture changes.

- **Real-time tempo with zero-lag keys:** Keys are generated continuously alongside operations. Mission data still flows over high-throughput classical networks; QKD only supplies the keys, so throughput and latency stay mission grade while key freshness increases.

- **Coalition trust at scale:** Standardized, quantum-safe authentication on the classical control plane plus QKD-supplied session keys enables federated re-key across task forces without exposing national ciphers critical for carrier groups, amphibious forces, and maritime patrol coalitions.

- **Harvest-now/decrypt-later immunity:** Traffic recorded today remains useless tomorrow because keys are short-lived, physics-protected, and discarded if touched. PQC hardens identities and software; QKD adds physics-grade secrecy for the most sensitive links, a layered defence against future quantum computers.

Planners should prioritize deployment where operational payoff is highest, space and free-space optical legs with clear line of sight, theatre fibre backbones, and gateway nodes that feed fleets and ISR hubs so that tamper-evident keys protect the links that most affect tempo and survivability. Enforce strict acceptance rules at all echelons: never load a degraded key, use quantum bit error rate thresholds to gate acceptance, and institutionalize automatic rollback to the last validated epoch; drill discard-and-re-key with the same rigor as EMCON so alarms translate into immediate, practiced action. Maintain a hybrid posture that pairs QKD for continuous key growth with post-quantum cryptography for identity and control-plane authentication, yielding immediate hardening against harvest-now/decrypt-later threats while securing the command chain against spoofing and man-in-the-middle, and preserving future-proof secrecy as quantum capabilities evolve.

QKD fits the network

What to do	*Why it matters*	*How to execute in practice*
Treat key rotation like manoeuvre	Syncs cryptographic freshness with decisive operations to frustrate enemy targeting cycles	Align QKD re-key windows to dispersal, deception, and massing events; plan re-key like refuelling or EMCON timing
Push security to the edge	Protects the unit's most exposed to jamming and targeting; HQ relies less on fragile "exquisite" links	Deliver quantum-seeded keys to USVs, UUVs, pickets, and forward logistics nodes; enable local re-key and rollback at the edge
Make alarms trigger action	Turns eavesdropping into an advantage by forcing adversary reveals and friendly resets	Tie QBER alarms to pre-briefed branches: aperture discipline, wavelength shifts, relay geometry, route flips; drill discard-and-re-key
Use crypto standards as coalition ROE	Speeds joint re-key, lowers liaison friction, and preserves trust under EW pressure	Share PQC suites, metadata schemas, acceptance thresholds; re-key allied formations on a common clock with provenance logs
Secure silent logistics	Closes a quiet vulnerability where spoofing and cyber ops can delay or misroute supplies	Enforce tamper-evident re-key at chokepoints (straits, UNREP zones, staging anchors) with pre-planned alternates
Hedge against spectrum denial	Maintains authentic orders when RF bands are jammed or emissions are limited	Use directional optical segments and buried fibre QKD to complement LPI/LPD; diversify paths to preserve C2
Stop fake orders, not just theft	Prevents command spoofing during EW and deception campaigns	Pair QKD keys with quantum-safe authentication; practice challenge-response re-key to expose deepfake C2 attempts
Measure comms readiness, not just "encrypted"	Predicts if a fleet can fight connected for days, not hours	Track mean time to re-key under attack, links under QBER thresholds, rollback latency, and allied PQC baseline compliance
Make industry speed a weapon	The fastest force to replace, re-certify, and roll keys holds initiative	Procure conformance-tested terminals/ detectors and signed orchestration updates; require red-team certification and spares-on-hand
Signal deterrence visibly	Discourages harvest-now/decrypt-later by showing physics-backed security at scale	Demonstrate space-pass synchronized re-key, cross-border partner links, and at-sea optical refresh on repeatable schedules

Quantum Computing: Accelerating War-Gaming, Simulation, and Threat Prediction

Outsmarting adversaries at light speed through predictive dominance

By the early 2030s, quantum computing is expected to move battlefield simulations from serial to massively parallel exploration, cutting course-of-action cycles to minutes and improving war-gaming and movement forecasting with faster optimization, richer scenario sampling, and earlier pattern discovery across noisy, contested data streams. Quantum processors will specialize in problems that dominate operational planning, routing, scheduling, sensor-tasking, and red–blue move/countermove simulations by exploring many branches simultaneously instead of one after another. In practice, this means rapidly generating and ranking viable plans under changing constraints such as weather, emissions limits, fuel states, and adversary jamming; then updating those plans on the fly as new intelligence arrives. This parallelism gives staffs more cycles of "plan–test–refine" inside the same decision window, increasing the chance of selecting a robust course of action before the opponent can react. Picture a chart room with hundreds of silent planners working in parallel: each tests a different convoy route, deception pattern, or strike sequence against live constraints and adversary counters. Within minutes, the best three options with risks, logistics loads, and timing windows are on the commander's desk. Quantum computing aims to deliver that effect in software, turning staff time from building a single plan to supervising many.

Operational relevance

- **What "exponential" means in practice**
    - ❑ **Parallel COA exploration:** Quantum and quantum-inspired solvers can evaluate thousands of routings, dispersal, and strike options simultaneously, pruning weak branches fast and elevating robust COAs for commander review inside a single briefing cycle. This compresses planning timelines and allows more red/blue iterations before an operation starts.
    - ❑ **Higher-fidelity simulations:** Quantum-accelerated models ingest live weather, EW conditions, and logistics states to re-compute outcomes in near real time, improving accuracy of kill-chain timing, deception payoffs, and UNREP windows as conditions change. This reduces brittle plans and improves survivability at sea.

- **War-gaming and decision support**
  - ❏ **Iterated war games at speed:** Forces can run more "what-if" branches per hour, testing different enemy doctrines and deception schemes, which improves the chance of identifying tipping points and timing windows for massing or dispersal. Staffs shift from hand tuning few scenarios to supervising many.
  - ❏ **Interpretable recommendations:** Tools return ranked COAs with constraints met/violated and sensitivity to intel changes, enabling commanders to trust outputs and convert them into orders quickly within ATO/OPTASK/NAVDATA cycles.

- **Enemy movement forecasting**
  - ❏ **Earlier pattern discovery:** Quantum-accelerated analytics sift vast maritime tracks, SIGINT, and OSINT to flag emergent behaviours rendezvous patterns, logistics staging, spoofing campaigns, hours to days sooner than classical baselines on select workloads. This advances left-of-boom posture and cueing.
  - ❏ **Adaptive re-plans:** As the adversary shifts, the system re-computes best responses in minutes, keeping the force inside the opponent's OODA loop even under EW and cyber pressure.

- **Fielding path and limits**
  - ❏ **Near-term path:** Start with hybrid stacks quantum-inspired solvers on classical hardware, then swap in quantum backends as hardware matures, preserving workflows while performance climbs. This de-risks adoption and yields early gains.
  - ❏ **Reality check:** Full-scale, fault-tolerant advantage will roll out unevenly by problem type and timeframe; expect optimization and sampling use cases to lead, with verification and security controls required in classified environments

- **Commander actions**
  - ❏ **Frame problems for speedups:** Pose routing, loading, strike sequencing and deception as optimization/simulation tasks to maximize quantum benefit and measurability of gains.
  - ❏ **Instrument outcomes:** Track time-to-COA, win-rate improvements in exercises, fuel/ton-mile savings, and re-plan latency under EW to justify scaling

What to deploy first

Step	Plain meaning	What to do now	Why it helps
Decision-support pilots	Use quantum methods to speed planning	Start with routing, loading, tanker/UNREP plans, and quick COA screening on limited scenarios	High payoff, low integration risk for early wins
Hybrid stacks	Use "quantum-inspired" on today's computers; plug in real quantum later	Keep current tools; add quantum-inspired solvers now; switch backends as hardware matures	Smooth path to better performance without breaking workflows
Data discipline	Clean data makes good outputs	Curate logistics, maritime tracks, and EW scenario data; define constraints clearly	Better, trustworthy recommendations for commanders

Commander's Checklist

Checklist item	What to ask for	Why it matters
Frame problems as optimization/simulation	"Given these constraints, which plan maximizes mission success?"	Translates ops problems into solvable math for quantum/quantum-inspired tools
Demand interpretable outputs	Ranked COAs, what constraints are met/violated, sensitivity to intel changes, time-to-execute	Let's staff act fast and explain trade-offs to seniors
Tie to battle rhythm	Align re-plans to ATO/OPTASK/NAVDATA cycles	Outputs arrive in time to influence actual orders
Measure utility	Track time-to-COA, exercise outcome deltas, logistics savings	Proof you can brief to justify scaling and resources

Quantum Sensing: Detecting the Undetectable in all Domains

Shattering stealth: bringing hidden threats into plain view

By 2030-2035, quantum sensing is expected to move from trials to mission use at sea, in the air, on land, and in orbit providing GPS-independent navigation, passive cues for anti-stealth detection, and infrastructure security at chokepoints while integrated AI fuses quantum cues with conventional ISR to deliver faster, lower-signature find–fix–finish chains in contested environments.

2030-2035 outlook

- **Navigation without GPS:** Ruggedized quantum inertial sensors and optical/atomic clocks provide reliable positioning, navigation, and timing on ships, submarines, and aircraft when satellites are jammed or spoofed, turning PNT into a sovereign onboard function instead of a space dependency.

- **Anti-stealth cueing:** Quantum magnetometers and gravimeters offer passive detection cues for low-observable submarines and aircraft, narrowing search volumes for ASW and air defence and reducing the need for high-power emissions that reveal friendly positions.

- **Base and chokepoint security:** Gravity-gradient and magnetic anomaly mapping help expose underground tunnels, UUVs, or tampering near ports and straits, adding a "silent tripwire" around critical infrastructure and sea lines of communication.

From sweep to cue, quantum sensing shifts the kill chain from loud, wide-area searches to precision cueing. Instead of broadcasting high-power emissions that expose friendly positions, forces employ quantum magnetometers, gravimeters, and quantum RF receivers to generate low-signature cues about where to look. These cues are then handed off to active radars and sonars for classification and targeting, reducing electromagnetic exposure, compressing sensor-to-shooter timelines, and conserving emissions for decisive moments. The result is a quieter, faster find-fix–finish cycle that relies on passive or low-power detection first, then applies active energy only where it counts.

Integrating ISR with AI turns quantum cues into actionable intelligence at scale. Machine learning models fuse quantum sensor alerts with SIGINT, radar, and sonar tracks to filter out clutter and flag emerging patterns such as anomalous rendezvous at sea, spoofed shipping lanes, or covert staging near chokepoints. By correlating weak quantum signals with multi-int source data, the system reduces false alarms and elevates high-confidence targets for commanders, improving left-of-boom posture and enabling earlier interdiction decisions while maintaining a lower overall signature.

Quantum sensing also enhances survivability under electronic warfare. Because many quantum sensors measure environmental fields (gravity, magnetic) and intrinsic timing (optical/atomic clocks) rather than relying on RF emissions or satellite signals, they continue to deliver useful navigation and surveillance data when jamming saturates the spectrum or GNSS is denied. Ships and aircraft retain position, timing, and cueing accuracy under heavy EW, preserving formation integrity, weapon timing, and ISR coverage even as adversaries attempt to blind or deceive conventional systems

Aspect	What it means	Where it starts	How it scales	Why it matters
Trials to tactics	Long, hands-off trials at sea show quantum navigation operating continuously without babysitting	Selected defence trials and exercises on naval vessels and test ranges	Move to high-value platforms and chokepoints first; broaden as devices get smaller and tougher	Proves real-world reliability before mass adoption
Initial fielding targets	Platforms with the most to gain from GPS resilience and passive detection	Submarines and large surface combatants integrate quantum inertial/gravity aids	Port security units and maritime patrol aviation adopt for passive cueing and GPS-denied ops	Boosts ASW cueing and keeps navigation accurate under jamming
Networks of sensors	Distributed quantum sensors provide shared "silent" cues	Coastal lines, straits, base perimeters, and key sea lanes	Federated arrays feed fleets, air defence, and coast guards with low-emission alerts	Builds a common operating picture with lower RF signature
Operational integration	Use quantum outputs to guide existing radars/sonars	Cue-to-classify workflows in watch floors and CICs	Standard playbooks for handoff, confidence scoring, and false-cue management	Speeds kill chains while preserving emissions control
Sustainment path	Ruggedize, miniaturize, reduce power and maintenance	Rack mount and strap down kits on capital ships	Compact packages for patrol craft, UAVs, and expeditionary forces	Expands coverage without heavy logistics burden
Readiness metrics	Prove value with measurable outcomes	Track mean time from cue to classification; false cue rates; nav error under GPS denial	Add MTBF, swap time for modules, and software update cadence	Ensures commanders can brief performance and plan resourcing

Battlefield Autonomy: Quantum AI Driving Decision Superiority

Letting algorithms lead when seconds define victory

Over the coming five years, "quantum AI" will primarily mean two things at the edge of combat: quantum-inspired algorithms running on today's hardware to speed up route planning, target assignment, and sensor tasking, and early quantum accelerators in secure cloud or shipboard racks tackling hard optimization and pattern-matching jobs during planning cycles. Together, they compress decision time by exploring many manoeuvre options at once and picking the best under fuel, risk, and emissions constraints, while remaining auditable for commanders. Picture your combat information centre staffed by dozens of expert navigators, tacticians, and logisticians working in parallel. In minutes, they hand you three viable plans with risks, timing, and contingencies. Quantum-powered AI aims to deliver that parallel brain testing many branches at once so the ship acts before the adversary finishes deciding.

Operational relevance

- **Autonomy with intent:** Quantum-accelerated planners rank manoeuvres for unmanned surface/subsurface vehicles and airborne scouts, balancing stealth, fuel, sea state, and EW exposure to maintain screen integrity around carriers and amphibs. Outputs are ranked COAs with constraint checks, enabling push button tasking and human-on-the-loop oversight.

- **Real-time adaption under EW:** When jamming or spoofing shifts constraints, quantum-inspired solvers re-compute routes, emissions plans, and decoy allocations in minutes; paired with quantum-resilient PNT and secure comms, this keeps the force inside the adversary's OODA loop.

- **Swarm coordination:** Autonomous swarms use fast assignment algorithms to distribute roles (sensor, decoy, striker) and maintain coverage; the planner updates roles as contacts, weather, and rules of engagement change, preserving formation goals without micromanagement.

- **Maintenance and sortie tempo:** Quantum-accelerated scheduling sequences repairs, spares, and crew rotations across ships and ports, sustaining sortie generation during surges while avoiding single-point failures that stall an operation.

5.2 QUANTUM-ENABLED AUTONOMOUS SYSTEMS

Strategic Use of Quantum-Enhanced AI in Autonomous Drones, Unmanned Submarines, and Intelligent Weapons Systems

Quantum Edge Dominance: Faster finds, smarter strikes, unstoppable autonomy

Quantum-AI Enabled Target Identification and Engagement

Instant threat recognition, precision strikes, zero hesitation

Quantum-AI pairs quantum-accelerated analytics with AI perception to help autonomous platforms "see through clutter" and decide faster. In practice, quantum-inspired algorithms run on today's hardware to speed up pattern-matching and target de-confliction, while early quantum accelerators (where available) assist with hard optimization tasks like sensor fusion and fire-control timing. The result is quicker, more reliable identification in jammed, deceptive, or camouflaged settings and faster handoff from detection to engagement with a human in the loop. Picture a strike package flying through fog and decoys. Classically, the lead navigator checks one path at a time while the weapons officer compares one contact at a time. With quantum-AI, it is like having many expert crews working in parallel: one team fuses radar, IR, and acoustic hints; another tests decoy hypotheses; a third optimizes attack geometry. Within moments, they converge on the true target and the safest, surest firing solution.

Operational relevance

- **Targeting in clutter:** AI will fuse weak, disparate signature-slow radar cross-section returns, faint magnetic and acoustic cues, and thermal micro-patterns into single, high-confidence tracks. Expect hybrid quantum–classical models to run in real time at the edge, using quantum-inspired compression and hypothesis testing to sift signals from noise even in heavy clutter such as busy ports, archipelagos, and urban littorals. This reduces false positives and fratricide risk while increasing the number of actionable tracks per watch cycle. Faster, cleaner track confirmation lets commanders reallocate sorties and munitions from decoys to true threats, preserving magazine depth and shortening the kill chain.

- **Rapid sensor-to-shooter:** Quantum-accelerated optimizers will collapse the time from initial detection to a validated firing solution by running many engagement geometries in parallel weapon–target pairing, timing windows, flight profiles and then returning ranked options with confidence and constraints. Platforms will re-verify, request authorization, and execute within tighter command windows, even during dynamic ROE. Minutes become seconds. Carrier groups, surface action groups, and UUV pickets maintain initiative, coordinating salvos and decoys with less human micromanagement but preserved human authority.

- **Resilience under EW and deception:** When GPS is spoofed and RF bands are saturated, quantum-aided perception leans on passive cues (gravity, magnetic, optical time) and robust multi-sensor fusion to sustain target ID quality and engagement timing. Quantum-enabled inertial/timing aids maintain navigation and timestamp integrity for legal/ROE compliance even under jamming. Forces keep fighting connected and authentic in RF-denied zones, preserving formation integrity and shot timing while adversaries waste jamming power against less relevant channels.

- **Precision engagement:** Fire-control optimization will incorporate more environmental and countermeasure variables: sea state, wind shear, ducting, hard/soft kill threats while aligning with ROE and collateral constraints. Quantum-accelerated evaluation of thousands of micro-scenarios yields pinpoint engagement plans with explicit confidence, fallback options, and abort thresholds. Higher single-shot effectiveness, fewer rounds expended, reduced collateral risk, and clearer audit trails supporting command accountability.

Dimension	Future state (5–10 years)	Operational payoff
Targeting in clutter	Hybrid quantum–classical fusion elevates weak, multi modal signatures to high-confidence tracks in ports, straits, and urban littorals	Fewer false positives; safer shots; faster track confirmation
Sensor-to-shooter speed	Parallel optimization returns ranked firing solutions within command windows; auto re-verify before release	Seconds instead of minutes; synchronized salvos and decoys
EW/deception resilience	Passive quantum cues + quantum-aided PNT sustain ID and timing when RF is denied	Authentic C2 and accurate timing under jamming/spoofing
Precision engagement	Environmental and countermeasure modelling at scale; ROE-aware fire plans with abort thresholds	Higher first-shot effects; lower collateral; preserved magazines

Long-Duration Autonomous Missions with Quantum-AI Adaptability

Sustained autonomy with mission-flexible intelligence

Quantum-AI combines quantum-inspired optimization (running on today's CPUs/GPUs) with early quantum accelerators in secure enclaves to let unmanned platforms plan, re-plan, and conserve resources over weeks, not hours. Practically, this means a drone or unmanned submarine can carry a library of pre-computed contingency plans, then adapt mid-mission as weather, threats, fuel, or orders change without constant reach back. Quantum-aided navigation and timing reduce dependence on GPS, while onboard perception fuses weak signals to maintain situational awareness with low emissions. Think of a long-range patrol that leaves port with a seasoned staff officer in the back room, constantly reshuffling routes, fuel, and risk as new reports arrive. Quantum-AI gives that "staff" to the vehicle itself: many options explored in parallel, the best one selected for current constraints, and the rest stored for fast switches if the situation turns.

Over the next five years, operational relevance of Quantum-AI for autonomous drones, unmanned submarines, and intelligent weapons will centre on four gains: higher-confidence targeting in clutter, faster sensor-to-shooter sequencing, resilience under EW and deception, and ROE-aware precision engagement delivered through hybrid quantum-inspired models at the edge and selective quantum acceleration in planning enclaves.

Operational Relevance

- **Targeting in clutter:** Quantum-AI will fuse weak, disparate signals, slow radar cross-section echoes, faint acoustic or magnetic cues, and subtle thermal patterns into single, high-confidence tracks that stand out from sea clutter, coastal background, and urban camouflage, reducing false positives and fratricide risk in complex littorals and ports. Systems will maintain rolling "track truth" by testing multiple hypotheses in parallel, discarding decoy-consistent explanations, and promoting targets that consistently fit multi-sensor evidence over time, which preserves magazine depth and shifts sorties from search to prosecution.

- **Rapid sensor-to-shooter:** Parallel engagement planning will compress the interval from first contact to validated shot options by evaluating many weapon target pairings, timing windows, and flight profiles at once, returning ranked solutions with confidence, constraints met/violated, and pre-briefed abort thresholds. Autonomous platforms will automatically re-verify tracks, request authorization inside command policy windows, and execute with minimal back and forth, enabling distributed groups, UUV pickets, USV screens, and UAV scouts to coordinate salvos and decoys at the speed of the fight while keeping humans in the loop.

- **Resilience under EW and deception:** When GPS is spoofed and RF bands are saturated, quantum-aided perception leans on passive cues (gravity, magnetic, celestial/optical timing) and robust multi-sensor fusion to keep identification quality and engagement timing within ROE limits. Paired with quantum-resilient inertial/timing aids, platforms maintain navigation, timestamp integrity, and track stability, allowing formations to hold station, weapons to meet timing gates, and decision logs to remain auditable even as adversaries attempt to blind or mislead classical channels.

- **Precision engagement:** Fire-control optimization will incorporate environmental factors: sea state, ducting, wind shear alongside countermeasures and collateral constraints to deliver pinpoint effects with minimal rounds expended and clear accountability. By evaluating thousands of micro-scenarios before release, systems will select shots that maximize probability of kill while honouring no-fire zones and confidence thresholds, with immediate fallback options if confidence drops or the target presents decoy behaviour mid-flight.

Mission area	*What it means (future view)*	*How it works in practice*	*Commander payoff*
Targeting in clutter	Fuses weak multi-sensor signals (low-RCS, faint acoustic/magnetic, subtle thermal) into one high-confidence track in ports, straits, and urban littorals	Hybrid quantum-inspired models test many hypotheses in parallel, discard decoy-consistent explanations, and promote tracks that fit evidence over time	Fewer false positives and fratricide; more sorties shifted from search to prosecution; better magazine discipline
Rapid sensor-to-shooter	Collapses time from first contact to approved firing solution with parallel evaluation of weapon–target pairings, timing gates, and flight profiles	Platforms auto re-verify, generate ranked, ROE-aware shot options, request authorization, and execute within policy windows	Seconds instead of minutes for coordinated salvos; tighter alignment to ROE; sustained initiative at sea
Resilience under EW and deception	Maintains identification and timing when GPS is spoofed and RF is saturated by leaning on passive cues and robust fusion	Quantum-aided PNT preserves clocks and position; fusion stabilizes tracks despite jamming or deceptive emissions	Authentic C2 and accurate timing under denial; formations hold station; decision logs remain auditable
Precision engagement	Optimizes fire-control against environment and countermeasures with explicit confidence and abort thresholds	Evaluates thousands of micro-scenarios for sea state, ducting, wind, hard/soft-kill threats, and collateral constraints	Higher single-shot effectiveness; lower collateral risk; preserved munitions and clear accountability
Autonomy governance	Keeps humans in the loop with interpretable outputs and policy guardrails	Outputs show confidence, constraints met/violated, reasons for rejection, and fallback options	Faster approvals, safer automation, smoother after-action review and compliance
Fielding path (0–5 years)	Edge-first with surge acceleration during planning windows	Quantum-inspired fusion/optimization on existing CPUs/GPUs; optional quantum accelerators in secure enclaves for periodic recompute	Early gains without disrupting workflows; avoids single points of failure
Readiness metrics	Measures that predict connected combat tempo	Time-to-ID, time-to-engage, re-plan latency under EW, and false cue rates tracked per watch cycle	Data-driven scaling decisions and credible readiness briefs

Integration of Quantum-AI Drones into Network-Centric Warfare

Linking machines into a hyper-intelligent combat web

Quantum-enhanced AI (largely quantum-inspired algorithms now, with selective quantum accelerators in secure enclaves) lets drones perceive, plan, and re-plan faster while sharing cues and tasks across a common network. When these autonomous air and maritime drones are tied into a network-centric architecture, they act as a distributed mesh of eyes, relays, and precision strikers feeding tamper-evident keys from quantum communications where feasible, and using quantum-aided navigation to hold timing and position when GPS is denied. The result is a force-wide "combat web" that keeps orders authentic, targeting accurate, and tempo high under electronic warfare pressure. Imagine a convoy with scouts on every ridge and cove, each scout able to whisper secure updates, swap roles instantly, and guide fires without revealing the convoy's position. Quantum-AI drones are those scouts at sea and in the air: low-signature, constantly coordinating, and resilient when radios are jammed.

Operational Relevance

- **Persistent ISR and cueing**
  - ❏ Quantum-AI fusion elevates weak, multi-modal signatures from UAVs/USVs/UUVs into high-confidence tracks, cueing ship and shore sensors only when and where needed to reduce emissions and speed classification. This aligns with NCW's shift to distributed sensing and decision-making across surface, subsurface, air, and land nodes.
  - ❏ Naval seminars and doctrine emphasize moving from platform-centric to network-centric operations for speed and adaptability, a posture that autonomous swarms realize by providing continuous, shareable ISR in contested seas

- **Strike and decoy orchestration**
  - ❏ Parallel planning assigns sensor, decoy, and striker roles across swarms and synchronizes salvos within ROE, allowing distributed lethality while preserving human authorization for release. This supports joint missions from counter-piracy to high-intensity conflict where swarms overwhelm defences and protect high-value units.

❑ Lessons from recent conflicts show AI-driven drone tactics evolving towards coordinated "hellscape" concepts at sea; networked autonomy brings that tempo into carrier screens, coastal defence, and undersea infrastructure protection.

- **Assured comms and timing under EW**

  ❑ Post-quantum cryptography hardens identity and software signing across fleets; where adopted, satellite/terrestrial QKD can provide tamper-evident keys to gateways for coalition inter-op, with policy bodies highlighting quantum-safe transitions for future networks.

  ❑ Field-validated quantum navigation trials demonstrate continuous, hands-off operation at sea, offering GPS-independent PNT that is immune to jamming/spoofing and keeps formation timing aligned with minimal emissions.

Integration of Quantum-AI Drones into NCW

Aspect	*What it means*	*How it works in practice*	*Commander payoff*
Persistent ISR and cueing	Drones create a low-signature sensor mesh that spotlights real targets	Hybrid quantum–classical fusion elevates weak signals; cues hand off to ship/shore sensors	Fewer false alarms, faster kill chains, lower emissions
Strike/decoy orchestration	Swarms coordinate roles to overwhelm defences	Parallel optimization assigns sensor/decoy/striker roles with ROE guardrails	Synchronized salvos, efficient magazine use, human-in-the-loop control
Assured comms and timing	Trustworthy orders and timestamps under EW	PQC for identity; QKD where feasible; quantum-aided inertial/timing for GPS-denied ops	Authentic C2, accurate windows for fires and manoeuvre
Logistics sustainment	Drones keep parts, data, and relays flowing	Optimized routes for re-supply and comms relays; preplanned alternates	Higher sortie generation, resilient connectivity
NCW integration	Drones plug into the common operational picture	Pub/sub data with provenance; rapid re-key and tasking via shared policies	Coalition tempo and inter-operability under pressure
Fielding path (0–5 yrs)	Hybrid now, quantum boost in planning	Edge AI on CPUs/GPUs; secure enclave access to accelerators during windows	Early gains without workflow disruption
Readiness metrics	Measures that predict tempo	Time-to-ID, time-to-engage, re-plan latency under EW, link integrity under re-key	Evidence-based scaling, credible briefs

- **Logistics, relays, and resilience**
  - ❑ Quantum-inspired optimization schedules drone relays for comms coverage and micro-resupply in chokepoints, sustaining NCW links through jamming and physical disruption; EW market growth underscores the need for resilient, redundant architectures.
  - ❑ National roadmaps and capability plans call for integrating AI, quantum, and unmanned systems into joint C2 and undersea infrastructure defence, aligning acquisition with swarm deployment and quantum-safe networking.

Quantum-Informed Decision Loops in Kill Chain Automation

Accelerating the OODA loop beyond human reach

Quantum-informed decision-making refers to the near-term use of quantum-inspired algorithms on existing computing hardware, complemented by selective access to early quantum accelerators during defined planning windows, to evaluate many tactical options in parallel rather than sequentially. In practical terms, these methods ingest multi-domain data for sensing, targeting, and routing, then return ranked courses of action with explicit confidence levels, constraint checks, and abort thresholds, enabling faster observe–orient–decide–act cycles without diluting commander authority or accountability. This approach treats computation as a tempo advantage: more viable branches are explored within the same time budget, while outputs remain interpretable and auditable for rapid approval in accordance with rules of engagement.

Within the kill chain, quantum-informed analytics tighten each link by accelerating detection cueing, speeding track validation, parallelizing target–weapon pairing, and optimizing fire-control with policy and collateral constraints front and centre. The effect is to move decision advantage "left of the kill chain": the force detects earlier with lower emissions, discriminates true targets from decoys more reliably, weighs multiple engagement geometries simultaneously, and presents the command team with top-tier options that are ready to brief, defend, and execute. Because these tools expose why options were promoted or rejected and how sensitive they are to shifting intelligence, they integrate cleanly into existing battle rhythms rather than replacing commander judgement.

A useful analogy is a combat operations centre staffed by many expert cells working at once: one cell fuses radar, electro-optical/infrared, and acoustic hints into coherent tracks; another tests decoy and spoofing hypotheses against the evidence; a third optimizes strike timing, routing, and weapon–target assignment under environmental and countermeasure constraints. Within minutes, the team converges on the top three, policy-compliant options complete with confidence, risks, and fallback plans. Quantum-informed decision loops aim to embed that "parallel brain" behind every autonomous patrol and distributed sensor–shooter team, while preserving human-in-the-loop control to approve, redirect, or abort as the tactical situation evolves.

Operational Relevance

- **Observe and orient at speed:** Autonomous pickets will lift weak multi-sensor cues (EO/IR, radar, acoustic, magnetic) out of clutter and hand them to commanders as high-confidence tracks, reducing emissions by cueing ship/shore sensors only when needed and enabling earlier discrimination of decoys and spoofing. This aligns with kill web concepts that favour many parallel find-fix paths over a single linear chain, improving survivability and tempo in contested seas.

- **Decide with interpretable options:** Quantum-informed optimizers will evaluate weapon-target pairings, flight profiles, and timing windows in parallel, returning ranked, ROE-compliant courses of action with constraints met/violated and pre-briefed abort thresholds. Trials and demonstrations across the force already show AI compressing targeting cycles; quantum-informed methods push this further while preserving human-in-the-loop authorization and audit trails.

- **Act under EW pressure:** Quantum-assured navigation and timing validated in recent maritime trials will keep formations on station and timestamps trustworthy when GPS is denied or spoofed, maintaining shot synchronization and legal compliance under jamming. Combined with post-quantum cryptography for identity and software signing, and QKD pilots where feasible, kill chains remain authentic and tamper-evident as they accelerate.

- **From chains to webs:** Joint programs aim to replace brittle linear chains with self-healing kill webs that re-route around damaged nodes, using machine-to-machine data exchange to sustain F2T2EA continuity. Quantum-informed analytics increase the number of viable pathways considered per minute, raising resilience and complicating adversary countermeasures in multi-domain operations.

Autonomous Threat Anticipation via Quantum-AI Fusion

Machines that predict conflict before it erupts

Autonomous threat anticipation via quantum-AI fusion represents a decisive evolution in the way future battles will be fought, where machines gain the capacity to detect, analyse, and react to hostile intent before conventional sensors or human commanders can respond. Over the next five years, rapid advances in quantum computing, combined with artificial intelligence, are projected to give autonomous defence and attack platforms the ability to process vast, uncertain, and complex battle space data streams at speeds that far exceed current capabilities. Unlike present-day machine learning systems constrained by processor limitations, quantum algorithms can assess multiple possibilities simultaneously, modelling adversary behaviour with unprecedented depth and speed. This enables unmanned systems not merely to track threats but to forecast engagements before shots are fired.

From a military analogy perspective, conventional early warning systems are akin to sentries watching the horizon for movement. Quantum-AI fused systems, however, act as seasoned battlefield strategists standing atop a command tower, observing subtle changes in posture, terrain usage, signal traffic, and resource positioning instantly correlating these patterns into probable enemy courses of action. These platforms will, in effect, be able to "read" the battle space with an intuition that today remains the domain of experienced human commanders, but applied at computational speeds and scales far surpassing human limits. A single autonomous UAV or maritime vessel could, through quantum-enhanced situational modelling, identify the risk trajectory of an enemy task force hours before a kinetic action occurs, allowing commanders to adjust tactics pre-emptively.

Operationally, this capability feeds directly into the emerging quantum arms race among technologically advanced militaries. In much the same way precision-guided munitions once shifted the balance of force, machines capable of predicting a conflict before it erupts will redefine deterrence and rapid decision-making. Quantum algorithms embedded in autonomous systems will allow mission priorities to be dynamically reassigned in real time and an unmanned strike drone tasked for reconnaissance could instantly switch to a counter-attack role if predictive modelling indicates a window of vulnerability in hostile air defences. Conversely, a naval defence platform could disengage from a surface threat and redirect resources towards protecting logistical convoys if quantum-AI analysis forecasts a high-probability ambush in a different sector. This anticipatory agility will be as critical to future combat success as stealth or firepower.

In practical terms, the battlefield of 2030 could be populated by fleets of unmanned combat aerial vehicles, submersible drones, and autonomous ground assets, all continuously exchanging encrypted, quantum-secured data while updating conflict prediction models in near real-time. The side that masters autonomous threat anticipation via quantum-AI fusion will not simply react faster; it will shape engagements by denying the adversary the element of surprise and by imposing decision dilemmas before hostilities escalate. This will transform the tempo of operations, giving commanders a predictive edge that compresses the Observe–Orient–Decide–Act (OODA) loop into fractions of current response times, redefining both defensive posturing and offensive manoeuvre in the quantum warfare era.

Operational Relevance

- **Deterrence and escalation control**
  - ❏ Predictive autonomy shifts deterrence from "punish after" to "pre-empt surprise," as machines flag converging indicators of impending conflict and push automated alerts and courses of action to commanders before an adversary commits to escalation, complicating enemy decision calculus and raising the cost of surprise.

  - ❏ Defence planners already frame deep-tech convergence quantum, AI, and space ISR as an integrated posture to see, decide, and signal earlier; near-term funding and programs indicate field experiments that tie predictive analytics to operational alerting and readiness levels by FY26–30.

- **Dynamic mission re-prioritization**
  - ❏ Hybrid quantum-classical solvers can re-optimize tasking across UAV swarms, maritime USVs, and ground UGVs as the battle space shifts,

turning reconnaissance sorties into immediate strike-support or screen-defence roles when predictive risk scores cross thresholds.

❑ Case work in large-scale scheduling shows quantum-enhanced optimization increases solvable problem size and reduces iteration cost, enabling faster re-tasking windows for convoys, patrols, and CAP orbits under hard constraints like fuel, airspace de-confliction, and EW threats.

- **ISR fusion and early warning**

  ❑ Quantum-accelerated analytics improve hyperspectral, RF and multi-INT processing, extracting weak signals of intent (logistics staging, emissions discipline shifts, anomalous comms patterns) to raise early warning fidelity and reduce false positives in contested environments.

  ❑ Programs oriented to rugged quantum sensing and navigation (e.g., RoQS) push resilient ISR and PNT under jamming, ensuring autonomous platforms keep collecting and cueing predictions even in GPS-denied or EW-intense fights.

- **Tempo and OODA compression**

  ❑ Quantum-AI fusion shortens planning cycles by rapidly simulating branches and sequels under uncertainty, letting autonomous elements propose and execute micro-adjustments (route changes, sensor pointing, decoy deployment) inside an adversary's OODA loop.

  ❑ This computational tempo advantage turns into operational tempo when paired with secure links and machine-to-machine C2, enabling swarms to exploit fleeting gaps in IADS or maritime screens identified by predictive models.

- **Resilience under contested C2**

  ❑ Quantum-secured communications and preparatory post-quantum measures protect anticipatory models and tasking data from adversary interception or tampering, preserving the integrity of autonomous decision aids during crisis.

  ❑ NATO, DARPA, and allied initiatives indicate a 5–10 year glide path to ruggedized quantum links and sensors; planning that assumes hybrid deployments (classical AI + quantum accelerators at the edge or in the loop) yields practical near-term gains without waiting for fault-tolerant systems.

Operational effect	What changes in practice (next 5 years)	Representative enablers	Command impact
OODA loop compression	Predictive analytics triage ISR feeds and pre-position effects; planners receive machine-generated COAs before adversary moves are finalized.	Quantum-accelerated analytics for multi-INT and hyperspectral; deep-tech convergence of quantum, AI, and space ISR.	Faster warnings, fewer false positives, earlier deterrent posturing and manoeuvre windows.
Dynamic mission re-tasking	Swarms re-optimize roles mid-mission (recon ↔ strike support; screen defence) as risk scores cross thresholds.	Quantum-inspired optimization for swarm tasking and real-time path/energy trade-offs.	Higher sortie effectiveness and resilience under EW/terrain disruptions.
ISR fidelity uplift	Weak indicators (logistics staging, emissions shifts) fused into higher-confidence early warning in cluttered theatres.	Quantum-enhanced processing for hyperspectral/RF; QML-assisted anomaly detection.	Raises detection probability while reducing analyst load and response latency.
Logistics continuity	Convoy, depot, and air task schedules re-computed in minutes under attack/disruption; sustainment aligns to predictive demand.	Quantum-inspired schedulers for multi-echelon inventory, depot MRO, threat-aware routing.	Higher mission availability; reduced turnaround and stock-out risk at the edge.
Contested C2 resilience	Anticipatory models and tasking data protected against interception/tamper; secure machine-to-machine control.	Post-quantum cryptography migration; quantum-secure comms roadmaps in defence.	Preserves integrity of autonomous decision aids during crisis and grey-zone pressure.
Swarm scale and agility	Coordination of 100+ UAVs with adaptive placement and load optimization in GPS/EW-contested zones.	Quantum-inspired evolutionary optimization; early quantum planning for swarms.	Broader coverage, faster massing/dispersal, fewer collisions/airspace conflicts.
Space-enabled warning	Space-based early warning architectures integrate quantum-ready analytics for faster cueing to shooters.	Space Force prototyping in missile warning/secure comms; quantum-AI embedding path.	Shorter sensor-to-shooter chains; improved cross-domain fires timing.
Fielding pathway	Hybrid deployments: classical edge AI + quantum-enhanced training/optimization via secure backhaul or theatre nodes.	Near-term TRL focus on logistics, ISR triage, mission de-confliction pilots FY26–30.	Measurable gains without waiting for fault-tolerant quantum hardware.
Deterrence posture	"Predict-to-deter": earlier signalling and posture shifts complicate adversary calculus pre-crisis.	Budgeted deep-tech convergence; irregular warfare use-cases for quantum-assisted analysis.	Raises cost of surprise; narrows adversary decision space.
Standards and migration	PQC adoption, interoperable radios/links, and data pipelines for quantum-assisted analytics.	Federal PQC summits and service-level interoperability pushes (e.g., SDR standards).	Secure integration of autonomous prediction into joint/combined C2.

- **Logistics and sustainment advantage**
  - ❑ Predict-then-optimize pipelines extend beyond fires and ISR to sustainment, where quantum-enhanced solvers demonstrated utility in convoy scheduling and large-scale routing; in combat, the same tools reprioritize fuel, munitions, and recovery tasks to keep forward autonomous assets mission-capable.
  - ❑ Faster recomputation under disruptions (bridge loss, port denial, EW corridor shifts) improves operational availability and reduces downtime, directly impacting sortie generation and persistent presence in contested zones.

Practical fielding stance (next 5 years)

  - ❑ Expect hybrid architectures: Edge AI on HPC/GPU nodes integrated with quantum-enhanced optimization/training pipelines via secure backhaul or theatre nodes, focusing initially on ISR triage, mission re-tasking, and convoy/air task de-confliction.
  - ❑ Prioritize TRL-7/8 sensing and post-quantum cryptography adoption, plus pilot programs that measure OODA compression and mission availability uplift, aligning with current defence investment signals and allied experimentation roadmaps.

Smart Payload Management Using Quantum Algorithms

Weapons that choose their own tactics

Over the next five years, hybrid quantum-classical optimization will begin to sit inside mission computers and shore-based/tender-side planners for unmanned systems, guiding what payload to carry, when to employ it, and in what sequence. In plain terms: instead of pre-scripted loadouts and fire plans, drones and UUVs will use quantum-inspired solvers to evaluate thousands of constraints and options in real time, rules of engagement, fuel/energy state, target defences, collateral risk, weather/sea state, EMCON posture and select the payload and firing pattern that best achieves the commander's intent. Early capability will come from quantum-inspired algorithms on GPUs/HPC, with selective routing to maturing quantum accelerators as they become available. Think of today's weapons officers doing "stores management" on a manned aircraft or a ship's combat system: they weigh threat type, distance, decoys, magazine depth, and risk of counter-fire before assigning munitions. Smart payload management is like having a tireless deputy weapons officer embedded in each unmanned platform and its control node, continuously re-computing the best combination of decoys, jammers, kinetic rounds, and sensors as the battle space shifts. Where a human might re-plan between sorties, the quantum-AI stack re-plans between seconds.

This shift is a pacing-race dynamic: forces that field autonomous payload managers which can select tactics faster than opponents can respond will dominate salvo exchanges and conserve magazines. Near-term defence programs are already funding quantum software for C2/ISR optimization and research into fault-tolerant pathways, signalling that "weapons that choose their own tactics" will emerge first as tightly bounded autonomy with human-in-the-loop oversight. The competitive edge will come from doctrine and data integration as much as hardware training commanders to set clear decision envelopes and integrating PQC/secure links so adversaries cannot spoof the planner's inputs.

Operational relevance (next 5 years)

- **Adaptive loadout and employment:** Quantum-inspired solvers improve pre-mission loadout selection (e.g., mix of sensors, decoys, and effectors) and in-mission sequencing (e.g., jam-decoy-strike) under uncertainty, raising probability of kill while conserving scarce munitions.

- **Magazine depth and sustainment:** By optimizing shot doctrine against layered defences, platforms expend fewer high-end rounds for the same effect, easing re-supply burden and extending on-station time, crucial for dispersed maritime and littoral ops.

- **Swarm coordination:** Coordinated payload choreography across UAV/USV/UUV swarms reduces fratricide and de-conflicts spectrum use; quantum-assisted tasking assigns who emits, who shoots, and who holds in reserve based on evolving threat lay-down.

- **Counter-air/air-defence breaching:** For penetrations against IADS, algorithms select standoff vs. close-in effects, order decoys and SEAD shots, and time salvos to saturate radars and interceptors, compressing the adversary OODA loop.

- **Maritime ASW/ASuW:** UUVs/USVs vary sensor modes, deploy decoys, or cue kinetic/non-kinetic payloads based on acoustic and EM anomalies; scheduling and routing are re-optimized when contact quality changes or counter-detection risk spikes.

- **C2 resilience:** If links are degraded, onboard planners fall back to commander's pre-authorized decision envelopes and continue optimized employment; when links recover, shore or mothership nodes can re-seed plans with larger quantum resources.

Dimension	FY26–FY27	FY28–FY29	FY30 objective
Pilot use cases	Limited-scope trials for ISR triage on unmanned platforms; initial convoy/air task de-confliction pilots using quantum-inspired schedulers; lab and range validation of WTAP-style salvo planners on representative scenarios.	Operational experiments integrating WTAP-style salvo scheduling with live-virtual-constructive exercises; UUV/UAV routing trials under EW stress with real-time re-planning; initial tactics, techniques, and procedures for predictive loadout sequencing.	Missionized playbooks for ISR triage, de-confliction, and salvo scheduling with human-in-the-loop sign-offs; routine UUV/UAV contested-EM routing in exercises and selected operations.
Architecture	Edge AI onboard for perception and classification; local quantum-inspired solvers for tasking and scheduling; secure backhaul prepared for PQC migration per NIST guidance.	Optional reach-back to theatre HPC/quantum nodes for batch re-training and heavier optimization; mixed classical-quantum pipelines aligned with SIPRI-noted maturity variance across compute, comms, and sensing.	Hardened PQC in links and enclaves; standardized interfaces for edge–theatre quantum acceleration; configuration baselines for joint/combined operations.
Governance	Human-in-the-loop control enforced in software; ROE-aware constraint sets embedded in planners; cryptographic audit logs for every autonomous recommendation.	Independent test and evaluation of decision envelopes; red team/blue team assessments for spoofing and harvest-now-; decrypt-later risk; staged PQC deployment across C2 segments.	Doctrine and training codify authority to proceed, override, and abort; continuous compliance auditing and after-action learning pipelines institutionalized.
Readiness metrics	OODA compression measured in seconds saved from detection to tasking; false-positive/negative rates in ISR triage; optimization latency for de-confliction.	Magazine efficiency (shots per effect) under WTAP-style scheduling; mission availability under EW-induced re-routes; cyber integrity under PQC pilots.	Standardized joint KPIs for anticipatory autonomy across services; interoperability pass rates in combined exercises.
Risk mitigations	Fail-safe reversion to human-directed modes; bounded autonomy with geo-fencing and ROE constraints; PQC migration plans initiated.	Model poisoning/spoofing drills; encrypted telemetry with hybrid-PQC; contingency procedures for quantum node unavailability.	Full PQC posture on critical paths; formal safety cases and certification for autonomous decision aids.

5.3 Quantum Cyber Security in Defence

Securing Military Assets, Infrastructure, and Operational Command Against Quantum Cyber Attacks

Quantum-hardened command, quantum-safe force: deny the breach, dominate the battle space

Shielding Autonomous Defence Systems from Quantum Hijacking

Ensuring our drones obey only one master: the Tricolour

Over the next five years, the most credible cyber risk to autonomous military systems is not science-fiction AI takeover but command spoofing, link hijacking, and model/data tampering that force a platform to misidentify, mis-navigate, or misfire. Quantum-grade defences harden these weak points by combining post-quantum cryptography (PQC) for all command, control, and update channels; quantum-secure keying where available (e.g., QKD/ MDI QKD on fixed links); verifiable execution proofs from compute nodes; and hardware-level tamper detection on sensors and radios. In simple terms: every command and every sensor byte must carry an unforgeable "military fingerprint," and every onboard decision must be provably authentic and auditable. Treat each drone, UGV, or UUV like a dispersed platoon post at night. The sentry challenges with a password, checks insignia under red light, and confirms orders via the company net before acting. Quantum authentication is the modern challenge-and-reply: PQC-signed orders (the password), quantum-safe session keys (the insignia), and verifiable computation receipts (the commander's confirmation) that together make it extraordinarily hard for an adversary to slip a fake order, replay an old one, or falsify sensor truth.

Deterrence in the quantum era will favour forces that can prove command authenticity at machine speed while denying an adversary any foothold to redirect swarms or falsify the picture. Embedding quantum authentication from PQC at the edge to MDI-QKD in backbone links turns every autonomous asset into a loyal node that rejects counterfeit orders, refuses tainted data, and reports tamper attempts in real time. In contested EM environments, that reliability preserves initiative: swarms keep formation, UUVs hold picket lines, and counter-UAS batteries fire on genuine threats, not on decoys. The side that fields verifiable autonomy with quantum-safe trust anchors will not only resist hijacking but also impose confidence shocks on adversary systems that cannot prove who is in command.

Operational relevance (next 5 years)

- **Command spoofing denial:** Replace RSA/ECC on C2 radios, telemetry, and swarm M2M protocols with NIST-profiled lattice schemes (e.g., Kyber KEM for key exchange, Dilithium signatures) and hybrid handshakes to maintain backward interoperability while phasing out vulnerable primitives against "harvest-now, decrypt-later" adversaries. PQC prevents future quantum decryption of captured command traffic and strengthens authenticity today; attackers already record traffic to decrypt later, so delaying migration leaves command streams exposed over their sensitivity lifetime. Use PQC-ready TLS/IPsec profiles on datalinks and gateways, test MTU/latency impact for control loops, and add MDI-QKD over fibered or fixed microwave backbones to remove detector side channel risks and rotate high-assurance keys for critical relays and command posts.

- **Anti-tamper sensing and data integrity:** Add optical isolators and narrowband filtering on quantum-adjacent photonic paths; monitor detector current/temperature to catch blinding or Trojan horse attempts; cryptographically bind every sensor frame to time, position, and platform ID with PQC signatures. Practical "quantum hacking" exploits often target device level physics rather than maths; authenticated sensing pipelines make injected or replayed frames detectable at ingestion and throughout the fusion chain. Implement signed sensor metadata and immutable logging at the edge; trigger safing states when physical telemetry deviates from envelopes (e.g., detector bias drift), and cross-validate with independent sensors to defeat single-point spoofing

- **Verifiable autonomy:** Require critical onboard planners and sensor-fusion modules to produce lightweight verifiable computation receipts that can be checked by a supervisory core or offboard node without re running the full computation. If an adversary tampers with models, libraries, or inputs, the proof fails, flagging compromise even when internals are opaque during flight; this preserves tempo while adding trust to machine decisions. Use succinct interactive proofs tailored for embedded/HPC co-processors, log proofs alongside inputs/parameters, and gate weapon-release or autonomous manoeuvres on proof validity within ROE-configured envelopes.

Line of effort	Concrete actions (FY26–FY30)	Integration notes	Operational effect
PQC first	Deploy lattice-based KEM for session keys and lattice-based signatures for command authenticity on C2 radios, datalinks, gateways, and update servers; enable hybrid handshakes during transition; baseline latency/MTU and tune control-loop timeouts	Inventory all crypto end-points; enforce crypto-agility (roll/rotate without platform re-flash); stage migration by mission criticality; include "harvest-now, decrypt-later" risk in classification and retention policies	Immediate uplift in command integrity and future-proofing against quantum decryption; sustained control even if adversaries capture traffic for later decryption
QKD where it fits	Pilot MDI-QKD between fixed command posts, service gateways, ship-to-shore relays, and satellite ground stations; automate key failover to PQC when quantum links degrade	Reserve QKD for stationary/high-value circuits; monitor QBER and link health; integrate key-orchestration that pre-ferences QKD keys but maintains mission with PQC	High-assurance key rotation on backbone links, removal of detector-side attack surface on trusted relays, and graceful continuity for mobile/tactical users
Device hardening	Add optical isolators, narrowband filters, detector current/ temperature monitors on photonic/quantum-adjacent sensors; EM shielding on radios; introduce randomized scheduling/ memory mapping to de-correlate side channel leakage	Tie physical telemetry to automated "safing" states; require dual-sensor corro-boration for critical cues; validate hardening in EW/ laser-illumination drills and side channel red-teams	Denial of blinding/Trojan-horse vectors and reduced exploitability of implementa-tions, preserving trustworthy sensing and autonomy inputs in contact
Verified execution	Require cryptographic attestations from autonomy kernels and sensor-fusion outputs before high-risk actions; log proofs with ROE constraints and input hashes for after-action audit	Use lightweight proofs checkable by supervisory cores; gate weapons release/ manoeuvres on proof validity; retain human-on-the-loop override paths	Tamper-evident autonomy that maintains tempo; "no proof, no fire" policy reduces risk of model/library compromise under pressure
Quantum-aware monitoring	Train ML detectors on QKD/C2 anomaly features (timing variance, decoy-state anomalies, error bursts); wire detectors to permission-reduction "safing" until human clearance	Feed telemetry from both classical and quantum links; define tiered responses (rate-limit, read-only, hold-fire) mapped to anomaly confidence; drill operators on clear/restore procedures	Early warning for quantum-specific attacks and safe degradation of autonomy, preventing escalatory misfires or adversary-triggered malfunctions

- **Supply chain and update hardening:** Enforce PQC-signed firmware and container images, randomized resource mapping to blur side channels, and strict sandboxing between perception, navigation, and fire-control domains to prevent lateral movement. Even quantum-safe algorithms can be broken by AI-assisted side channel attacks against weak implementations; strong code provenance and isolation reduce exploitability and blast radius. Integrate PQC attestation in CI/CD, require hardware roots of trust for boot and module loading, and continuously fuzz/test PQC stacks against AI-driven leakage analysis before fleet rollout.

Building Quantum-Defensive Layers into Smart Military Infrastructure

Fortresses of the future are secured by entanglements

Over the next five years, smart bases and mission-critical installations will add quantum-defensive layers that harden both the digital perimeter and the physical estate. Practically, this means deploying post-quantum cryptography (PQC) across command networks; introducing quantum-secure backbones (where feasible) for key distribution; and integrating quantum-enhanced sensors for tamper, intrusion, timing, and navigation resilience. The goal is simple for commanders: even if an adversary records your traffic today or degrades GNSS tomorrow, your bases keep communicating securely, your sensors keep telling the truth, and your decisions remain trustworthy under pressure. Governments and alliances are already signalling this trajectory, treating quantum as a convergence layer spanning encryption, sensing, and timing for defence infrastructure. Think of a fortified airbase with layered defences: outer fences and patrols, hardened blast walls, guarded gates with challenge-and-reply, and an inner citadel for command. Quantum-defensive layers are the digital-physical equivalent. PQC is the reinforced gate key that won't be copied by tomorrow's master locksmith; quantum key distribution on fixed backbones is the guarded courier route that can't be ambushed without detection; quantum sensors are the silent ground microphones and tripwires that reveal tunnelling or spoofing attempts; and quantum-stable clocks are the base's synchronized watches that keep every sortie and interceptor aligned even when GPS is jammed.

Fortresses of the future are secured by entanglement. Strategic competitors are already trailing quantum-secure networks and quantum sensing alongside AI and space systems. The race favours forces that can prove command authenticity, maintain secure timing, and detect stealthy physical intrusions even in contested EM environments. Building quantum-defensive layers into smart infrastructure denies adversaries quick wins no easy uplink spoof, no silent perimeter breach, no time de-synchronization gambit and preserves the initiative for commanders. Budget signals and national programs indicate this convergence is moving from concept to field pilots within the five-year window, with uneven maturity across compute, comms, and sensing best addressed through hybrid deployments (PQC everywhere; QKD and specialized sensors where feasible).

Operational relevance (next 5 years)

- **Harden command authenticity and continuity:** Migrate base C2, facility control systems, and cross-domain gateways to standardized PQC with crypto-agile designs to defeat harvest-now, decrypt-later risk and preserve interoperability during transition. Where fibered or fixed microwave backbones exist (e.g., between ops centres, radar sites, SATCOM ground stations), layer measurement-device-independent QKD (MDI-QKD) segments to raise key assurance and detect tapping attempts.

- **Secure edge for critical infrastructure:** Field PQC-hardened edge controllers for power, fuel farms, magazines, and runway systems; bind sensor data (video/RF/SCADA) with quantum-safe signatures at source so forged frames can't enter the fusion chain unnoticed.

- **Quantum sensing for facility protection:** Pilot quantum-enhanced intrusion and tamper detection (e.g., distributed fibre sensing, magnetometry, and gravimetry) around perimeters, piers, and magazines to reveal tunnelling, underwater approach, or concealed movement with higher sensitivity and lower false positives in cluttered environments.

- **Timing and navigation resilience:** Introduce quantum-grade timing holdover (optical/atomic references) at bases, radar stations, and uplinks so air defence and sortie scheduling remain synchronized under GNSS denial; pair with quantum-aware navigation aids for harbour and airfield ops.

- **Space and uplink assurance:** For satellite gateways and long-haul relays, combine PQC for mobile segments with QKD pilots on fixed ground-to-ground trunks; monitor quantum bit error rates and anomaly signatures to trigger key failover without interrupting mission traffic.

- **Doctrine and readiness:** Update base-defence SOPs to include quantum anomaly drills (QKD tamper alerts, PQC key rotations, sensor-authentication failures) and measure recovery time objectives; align procurement to require PQC readiness and quantum-sensor interfaces by default.

Quantum Forensics for Cyber Attack Attribution and Response

Detecting the undetectable, tracking the untraceable

Over the next five years, quantum-enhanced forensics will augment classical cyber investigations by hardening evidence integrity and accelerating pattern correlation across vast, noisy data. In plain terms, post-quantum cryptography (PQC) preserves chain of custody against future decryption; quantum-safe hashing and time-stamping seal logs and AI-generated artefacts; and quantum-inspired algorithms speed correlation of indicators of compromise (IOCs) across multi-jurisdictional infrastructures. Together, these tools raise attribution confidence and compress response timelines from days to near real time, even when adversaries use deception, false flags, or anonymization networks to obscure their origin. Imagine a joint task force conducting a maritime interdiction. Classical tools are your binoculars and ship logs useful but spoof able. Quantum forensics adds tamper-evident seals to every logbook entry, synchronized chronometers that cannot be skewed by jamming, and reconnaissance that can correlate faint wakes across open ocean to a single vessel. Even if an adversary changes flags mid-sea (false flags in cyber), the sealed logs and cross-domain correlations reveal the true course and ownership, enabling lawful interception.

Strategic competitors are racing to pair AI-enabled campaigns with quantum-era anti-forensics. Forces that field quantum-forensic chains, PQC for authenticity, quantum-resistant time-stamping for chronology, and accelerated multi-domain correlation will expose covert operations quickly, deny plausible deniability, and underwrite decisive, legally defensible responses. This is less about exotic hardware and more about disciplined, quantum-ready tradecraft embedded across sensors, networks, and command posts.

Operational relevance (next 5 years)

- **Evidence integrity under quantum threat:** Adopt PQC-backed signatures and quantum-resistant hashing for logs, telemetry, and AI artefacts to preserve evidentiary value across the full lifecycle defeating "harvest-now, decrypt-later" strategies and enabling credible, shareable records for coalition response and proportional countermeasures.

- **Faster, deeper correlation:** Use quantum-inspired optimization to link IOCs across fragmented infrastructures (cloud, proxies, botnets), improving detection of campaign-level patterns that traditional point tools miss; pair with AI that maps observed TTPs to known actors while keeping proofs and hashes quantum-safe.

- **Deception-aware analytics:** Design attribution workflows that treat obfuscation, false flags, and jurisdictional hops as expected noise; quantum-secure evidence chains plus multi-source correlation reduce the risk of misattribution and the political costs of premature public statements.

Line of effort	*Concrete actions (FY26–FY30)*	*Integration notes*	*Validation metrics*
Seal the chain	Apply PQC signatures and quantum-resistant hashes to all high-value logs, model outputs, and evidence packages; deploy tamper-evident, synchronized time sources at ops centres and gateways	Use crypto-agile logging pipelines and standardized evidence formats; ensure time sources are independently auditable and resistant to desync/spoof	Chain-of-custody verification rate; time drift under attack; admissibility/readiness for coalition sharing.
Correlate at scale	Stand up quantum-inspired graph/ optimization analytics to stitch IOCs across clouds, proxies, and botnets; tag events with ATT&CK and Diamond-model metadata to generate actor hypotheses with confidence scores	Integrate with TI platforms (MISP, ATT&CK) and SIEM/SOAR; calibrate confidence thresholds for operational vs. public attribution	Time-to-linkage for multi-infrastructure campaigns; correlation precision/recall; actor-hypothesis confidence calibration curves.
Grade confidence, act proportionally	Formalize tactical/operational/ strategic attribution tiers with decision thresholds tied to response playbooks (isolation, sinkholes, sanctions, public statements)	Embed thresholds in CONOPS; require red team challenge reviews before escalation; maintain rollback paths if confidence degrades	Decision latency by tier; rate of correct/contested attributions; proportionality alignment in after-action reviews
Train and exercise	Run joint drills featuring false flags, fragmented evidence, and coalition sharing over PQC-secured exchanges; include legal/policy cells for release decisions	Use realistic datasets and live traffic captures; rehearse cross-border information-sharing mechanisms and evidentiary standards	Mean time-to-attribution; confidence vs. ground truth; inter-ally data-fusion success and friction points.

- **Near-real-time response:** With sealed telemetry and accelerated analytics, defenders can publish confidence-graded attribution assessments faster, cue technical countermeasures (isolation, key rotation), and brief leadership with defensible evidence, aligning response proportionality to campaign scope.

- **Coalition interoperability:** Standardize PQC-capable logging, exchange formats, and evidentiary thresholds so allied commands can fuse data without downgrading security, enabling joint attribution statements and synchronized actions.

Chapter Six

Global Quantum Military Landscape and India's Strategic Quantum Roadmap

This chapter, "Global Quantum Military Landscape and India's Strategic Quantum Roadmap," opens with a clear assessment of the emerging quantum arms race and how it is reshaping power balances, deterrence models, and escalation dynamics across land, sea, air, space, and cyber. It explains why quantum communications, sensing, and computing are moving from research programs to force planning assumptions, compressing decision cycles, undermining legacy encryption, and enabling new forms of resilience under electronic and cyber-attack. It provides an intelligible framework to read competitor capability statements, separate engineering reality from hype, and translate global breakthroughs into operational risk and opportunity.

A comparative survey then maps the trajectories of major powers and coalitions, comparing national quantum initiatives, space based quantum links, city scale and backbone networks, precision timing and navigation, and dual use industrial ecosystems. The focus is on what these investments mean in practice: survivable C2 control in GPS denied environments, low probability of intercept communications, stealth detection through quantum sensing, and quantum accelerated planning for complex operations. The analysis highlights likely timelines, dependencies such as photonics supply chains and cryogenic infrastructure, and the indicators and warnings that should trigger doctrine reviews, procurement decisions, and coalition interoperability work.

Building on this global picture, the chapter articulates India's Quantum Military Defence Strategy as a coherent, phased roadmap. Near term actions prioritize crypto agility and the transition to post quantum cryptography; pilot projects of quantum key distribution across strategic corridors; and operational test campaigns for quantum assisted navigation, timing, and passive sensing in contested electromagnetic environments. Mid term goals include integrating quantum secure command paths into joint C2 architectures, deploying hybrid QKD/PQC at scale, and fielding quantum enhanced decision tools for logistics, maritime air operations, and integrated air/ missile defence. Long term objectives focus on sovereign manufacturing of critical components, space based quantum links, and mission tailored quantum processors aligned to defence problem sets.

The chapter concludes with governance, people, and procurement levers to make the roadmap executable: a joint quantum requirements board linking the three Services, DRDO, ISRO, academia, and industry; red teamc and validation ranges for quantum communications and sensing; workforce continuums for operators, engineers, and acquisition professionals; and standards for interoperability with trusted partners. It sets measurable milestones such as coverage targets for quantum secure links, certification gates for post quantum migration, readiness levels for quantum sensors, and decision support performance thresholds, so that leaders can track progress, contain risk, and ensure that the Indian armed forces retain strategic advantage as quantum capabilities move from laboratories into the battlespace.

6.1 THE QUANTUM ARMS RACE

From qubits to command, whoever secures trust, time, and truth will set the tempo of power

Changing Balance of Power: From Nuclear Deterrence to Quantum Deterrence

The next Cold War may be fought in Qubits, not kilotons

Quantum deterrence shifts credibility from destructive payloads to information superiority, or the ability to secure command and control, see first in denied environments, and act fastest with trustworthy data. In practice, three quantum pillars underpin this shift. First, *quantum-safe communications* (post-quantum cryptography today, with selective quantum keying pilots where feasible) make orders tamper-evident and deny "harvest-now, decrypt-later" payoffs, preserving C2 secrecy during prolonged crises. Second, *quantum-aided sensing and timing* (quantum inertial navigation, precision clocks, magnetometry/ gravimetry) erode the sanctuary of stealth and spoofing, making covert movements riskier and GPS denial less decisive. Third, *quantum-informed computing* (largely quantum-inspired algorithms in the near term, with emerging accelerators for planning windows) compresses the observe-orient-decide–act cycle by evaluating many courses of action in parallel. The combined effect is deterrence by assured command integrity, persistent sensing, and machine-speed decision tempo factors that complicate brinkmanship by reducing the value of surprise, disruption, and deception. In a traditional sense, the biggest arsenal often discourages attack, but in the quantum era, the best shield of information and timing deters even coercion. Imagine two adversaries at sea: one sails with unforgeable orders, passive sensors that reveal hidden threats, and a staff that can test dozens of operational plans at once; whilst the other must signal loudly, trust information that could be forged, and can only check one plan at a time. The quantum-enabled force is at an advantage, since it cannot be blinded, spoofed, or delayed, making coercion costly and escalation unattractive. The opponent's Course Of Action generation is made that much more difficult.

Credible command and control under pressure begins with quantum-safe communications and disciplined key management. By mandating post-quantum cryptography for identity, code signing, and mission messaging and piloting quantum key distribution on select, high-value links, orders remain authentic and confidential throughout crises, enabling rapid, defensible directives and coherent messaging. In parallel, quantum-aided navigation and precision timing keep formations synchronized and information auditable in GPS-denied environments, while quantum magnetometry, gravimetry, and low-signature optical techniques raise the probability of detecting stealth platforms, increasing their cost of remaining hidden and narrowing windows for surprise.

Deterrence at decision tempo follows from quantum-informed planning engines that generate ranked, ROE-compliant options with confidence and constraint checks in minutes rather than hours, making proportionate responses credible before an adversary can consolidate gains. To manage escalation during joint/ multi-agency operations, tamper-evident evidence packages, signed telemetry, authenticated timestamps, and sensor attribution enable rapid and trusted sharing, aligning thresholds for public attribution and joint action, while reducing miscalculation in high-tempo standoffs. Routine but non-provocative demonstrations, GPS-denied navigation trials, low-emission ISR cueing, rapid re-key and restoration under electronic warfare, signal resilience and readiness, deterring coercion without the escalatory optics of overt nuclear posturing.

A credible quantum posture starts with a PQC-first migration discipline and rehearsed emergency re-key. There should be an inventory of all cryptographic dependencies across platforms, gateways, and applications; move identities and software signing to standardized post-quantum suites; and pre-stage automated certificate lifecycles with "break-glass" re-key paths for crisis rotation. In parallel, field quantum-aided PNT on high-value platforms, precision timing and inertial aids to validate GPS-denied battle phases/ stages, synchronized fires, and legally robust time-stamping during exercises. Inside operations centres, hybrid decision engines use quantum-inspired optimization for routeing, sensor-tasking, and weapon to target matching, thus reserving emerging accelerators for scheduled planning windows to avoid single points of failure. Finally, the institutionalisation of deterrence metrics and SOPs: track time-to-identify, time-to-engage, re-plan latency under contested electromagnetic conditions, cryptographic link integrity, and navigation error without GPS, so that thresholds can be linked to proportionate, pre-approved responses executable rapidly across the task organisation.

However, since new technologies often take time to be absorbed and peacetime exercises may generate the wrong lessons, the integration of quantum capabilities must be governed by realistic risk controls. Perception gaps should be avoided, as should overstating capability internally, by emphasizing layered assurance PQC, selective quantum keying where feasible, backed by continuous testing and red teaming against deception and data poisoning. Since the international legalities of using lethal autonomous weapon systems are yet to be codified, escalation should be managed by retaining human-in-the-loop, enforcing interpretable outputs (ranked options, confidence, constraints), and publishing incident-response norms for communications anomalies. Sustain readiness by prioritizing conformance-tested quantum communications and sensing kits with signed updates, and by building training pipelines in crypto stewardship, AI oversight, EW/EMCON integration, and quantum-aided navigation. This combination of disciplined migration, GPS-independent timing, hybrid planning, hard metrics, and sober risk controls delivers fast, accountable decision cycles without trading stability or trust.

Strategic Approaches Towards Pre-emptive Tech-Disruption Operations

The first strike may now be digital, silent, and instantaneous
Pre-emptive tech-disruption plans can now shift deterrence from kinetic overmatch to disabling the adversary's decision-making capability even before

conflict breaks out. In practice, this means prioritizing the neutralization of adversary quantum assets, communications, sensing, navigation, and emerging compute accelerators, so that their command cannot issue orders with assured opsec, forces cannot synchronize timing, and their stealth advantages are lost. The strategic logic is straightforward: if crises are decided at machine speed, the side that undermines the other's data integrity, timing, and sensor trust early, gains decisive advantage while staying below the threshold of overt escalation. This is not "first strike" in the nuclear sense; it is pre-emption in the information domain, reversibly degrading the opponent's trust architecture while preserving one's own. Think of two fleets preparing to sail. Rather than firing the first shot, one fleet quietly swaps the enemy's charts, desynchronizes clocks, and generates false data along key sea areas, while securing its own maps and clocks against tampering. When operations commence, one fleet moves as a coordinated whole, whilst the other struggles to align systems, validate orders, or distinguish real signals from decoys. Quantum denial and disruption aims to deliver that asymmetry, assuring friendly perception, timing, and command integrity while injecting calibrated doubt into the adversary's.

To degrade an adversary's decision-making architecture without triggering premature escalation, tactical doctrines should prioritize systematic pressure on "trust anchors" while preserving plausible deniability and legal reversibility. The immediate objective is to stress, and selectively neutralize quantum-safe identity systems, key-distribution gateways, software-signing chains, precision timing sources, and quantum-aided navigation nodes through measures that detect compromise, force re-key at tactically unfavourable moments, or induce controlled de-synchronization within pre-authorized bounds. Options must be architected as observable to national leadership, yet deniable to broader publics, sustaining escalatory control and diplomatic manoeuvre space.

Concurrently, strategy should expose deception and stretch adversary decision cycles at machine speed while hardening one's own C4ISR beyond dispute. Quantum-aided sensing with emission-aware ISR should cue suspected stealth corridors, undersea staging, and spoofed maritime/air traffic, using tailored disclosures—public or allies-only to narrow the adversary's operational window and raise operational risk for covert deployments. Quantum-informed optimization should inject calibrated friction logistics re-routes, decoy saturation, forced communications, re-key timing while safeguarding friendly clock and key integrity. On the friendly side, mandate post-quantum cryptography for identities, code signing, and mission messaging; field quantum-aided timing and inertial navigation on high-value platforms; and standardize interpretable autonomy so rapidly generated COAs remain auditable, ROE-compliant, and coalition-shareable. This blend of denial, exposure, friction, and assurance shifts initiative without firing a shot, positioning senior leaders to coerce, deter, or de-escalate on timelines measured in minutes, not days.

Adopt effects that are deliberately reversible and tightly attributed to preserve coercive leverage without foreclosing diplomacy: design operations that can be paused or rolled back on command, and wrap every action in tamper-evident telemetry with authenticated timestamps and provenance so national leaders can be briefed with confidence while parallel, classified attribution lines mature before any public claim. In parallel, lock doctrine to law and partnerships by pre-approving playbooks with legal review and partner consultation, specifying thresholds, authorities, and notification protocols for tech-disruption moves, and rehearsing cross-domain crisis communications to prevent misreads when timelines compress.

Signal strength without provocation by institutionalizing routine, non-escalatory demonstrations of resilience, GPS-denied navigation exercises, rapid cryptographic re-key under electronic warfare, and low-emission ISR cueing, making clear that coercion via denial or deception will not yield advantage. This triad of reversibility, legally grounded partner alignment, and measured signalling preserves escalation control, sustains strategic ambiguity where useful, and underwrites credible deterrence in arenas where decision speed and trust in data, not kinetic fires define advantage.

Changing Definitions of Sovereignty and Control in the Information Battle Space

Owning the information sphere will mean owning the battlefield

Quantum technologies will expand practical sovereignty from physical territory into the information battle space by hardening command authenticity, extending sensor reach, and decoupling navigation and timing from vulnerable satellite services. In effect, quantum-safe communications make orders tamper-evident; quantum-aided sensing and timing allow persistent, low-emission awareness beyond conventional airspace and maritime limits; and quantum-informed analytics compress decision cycles to machine speed. This confers de facto control over flows of truth, origin, time, and trust redefining where national power can be projected and defended without crossing a kinetic threshold.

Operational Relevance

- **Quantum-assured command integrity:** A PQC-first posture for identity, software signing, and mission messaging, with selective quantum keying on high-value legs, establishes a sovereign "trust corridor" that remains authentic even when traffic crosses foreign clouds, undersea cables, or coalition relays. This denies adversaries leverage from "harvest-now, decrypt-later" schemes and preserves lawful, rapid authority to act during crises.

- **Borders extended by sensing and time:** Quantum-aided navigation and precision timing keep formations synchronized and logs auditable when GNSS is jammed or spoofed; quantum magnetometry, gravimetry, and low-signature optical techniques raise the probability of cueing stealth platforms, protecting sea routes, undersea infrastructure, and air corridors without overt emissions. This extends effective control into choke points and grey zones while maintaining escalation control.

- **Decision-tempo sovereignty:** Quantum-informed planning engines evaluate many courses of action in parallel and output interpretable, ROE-compliant options within minutes, not hours. Predictable speed with accountability lets states impose timelines on adversaries pre-empting fait accompli gambits, coordinating allied action faster, and shaping the narrative before misinformation hardens.

- **Digital sovereignty and coalition interoperability:** Standardized evidence schemas, signed telemetry, authenticated timestamps, sensor provenance enable fast, trusted sharing and joint action without surrendering data ownership. This balances national control of critical truth artefacts with credible allied deterrence and reduces miscalculation in high-tempo standoffs.

Quantum sovereignty must be codified in doctrine and practice, not just by rhetoric. First, define explicit thresholds, authorities, and notification protocols for quantum denial and disruption, and pre-approve reversible, legally reviewed measures that can be paused or rolled back on command to preserve escalatory control and diplomatic manoeuvre space. Second, build sovereign technological depth by prioritizing conformance-tested quantum communications, sensing, and timing kits with signed update pipelines, while developing domestic capabilities for key-management, precision time-transfer, and evidence-provenance to avoid single-vendor chokepoints and assure verifiable control of critical truth artefacts. Third, normalize resilience through training: institutionalize GPS-denied and comms-contested exercises that combine quantum-aided navigation, rapid cryptographic re-key under electronic warfare, and low-emission ISR cueing so units default to disciplined, low-signature operations without provocation. Finally, maintain human oversight where it matters most: keep human-in-the-loop for lethality; mandate interpretable autonomy that presents confidence, constraints, and sensitivity to new intelligence for rapid, lawful approval; and preserve parallel, classified attribution channels until public claims are strategically warranted, ensuring speed never outruns legitimacy.

Militarization of Space as Quantum Infrastructure Expands Off-Earth

From sea beds to satellites, battle space becomes omni domain

As quantum satellites and space-based payloads transition from experiments to operational backbones, space control becomes central to deterrence and day-to-day C4ISR. Quantum key distribution constellations will provide tamper-evident keys across theatres, anchoring sovereign command integrity even when terrestrial links are compromised. Quantum-aided timing and navigation from space will harden force synchronization in GNSS-denied conditions, while quantum-enhanced sensing from orbit spanning magnetic, gravimetric, and RF modalities will extend low-emission ISR across maritime choke points, undersea infrastructure, and polar approaches. The strategic effect is a shift from space as a support domain to space as the trust layer of joint operations.

Operational Relevance

- **Space-anchored C2 assurance:** Treat quantum-secured satellite links as the sovereign backbone for key exchange, identity validation, and software signing, enabling rapid, defensible orders under cyber pressure and denying adversaries "harvest-now, decrypt-later" leverage. Pair with post-quantum cryptography end-to-end so space keys harden, rather than replace, terrestrial security.

- **Hardened timing and navigation:** Field precision time-transfer and quantum-aided inertial updates from space to keep formations synchronized, fire windows aligned and mission logs auditable when GNSS is jammed or spoofed. Prioritize high-value platforms, patrol aircraft, SSBN/SSN bastions, and carrier groups for early integration.

- **Low-emission orbital ISR:** Deploy quantum-enhanced sensing concepts from LEO/MEO to cue stealth corridors, map magnetic/gravimetric anomalies around subsea cables and ports, and detect deceptive RF behaviours with minimal electromagnetic signature. Fuse orbital cues with unmanned pickets to accelerate find-fix without revealing intentions.

- **Space control as deterrence:** Protect quantum satellites with layered defence manoeuvre, hardening, custody by patrol assets, and rapid reconstitution and pre-plan proportionate, reversible responses to hostile acts on quantum infrastructure. Make clear in doctrine that interference with quantum trust services is a strategic threshold.

- **Coalition interoperability and sovereignty:** Standardize evidence-provenance (signed telemetry, authenticated timestamps) and key-exchange schemas so allies can act quickly without surrendering data ownership. Maintain national control of root keys, time sources, and update pipelines to avoid single-vendor chokepoints.

A near-term space posture should begin with a PQC-first transition anchored by tightly scoped space pilots: migrate identities, software signing, and mission messaging to standardized post-quantum suites, while trailing quantum keying on a limited set of satellite links that serve strategic commands and exercising rapid rekey under electronic-warfare conditions to validate continuity. In parallel, institutionalize space-to-edge timing drills, routine GPS-denied exercises that rely on space-anchored precision time to keep maritime formations and joint fires synchronized and to maintain legally robust time stamping across services. Build an orbital cue-to-edge prosecution chain in which low-emission, space-borne cues trigger emission-disciplined UAV/USV/UUV investigations, with interpretable autonomy producing ranked, ROE-compliant courses of action for fast approval. Finally, harden resilience and reconstitution by pre-contracting launch capacity, maintaining on-orbit spares or hosted payload options, and rehearsing rapid reassignment of keys and time sources to surviving nodes in the event of satellite degradation or loss, ensuring that keys, clocks, and cues remain sovereign even under sustained contestation.

Risk controls and norms for quantum space infrastructure should be codified to preserve escalation control, safety, and sovereign assurance across contested orbits. Escalation management must define precise thresholds, authorities, and notification protocols for responses to interference with quantum-enabled satellites or ground segments, favouring reversible, attributable measures, and retaining human—in-the-loop for any action with kinetic spill over risks. Debris and safety imperatives require manoeuvrable platforms, rigorous conjunction awareness, and debris-minimizing designs, underpinned by bilateral and multilateral understandings that recognize quantum trust services (keying, time transfer, provenance) as critical infrastructure warranting special protections. Supply chain assurance demands conformance-tested quantum communications and sensing kits with cryptographically signed update pipelines, deliberate supplier diversification to avoid single points of failure, and retained domestic capacity for key management, precision time transfer, and payload security modules ensuring that keys, clocks, and software provenance remain sovereign and recoverable even under sustained contestation.

Weaponization of Quantum Intelligence: Strategic Superiority through Predictive Warfare

Predicting moves before they're made rewrites the rules of combat

Quantum intelligence marries quantum-informed computation with operational AI to forecast adversary behaviour, logistics, and deception faster and more reliably than today's methods. In near-term practice, this will rely on quantum-inspired algorithms at scale, complemented by access to early quantum accelerators during planning windows, to evaluate vast "what-if" branches in parallel courses of action, routeing, weapon-target pairing, decoy and spoof detection, then return interpretable, confidence-graded recommendations. The shift is from reacting to events, towards accurate pre-calculation: shaping the battle space, pre-positioning assets, and queuing coalition actions before the adversary's plans mature, while maintaining human-in-the-loop authority. Instead of playing chess one move at a time, commanders are handed the top three winning continuations after testing thousands of lines, each with risks, logistics demands, and deception flags explained. The game still requires judgement, but the clock and the board now favour the side that can explore the future at scale, and then act with discipline. COA development will be more refined, as the requirement to make educated guesses is reduced.

Objective	Actions	Outputs/ Standards	Metrics	Risks	Controls
Rapid capability without fragility	Deploy quantum-inspired planners on existing compute for COA generation, routing, sensor-tasking; use quantum accelerators only as surge capacity in scheduled planning windows	No single points of failure; edge-executable workloads; surge pathways documented	Time-to-plan; compute utilization; failover success rate	Architectural lock-in; dependency on scarce accelerators	Modular orchestration; fallback COAs; periodic surge drills
Accountable, lawful speed	Require ranked options with confidence, constraints met/violated, and sensitivity to new intel; keep human-in-the-loop for lethality; codify abort thresholds	Standardized decision packets; ROE-aligned artefacts; audit trails	% lethal COAs with human authorization; decision packet completeness	Opaque recommendations; decision creep	Interpretable ML mandates; ROE guardrails; pre-briefed abort triggers
Resource to results, not hype	Track Brier scores, false-flag catches, time-to-decision, and downstream effects on time-to-engage and sustainment readiness; tie funding to measured gains	Quarterly scorecards; AAR-integrated metrics; budgeting linked to outcomes	Brier score; false-flag detection rate; TTD/TTE deltas	Metric gaming; over-fitting	Independent validation; red teaming; hold-out datasets
Robustness under deception	Red team against novel TTPs and adversarial data; enforce multi-sensor corroboration and uncertainty quantification	Deception-aware scoring; uncertainty bounds	Deception detection rate; misclassification rate	Spurious certainty; brittleness in edge cases	Diverse sensors; adversarial training; uncertainty thresholds
Prevent inadvertent escalation	Bind pre-calculated actions to ROE, legal review, and political authority; favour reversible, observable measures until thresholds are crossed	Authority matrices; reversibility checklists	% actions with legal sign-off; reversibility compliance	Premature or escalatory actions	Legal playbooks; phased effects; notification protocols
Preserve control and credibility	Maintain PQC-secured identities, signed telemetry, authenticated timestamps; keep parallel classified attribution channels until public claims are warranted	Provenance-complete evidence packages	Evidence acceptance rate by allies; integrity check pass rate	Compromised provenance; premature attribution	PQC-first policy; dual-track attribution; info-release governance

Operational Relevance

- **Campaign design and posture:** Use quantum-informed COA generators to war game multi-domain contingencies in hours, not weeks optimizing pre-positioning, replenishment, and escalation ladders with interpretable outputs that staffs can brief and approve.

- **Deception detection and target vetting:** Parallel hypothesis testing compares sensor signatures, C2 patterns, and historical TTPs to elevate true targets and mark false-flags early, reducing effort and lowering fratricide risk in cluttered seas and urban littorals.

- **Tempo and logistics synchronization:** Predict adversary surge windows and chokepoint usage, align sealift/air bridge schedules and refuel tracks, and time swarm operations to arrive just before consolidation phases.

- **Joint Force/Coalition coherence:** Package evidence and recommendations with signed provenance, timestamps, and rationale so partners can adopt shared COAs quickly without surrendering data ownership, keeping joint tempo ahead of misinformation cycles.

6.2   INDIA'S QUANTUM MILITARY DEFENCE STRATEGY

Assessment of Current Defence Capabilities and Quantum Readiness

Mapping today's forces to meet tomorrow's quantum challenges

A credible assessment begins with a whole-of-enterprise inventory of "trust surfaces" across command, communications, sensing, navigation, space, cyber, and logistics systems, because quantum readiness is fundamentally about securing authenticity, time, and provenance/origin at operational tempo. Practically, this means cataloguing cryptographic dependencies (identities, software signing, link/protocol suites), timing and navigation anchors (GNSS reliance, atomic clocks, inertial baselines), sensing portfolios (EO/IR/radar/ acoustic and their counter-spoof measures), space dependencies (satcom, ISR, ground segment), and data pipelines from edge to fusion centres. Each element should be scored for PQC maturity, crypto-agility, key-management hygiene, evidence-provenance (signed telemetry and authenticated timestamps), EW resilience, and ability to operate in GPS-denied and comms-contested conditions. The output is a heat-map of mission threads, air defence, maritime security, integrated fires, border surveillance, nuclear C2 support showing where quantum-safe upgrades and quantum-aided capabilities would most reduce operational risk.

From that baseline, define near-term integration vectors that deliver measurable gains without waiting on uncertain hardware timelines. First, mandate a PQC-first posture: transition identities, code-signing, and mission messaging to standardized post-quantum suites; establish automated certificate lifecycles, crypto-agility policies, and "break-glass" re-key procedures for crises. Second, harden timing and navigation: field precision timing and quantum-aided inertial kits on high-value platforms, and institutionalize GPS-denied drills for synchronized fires, maritime deployments, and legally robust time-stamping. Third, deploy quantum-inspired optimization where it matters now: sensor-tasking, routeing, weapon to target matching, deception adjudication on existing compute, reserving access to early accelerators for scheduled planning windows to avoid single points of failure. Fourth, prepare space as the trust layer: pilot quantum keying or enhanced secure key distribution on select satellite links serving strategic commands; exercise rapid re-key and time re-assignment under EW to validate continuity.

To convert readiness into doctrine and budgets, couple technology plans to mission metrics and governance. Establish a quantum readiness scorecard per mission thread that tracks time-to-identify, time-to-engage, re-plan latency under EW, navigation error without GPS, integrity pass rates for signed telemetry and time stamps, certificate/re-key MTTR, and coalition acceptance of evidence packages. Tie resource allocation to measured improvements against these metrics, not hype, and require interpretable outputs from any decision engine: ranked COAs, confidence levels, constraints met/violated, and sensitivity to new intelligence, with human-in-the-loop for lethality and codified abort thresholds. Build sovereign depth where dependence is riskiest by prioritizing conformance-tested quantum comms/sensing/timing kits with signed update pipelines, diversifying suppliers, and developing domestic competencies in key management, precision time transfer, and evidence-provenance services.

Finally, integrate policy, training, and alliances so readiness survives contact with crisis timelines. Pre-approve legally reviewed SOPs for quantum denial/disruption responses with thresholds, authorities, and notification protocols. Institutionalize routine, non-escalatory demonstrations, GPS-denied navigation, low-emission ISR cueing, rapid cryptographic restoration under

EW to normalize resilient behaviour and signal deterrence without provocation. In parallel, mature coalition interoperability via standardized evidence schemas and key-exchange processes that protect national data sovereignty while enabling fast joint action. The outcome of this assessment-driven approach is a prioritized roadmap that secures today's C4ISR fabric, inserts quantum-aided capabilities where they yield immediate operational advantage, and builds the industrial and doctrinal foundations for the quantum decade ahead.

Assessment domain	What to evaluate	Immediate actions (0–12 months)	2–5-year objectives	Readiness metrics	Key risks	Governance and safeguards
Command and identity (C2 trust)	Identity, access, code signing, mission messaging	Inventory crypto dependencies; mandate post-quantum cryptography (PQC) for identities and software signing; automate certificate lifecycles; pre-stage emergency re-key	PQC across mission threads; coalition-ready trust schemas; selective quantum key pilots on high-value links	% assets on PQC; mean time to re-key; signed-artefact integrity rate; coalition acceptance rate	Legacy crypto exposure; key sprawl; mis-configurations	Crypto-agility policy; centralized key governance; conformance testing; continuous audits
Communications transport	Satcom, terrestrial, HF/VLF, tactical radios, cross-domain guards	Harden link protocols; deploy PQC for data-in-motion; segment and label traffic by mission criticality	Space-anchored key distribution pilots; resilient routing with auto-failover	Link integrity under EW; packet loss vs. baseline; time-to-failover	EW denial; QoS collapse; single points of failure	Redundant paths; emissions control SOPs; chaos-testing
Evidence provenance	Sensor telemetry, logs, chain of custody, time stamps	Enforce signed telemetry and authenticated time stamps at point of creation	Standardized evidence packages for inter-service and allied sharing	Integrity check pass rate; audit completeness; time-to-share	Evidence contamination; attribution disputes	Provenance policy; tamper-evident storage; parallel classified attribution lines
Sensing portfolio	EO/IR, radar, acoustic, passive RF, magnetic/ gravimetric cues	Map spoofing/decoy countermeasures; fuse multi-sensor corroboration	Introduce quantum-aided sensing pilots where feasible; deception-aware scoring	True-positive rate; false-flag catch rate; sensor fusion latency	Deception brittleness; saturated ISR	Red team deception; calibrated confidence thresholds
Timing and navigation (PNT)	GNSS reliance, atomic clocks, INS baselines	Field precision timing and enhanced INS on high-value platforms; exercise GPS-denied operations	Space-to-edge precision time transfer; fleet-wide GPS-denied proficiency	Navigation error without GNSS; time-sync error; PNT continuity under EW	Timing drift; legal time stamp gaps	PNT drills; signed time sources; dual redundant clocks

Assessment domain	What to evaluate	Immediate actions (0–12 months)	2–5-year objectives	Readiness metrics	Key risks	Governance and safeguards
Mission data pipelines	Edge collection to fusion centres and COPs	Map data flows; remove bottlenecks; enforce schema validation	Secure, low-latency pipelines with provenance; schema harmonization	Ingest latency; drop rate; schema conformance	Data silos; version drift	API governance; schema registries; SLOs
Decision engines	COA generation, routing, tasking, pairing	Deploy quantum-inspired optimization on existing compute; require interpretable outputs	Integrate accelerators as surge capacity for planning windows	Time-to-decision; time-to-engage; re-plan latency under EW	Opaque recommenda-tions; overfitting	Interpretability mandates; human-in-the-loop; abort thresholds
Space dependence	Satcom, ISR, ground segment, reconstitution	Identify trust services; simulate rapid re-key/ time reassignment	On-orbit spares/hosted payloads; pre-contracted launch; orbital cue-to-edge chains	Re-key time; cue-to-decision latency; reconstitution MTTR	ASAT disruption; debris constraints	Reversibility ladders; manoeuvrable platforms; space norms
Cyber defence	SOC pipelines, hunt, incident response	PQC for SOC tooling; parallel attribution cells; machine-speed triage	Coalition-ready incident exchange; automated containment playbooks	Time-to-detect; time-to-contain; attribution confidence	Lateral spread; "harvest-now, decrypt-later"	Segmentation; zero trust; tabletop with allies
Logistics and sustainment	Supply, fuel, spares, maintenance	Use optimization for routing and stock pre-positioning	Predictive sustainment linked to operational tempo	Sortie genera-tion; mean logistics delay; readiness vs. plan	Chokepoints; demand shock	Multi-node stockpiles; alternates and hedges
Industrial base	Vendors, labs, standards, talent	Qualify conformance-tested kits; sign update pipelines	Domestic capacity for keys, time transfer, provenance	% kits con-formant; signed update compliance; staffing levels	Vendor lock-in; skill shortages	Multi-vendor sourcing; train-ing pipelines; export/standards alignment
Doctrine and ROE	Delegation, thresholds, escalation	Pre-approve playbooks; legal review; notification protocols	Routine non-escalatory demos (GPS-denied, rapid re-key, low-emission ISR)	Playbook adherence; incident comms accuracy	Misreads in crisis	Cross-domain comms drills; reversible effects

Prioritization of Quantum Technologies for Defence Applications

Targeted investments where quantum advantage delivers battlefield supremacy

The strategic focus for the next five years should concentrate on three pillars that directly improve operational security and decision tempo: quantum communication, quantum sensing, and quantum computing. Quantum communication must be prioritized to harden command integrity and coalition interoperability through a post-quantum cryptography baseline for identities, software signing, and mission messaging, with targeted pilots of quantum keying on high-value links that support strategic commands. This combination preserves authenticity and confidentiality under cyber pressure, denies "harvest-now, decrypt-later" payoffs, and enables rapid, defensible orders and evidence-provenance sharing during crises. Procurement should emphasize conformance-tested systems with cryptographically signed update pipelines, crypto-agility policies, automated certificate lifecycles, and pre-staged emergency re-key procedures, ensuring that keys and provenance remain sovereign, auditable, and recoverable.

Quantum sensing should be treated as the near-term force multiplier for contested domains. Prioritize quantum-aided navigation and timing (precision clocks and enhanced inertial systems) for high-value platforms to sustain synchronized fires, formation keeping, and legally robust time stamping in GNSS-denied environments. In parallel, invest in low-emission sensing concepts, magnetic/gravimetric cueing, passive RF anomaly detection, and space-borne quantum-enhanced modalities to expose stealth corridors, protect subsea infrastructure, and accelerate find-fix without inviting escalation. Doctrine and training must institutionalize GPS-denied and comms-contested exercises, with clear emissions control and deception-aware procedures, so units default to resilient behaviours in real operations. Success should be measured by navigation error without GNSS, time-synchronization accuracy, false-flag detection rates, and cue-to-decision latency.

Quantum computing should be integrated through a pragmatic, "hybrid engines" approach that delivers value now while hedging technology risk. Field quantum-inspired optimization on existing compute for course-of-action generation, sensor-tasking, routing, and weapon-target pairing, reserving access to emerging accelerators as surge capacity during scheduled planning windows rather than as operational linchpins. Mandate interpretability ranked options, confidence, constraints met/violated, and sensitivity to new intelligence with human-in-the-loop for lethality and codified abort thresholds. Tie scaling and budgets to demonstrated operational metrics, not hype: Brier scores for predictive quality, time-to-decision and time-to-engage under electronic warfare, re-plan latency, sustainment impacts, and coalition acceptance of evidence packages. Finally, anchor all three pillars in governance: legally reviewed playbooks and thresholds for quantum denial/disruption actions, multi-vendor sourcing to avoid chokepoints, domestic capacity for key management, precision time transfer, and provenance services, and routine, non-escalatory demonstrations (GPS-denied navigation, rapid re-key, low-emission ISR cueing) to signal resilience without provocation.

Phased Integration Plan across Land, Sea, Air, and Space Domains

Synchronizing multi-domain quantum deployment for unified military power

A phased integration plan should begin with tightly scoped pilot programmes that prove operational value under contested conditions, then scale into strategic assets through metrics-driven governance and coalition-ready standards. Phase 1 (0–12 months) should establish a post-quantum cryptography baseline for identities, software signing, and mission messaging across selected formations, accompanied by GPS-denied exercises that validate quantum-aided timing and enhanced inertial navigation on high-value platforms. In parallel, field quantum-inspired decision engines on existing compute for course-of-action generation, sensor-tasking, routing, and weapon-target pairing mandating interpretable outputs with confidence levels, constraints met/violated, and sensitivity to new intelligence, while retaining human-in-the-loop for lethality and codifying abort thresholds. Pilot space-to-edge trust services by trailing secure key exchange and precision time transfer on limited satellite links serving strategic commands, and conduct rapid re-key and time reassignment drills under electronic warfare to validate continuity. Phase 1 success should be judged by reductions in time-to-identify and time-to-engage, re-plan latency under EW, navigation error without GNSS, integrity pass rates for signed telemetry and time stamps, and coalition acceptance of evidence packages.

Phase 2 (12–36 months) should expand targeted adoption in mission-critical threads integrated air and missile defence, maritime security and chokepoint control, joint fires, border surveillance, and nuclear C2 support. Land forces should extend PQC and provenance to sensor edges, embed hybrid decision engines at brigade/division HQs, and institutionalize deception-aware fusion; naval forces should combine low-emission ISR with orbital cues, exercise rapid rekey at sea, and enforce EMCON discipline tied to deception metrics; air forces should optimize sortie pairing, tanker routing, and SEAD packages with rapid, interpretable re-plans under EW; space should integrate on-orbit spares/hosted payload options, orbital cue-to-edge tasking chains, and conjunction awareness for manoeuvrable platforms. Jointly, standardize evidence schemas (signed telemetry, authenticated time stamps, sensor provenance) to accelerate allied decision-making without surrendering national data sovereignty. Scale only where Phase 1 metrics improve: Brier scores for predictive quality, cue-to-decision latency, time-to-engage under EW, GNSS-out synchronization, and chain-of-custody integrity.

Phase 3 (36–60 months) should elevate quantum-enabled capabilities into the strategic backbone of C4ISR and deterrence. Treat space as the trust and timing layer for joint operations codifying reversible, proportionate responses to interference with quantum trust services in doctrine and operate a resilient, self-healing "kill web" that re-routes around degraded nodes across land, sea, air, and space. Ensure accelerators are surge capacity for scheduled planning windows, never single points of failure, and maintain quantum-inspired fallbacks for continuity. Harden supply chains with conformance-tested kits, cryptographically signed update pipelines, multi-vendor sourcing, and domestic capacity for key management, precision time transfer, and evidence-provenance. Institutionalize routine, non-escalatory demonstrations GPS-denied navigation trials, low-emission ISR cueing, rapid cryptographic restoration under EW as strategic signalling that deters coercion without provocation. Across all phases, anchor governance in pre-approved playbooks with legal review and allied notification protocols, and tie budgets to mission outcomes, not technology hype, prioritizing deployments that measurably compress decision cycles, preserve command authenticity, and maintain synchronized operations in the face of denial and deception.

Domain	Phase 1: Pilot focus (0–12 months)	Phase 2: Targeted adoption (12–36 months)	Phase 3: Strategic rollout (36–60 months)	Core capabilities introduced	Success metrics	Key risks	Controls and safeguards
Land	PQC-first for identities, code signing, and mission messaging in selected corps; GPS-denied drills with quantum-aided timing/INS on high-value platforms	Hybrid decision engines for COA generation, routing, and sensor-tasking at brigade/division HQs; signed telemetry and authenticated time stamps at sensor edge	Corps-level kill web orchestration with interpretable autonomy; coalition-ready evidence packages and rapid re-key SOPs	Post-quantum comms, quantum-aided PNT, quantum-inspired optimization	Time-to-identify; time-to-engage; nav error without GNSS; integrity pass rate	EW denial; legacy crypto exposure; model opacity	Crypto-agility policy; EW playbooks; interpretability mandates; human-in-the-loop for lethality
Naval	PQC for afloat networks and shore gateways; space-to-edge timing pilots for carrier/amphibious groups	Low-emission ISR cueing fused with orbital cues; deception-aware targeting; rapid re-key under EW exercises	Fleet-wide GPS-denied formation keeping; resilient, multi-path maritime kill web with autonomous pickets	PQC comms, space-anchored timing, deception-aware fusion	Formation time-sync error; cue-to-decision latency; re-key MTTR	Emissions detection; comms chokepoints; AIS spoofing	EMCON discipline; multi-path routing; signed provenance and adjudication protocols
Air	PQC for mission planning and datalinks; quantum-aided timing on AWACS/tanker nodes; ROE-aligned interpretable autonomy for sensor-tasking	Quantum-inspired sortie pairing, tanker routing, and SEAD plan optimization; coalition evidence-sharing schemas	Theatre-level dynamic targeting with rapid re-plan under EW and GNSS denial	PQC comms, quantum-aided timing, hybrid optimization	Re-plan latency; sortie generation rate; ROE compliance rate	Airspace congestion; brittle plans under deception	Deception red-teaming; abort thresholds; layered PNT sources
Space	Pilot quantum keying for select satellite links; simulate rapid time/key reassignment	On-orbit spares/hosted payloads; orbital cue-to-edge tasking chains; conjunction awareness integration	Treated as trust layer for joint ops; reversible responses to interference codified in doctrine	Space-anchored keying/timing, orbital cueing	Re-key time; reconstitution MTTR; cue validation rate	ASAT/interference; debris; supply chain choke points	Reversibility ladders; manoeuvrable platforms; signed update pipelines; multi-vendor sourcing
Joint C2	Standardize evidence provenance (signed telemetry, time stamps); establish delegation thresholds and notification protocols	Coalition inter-operability drills; cross-domain crisis comms rehearsals; parallel classified attribution	Routine, non-escalatory demonstrations (GPS-denied ops, rapid re-key, low-emission ISR) as deterrence signalling	Provenance, governance, escalation control	Evidence acceptance by allies; decision packet completeness; comms accuracy	Misreads/escalation; policy gaps	Pre-approved playbooks; legal review; transparency channels

Domain	Phase 1: Pilot focus (0–12 months)	Phase 2: Targeted adoption (12–36 months)	Phase 3: Strategic rollout (36–60 months)	Core capabilities introduced	Success metrics	Key risks	Controls and safeguards
Cyber/SOC	PQC in SOC tooling; machine-speed triage with interpretable outputs; tamper-evident chains of custody	Parallel attribution cells testing multiple hypotheses; automated containment playbooks	Coalition-ready incident exchange; sector-wide crypto hygiene enforcement	PQC, provenance, hybrid analytics	Time-to-detect; time-to-contain; attribution confidence	False attribution; lateral spread	Multi-sensor corroboration; uncertainty thresholds; segmentation
Logistics	Quantum-inspired routing and inventory pre-positioning pilots; signed maintenance logs	Predictive sustainment aligned to operational tempo; alternate corridors and hedges	Integrated logistics-kill web synchronization across theatres	Optimization, provenance	Mean logistics delay; readiness vs. plan; on-time replenishment	Chokepoints; demand shock	Multi-node stockpiles; surge contracts; stress tests
Work force	Core teams for crypto stewardship, PNT, model assurance, EW/EMCON; officer education modules	Certification tracks; embedded data engineering at formation HQs	Career pathways and retention incentives; joint standards board	Skilled operators and stewards	Certification completion; retention; exercise performance	Skill shortages; turnover	Incentives; continuous training; joint curriculum
Governance	Quantum readiness scorecard per mission thread; budget tied to metrics	Quarterly reviews and external validation; tech horizon scanning	Doctrine captured in TPCRs and service pubs; audit and accountability loops	Metrics-driven governance	KPI adherence; audit findings	Metric gaming; lagging doctrine	Independent validation; AAR loops; doctrine refresh cadence

Strengthening Indigenous R&D and Manufacturing Ecosystems

Building India's sovereign quantum defence industrial base

Over the next five years, the priority is to convert India's scientific strengths into deployable, sovereign quantum capabilities that harden C4ISR, accelerate decision tempo, and deter coercion without dependence on foreign supply chains. This requires a mission-oriented industrial strategy that links laboratory breakthroughs to fielded systems through assured demand, conformance testing, and secure update pipelines. Defence laboratories should focus on a defined portfolio with near-term operational payoff: post-quantum cryptography for identities and software signing; quantum-secure keying for selected high-value links; quantum-aided timing and inertial navigation for GPS-denied operations; and low-emission sensing concepts (magnetometry, gravimetry, passive RF anomaly detection) to expose stealth corridors and protect subsea infrastructure. Each line of effort must come with reference architectures, open interface standards, and signed firmware/update pathways that industry can adopt and scale, ensuring provenance and cyber hardening from design to depot.

Partnerships with private industry and academia should be structured as outcome-based programs, not one-off grants, with assured orders tied to measurable operational metrics. For communications and cyber security, incentivize companies to deliver PQC-first identity and code-signing stacks, automated certificate lifecycles, and "break-glass" re-key procedures demonstrably executable under electronic warfare. For PNT, co-develop precision clocks, ruggedized quantum-aided INS modules, and secure time-transfer services with clear qualification routes on high-value platforms. For sensing, fund dual-use components, single-photon sources/detectors, low-SWaP lasers, vapour cells, photonic chips that can feed both military and commercial demand, driving volume and learning curves. Establish national conformance labs to validate cryptographic suites, timing accuracy, sensor performance, electromagnetic compatibility, and update integrity, issuing certifications that procurement mandates as gatekeepers for fielding.

To sustain speed and scale, align procurement and policy with the realities of frontier tech maturation. Use multi-year, production-linked orders and spiral development plans that move prototypes into limited-rate initial production with embedded user trials, reducing the valley of death between lab and line unit. Require interpretable outputs from any decision-support software (ranked options, confidence, constraints, sensitivity to new intelligence) and enforce human-in-the-loop for lethality with codified abort thresholds, so legal and operational accountability scales with deployment. Build sovereign depth where chokepoints are most dangerous: domestic capacity for key management, precision time transfer, evidence-provenance services, and secure payload modules; a trusted foundry and packaging pipeline for photonic and quantum-sensing components; and a signed-update infrastructure that assures supply chain integrity throughout platform life. Finally, invest in people—crypto stewards, PNT engineers, photonics and laser specialists, model-assurance teams, and EW/EMCON practitioners through joint curricula, certification tracks, and rotational billets between labs, formations, and industry. The outcome is an indigenous quantum ecosystem that delivers operationally relevant capability on reliable schedules, protects national sovereignty over keys, clocks, and truth artefacts, and positions India to lead, not follow, in the quantum decade.

Strategic governance and sovereignty

- Establish a national "Trust Authority" for keys, time, and provenance: Operate sovereign root key management, precision time-transfer, and evidence-provenance services with 24/7 re-key drills, cryptographic time-stamping, and tamper-evident telemetry standards.

- Codify reversible, legally reviewed actions: Pre-approve playbooks for quantum denial/disruption with thresholds, authorities, audit trails, and notification protocols; keep human-in-the-loop for lethality and define abort thresholds to maintain political control at machine speed.

- Protect sovereignty and avoid lock in: Use open interface standards, conformance testing, and multi-vendor sourcing; retain domestic capacity for key management, time transfer, secure payload modules, and signed-update pipelines.

Programmatic controls and conformance

- Spiral development with measurable gates: Move from pilots to LRIP with field exercises; scale only when mission metrics improve (time-to-identify/engage, EW re-plan latency, GNSS-out navigation error, integrity pass rates, cue-to-decision latency, predictive Brier scores).

- National conformance lab network: Certify PQC suites, timing accuracy, sensor performance, electromagnetic compatibility, and update integrity; make certifications mandatory for procurement and fielding.

- Signed supply chain: Mandate reproducible builds, signed firmware with threshold approvals, SBOMs, and continuous CI/CD attestation for all mission software and device updates.

Cyber resilience and provenance

- PQC-first policy: Migrate identities, code signing, and mission messaging to standardized post-quantum suites; automate certificate lifecycles and practice "break-glass" re-key under EW; enforce crypto-agility policy for rapid algorithm rotation.

- Evidence packages by default: Require signed telemetry, authenticated time stamps, and sensor provenance at the point of creation; publish standardized evidence schemas compatible with allied systems while retaining national data sovereignty.

- Parallel attribution tracks: Maintain classified attribution lines until public release is politically warranted; rehearse cross-domain crisis communications to avoid misreads.

Operational robustness and escalation control

- Deception-resilient targeting: Red team models against false flags and adversarial inputs; enforce multi-sensor corroboration, uncertainty quantification, and interpretable outputs (ranked COAs, confidence, constraints, sensitivity to new intel).

- PNT hardening: Field dual-redundant precision clocks and quantum-aided INS on high-value platforms; conduct routine GPS-denied drills and space-to-edge time-transfer exercises for synchronized fires and legally robust logs.

- Space security norms and reconstitution: Treat quantum trust services in orbit as critical infrastructure; codify reversible responses to interference, ensure manoeuvrable platforms, on-orbit spares/hosted payload options, and pre-contracted launch for rapid recovery.

Industrial base and talent

- Domestic capacity where chokepoints bite: Incentivize Indian production of photonics, single-photon detectors, precision timing modules, vapour cells/lasers, and secure payload hardware; maintain critical spares stockpiles with lifecycle plans.

- Workforce pipelines: Create joint curricula and certification tracks for crypto stewardship, PNT engineering, photonics/laser safety, model assurance, and EW/EMCON; use rotational billets between labs, formations, and industry to retain skills.

- Outcome-based procurement: Tie multi-year orders to operational metrics and conformance certifications; use milestone-based disbursements and independent validation to prevent hype-driven allocation.

Coalition interoperability without surrendering control

- Standards that travel, keys that don't: Share evidence schemas and decision-packet formats (confidence, constraints, sensitivity), not raw keys; ensure rapid allied acceptance of Indian evidence while preserving sovereign control of roots and logs.

- Joint rehearsals: Conduct allied exercises in GPS-denied navigation, rapid cryptographic restoration under EW, and low-emission ISR cueing to normalize resilient behaviour and coordinated signalling.

Policy Frameworks for Quantum Technology Adoption and Security for Military

Crafting robust policies to govern quantum warfare readiness

A defence policy framework for quantum technologies must do three things: protect sovereignty over keys, clocks, and truth; accelerate fielding where quantum advantage is real; and maintain escalation control under legal and allied constraints. The scope spans procurement, classified R&D, cyber and space security, data governance, training, and ethical use. The governing principle is assurance before ambition: PQC-first for today's systems, selective quantum keying pilots where they add resilience, quantum-aided PNT for GPS-denied operations, and hybrid decision engines that are interpretable, auditable, and under human authority.

Procurement and accreditation

- Standards and conformance: Mandate open interface standards, national conformance certifications for PQC suites, timing accuracy, sensor performance, electromagnetic compatibility, and signed update integrity; make certification a prerequisite for procurement and fielding.

- Spiral acquisition: Use pilot-LRIP-scale pathways with field exercises; tie disbursements to mission metrics (time-to-identify/engage, EW re-plan latency, GNSS-out navigation error, evidence integrity pass rates, cue-to-decision latency, predictive Brier scores).

- Supply chain sovereignty: Require SBOMs, reproducible builds, threshold-signed firmware, and domestic second-source plans for photonics, timing modules, secure payload hardware; prioritize vendors that support crypto-agility and secure update pipelines.

- Space as trust layer: Treat quantum trust services in orbit (keying, time) as protected infrastructure; procure manoeuvrable platforms, on-orbit spares/hosted payloads, and pre-contract launch capacity with debris-minimizing designs.

Classified R&D and technology protection

- Portfolio focus: Prioritize R&D with near-term operational payoff PQC stacks, key distribution gateways, quantum-aided INS/clock modules, low-emission sensing, and hybrid optimization engines.

- Secure labs: Enforce cleared facilities, secure enclaves for signing keys, air-gapped test beds, and hardware security modules; require red team and side channel testing for high-value components.

- IP and export controls: Use sovereign-use IP clauses, design escrow, and export-control compliance plans; prefer designs that can be domestically fabricated, integrated, and sustained.

Data security, provenance, and attribution

- PQC-first governance: Migrate identities, code signing, and mission messaging to standardized post-quantum suites; automate certificate lifecycles; conduct "break glass" re-key drills under EW.

- Evidence packages by default: Require signed telemetry, authenticated time stamps, and sensor provenance at creation; standardize evidence schemas for coalition acceptance while maintaining national data sovereignty.

- Parallel attribution tracks: Maintain classified attribution lines until public release is strategically warranted; publish incident-response norms for communications anomalies and suspected deception.

Operational ethics and escalation control

- Human authority and interpretability: Keep human-in-the-loop for lethality; require interpretable outputs (ranked options, confidence, constraints, sensitivity to new intel) and auditable decision trails.

- Reversibility and proportionality: Favour reversible, observable effects for pre-hostility tech disruption; codify thresholds, authorities, and notification protocols to preserve diplomatic manoeuvre space.

- Non-provocative signalling: Institutionalize routine GPS-denied navigation, rapid cryptographic restoration under EW, and low-emission ISR cueing as demonstrations of resilience without escalatory optics.

Coalition interoperability and sovereignty

- Standards that travel, keys that don't: Share evidence schemas, decision-packet formats, and incident-reporting protocols; retain sovereign control of roots (keys, time, logs).

- Joint rehearsals and transparency: Conduct allied exercises in GPS-denied navigation, rapid re-key, and low-emission ISR; pre-brief doctrines for responses to interference with quantum trust services.

Continuous Evaluation and Adaptation to Emerging Quantum Threats for Military

Staying agile in a rapidly evolving quantum battlefield landscape

Continuous evaluation must be institutionalized as a command practice, not a periodic audit. The core objective is to detect adversary inflection points early, new post-quantum cryptography breakage paths, practical quantum-aided sensing against stealth corridors, space-based keying disruptions, or accelerators that compress decision cycles and to translate these signals into timely doctrine, procurement, and training updates. Practically, this requires a standing "Quantum Readiness Board" at the strategic level that runs quarterly threat sprints and annual deep reviews across C4ISR, space, cyber, EW, and logistics mission threads. The board should own a living risk register tied to mission metrics time-to-identify, time-to-engage under EW, re-plan latency, GNSS-out navigation error, cryptographic link integrity, evidence-integrity pass rates, and cue-to-decision latency, so that shifts in adversary capability produce immediate changes in thresholds, authorities, playbooks, and budget allocations. Continuous red-teaming against deception (false flags, data poisoning, sensor spoofing), model brittleness (domain shift, adversarial inputs), and chain-of-custody attacks (tampered logs, forged time stamps) should feed after-action reviews that update doctrine and technical baselines on a fixed cadence.

Adaptive safeguards must make reversibility, interpretability, and sovereignty default, so responses remain fast yet controlled. A PQC-first posture should be maintained with crypto-agility policies, automated certificate lifecycles, and rehearsed "break-glass" re-key under electronic warfare, while selective quantum keying pilots on high-value links are expanded only when they reduce measured risk and do not introduce single points of failure. For PNT, dual-redundant precision clocks, quantum-aided INS on high-value platforms, and space-to-edge time-transfer drills should be exercised routinely to sustain synchronized fires and legally robust logs in GNSS-denied conditions. Decision engines must output ranked, ROE-compliant options with confidence, constraints met/violated, and sensitivity to new intelligence; human-in-the-loop remains mandatory for lethality with codified abort thresholds and audit trails. In space, treat quantum trust services (keys, time) as protected infrastructure, with manoeuvrable platforms, on-orbit spares or hosted payloads, and pre-contracted launch capacity for rapid reconstitution, alongside doctrinal thresholds for proportionate, reversible responses to interference. Coalition interoperability should be updated through standardized evidence packages, signed telemetry, authenticated timestamps, and sensor provenance that are accepted by allies without surrendering national keys or logs. Finally, align budgets to measured effect: scale what shortens time-to-decision and time-to-engage, lowers GNSS-out navigation error, improves integrity pass rates, and reduces false-flag susceptibility; de-fund what cannot show gains. This make-adapt-measure loop ensures India's quantum posture stays ahead of adversaries, preserving deterrence and escalation control even as the technology landscape shifts

Strategic approach to strengthen continuous evaluation and adaptation against emerging quantum threats must adhere to:

- Adversary quantum watch floor: Establish a 24/7 "quantum watch" cell integrating intel, space, cyber, and EW to track rival deployments (QKD links, precision timing beacons, stealth-cue sensors, accelerator access), with weekly briefs that update risk posture and alert thresholds.

- Mission-thread kill-chain reviews: Run quarterly "quantum stress tests" on priority mission threads—integrated air and missile defence, maritime choke-point control, joint fires, border ISR, nuclear C2 support—injecting denial, deception, and timing faults to expose brittle nodes and update playbooks.

- Crypto-agility as a standing order: Enforce rolling PQC algorithm agility (NIST-class families plus national variants) with time-boxed rotations and compatibility matrices; evaluate emergent cryptanalytic results monthly to pre-authorize hot-swaps without service interruption.

- Evidence-first coalition readiness: Standardize evidence packages (signed telemetry, authenticated timestamps, sensor provenance) that allies can validate in minutes; exercise crosswalks to partner schemas so joint action is not delayed by format disputes.

- Space trust-layer guardianship: Treat space-based keying and time-transfer as a critical trust layer; define proportionate, reversible responses to interference; preposition on-orbit spares/hosted payloads and contract rapid-launch options; routinely drill re-key and time reassignment.

- Decision engine assurance: Mandate interpretable autonomy: all recommendations must present ranked COAs, confidence, constraints met/violated, and sensitivity to fresh intel; certify models through deception red-teaming, uncertainty quantification, and domain-shift trials before fielding.

- GPS-denied battle rhythm: Institutionalize GPS-out exercises with quantum-aided INS and precision timing on high-value platforms; score units on synchronization under EW (fires windows, formation keeping, legal time stamping), and tie readiness to performance.

- EW/EMCON-integrated sensing: Expand low-emission ISR cueing (magnetic/gravimetric, passive RF, space cues) to expose stealth corridors without revealing intent; set emissions budgets and deception-aware fusion thresholds that adjust automatically under contested conditions.

- Industrial base shock absorbers: Secure domestic capacity for keys, clocks, and provenance (KCP): key management, precision time-transfer, and signed-update services as national utilities; diversify photonics and timing supply lines; maintain critical spares and tooling.

- Red-teaming and war gaming cadence: Run biannual red-team campaigns focused on false flags, data poisoning, firmware supply chain attacks, and space custody breaches; feed findings into doctrine refresh and acquisition gate reviews.

- Budget-to-metrics discipline: Tie funding to mission outcomes: shorten time-to-identify and time-to-engage, reduce re-plan latency under EW, lower GNSS-out navigation error, improve integrity pass rates, and raise false-flag catch rates; sunset programs lacking measured effect.

- Legal and escalation guardrails: Keep human-in-the-loop for lethality; codify abort thresholds; maintain parallel classified attribution until public claims are strategically warranted; publish incident-response norms for comms anomalies to prevent misreads.

- Coalition signalling without provocation: Normalize non-escalatory demonstrations—rapid re-key/restoration under EW, GPS-denied navigation trials, low-emission ISR cueing—to deter coercion while avoiding optics of overt escalation.

- Surge planning with accelerators: Use quantum accelerators as planning-window surge capacity, never as operational linchpins; maintain quantum-inspired fallbacks on conventional compute; drill failover to avoid single points of failure.

- Doctrine refresh cycle: Institutionalize a 12–18-month doctrine update cycle for quantum denial/disruption, evidence sharing, and space trust-layer protection, synchronized with conformance lab findings and allied standards evolution.

Index